T0299203

Apples

Due to polymorphism, apples have extraordinary diversity. Depending on variety, apple fruits can differ in color, shade or size; apples even can be oval or pear-shaped. There are more than 10,000 varieties of apple, which vary in taste, shape, juiciness, texture, color, firmness and other qualities. For these reasons, apples have been diversely studied, and many improvements have been made such as the introduction of high density cropping; rootstock breeding; or varietal development. Therefore it is important to understand and document the production methods adopted and implemented in recent times for harvesting maximum benefits of the crop. *Apples: Preharvest and Postharvest Technology* documents production practices along with detailed illustration on varieties, rootstocks, important cultural practices and post-harvest management. This book will serve as a complete guide for apple production from farm to fork and will help students, scholars, researchers and scientists working in this domain. The book will also help growers all over the world to understand best practices for apple production, to harvest maximum yields, and in turn, to increase their returns.

Rafiya Mushtaq, PhD
Division of Fruit Science, Sher-e-Kashmir
University of Agricultural Sciences & Technology of Kashmir, Shalimar, Srinagar, Jammu and Kashmir, India

Gulzar Ahmad Nayik, PhD
Department of Food Science & Technology
Govt. Degree College Shopian, Jammu and Kashmir, India

Ab Raouf Malik, PhD
Division of Fruit Science
Sher-e-Kashmir University of Agricultural Sciences & Technology of Kashmir, Shalimar, Srinagar, Jammu and Kashmir, India

Apples

Preharvest and Postharvest Technology

Edited by Rafiya Mushtaq,
Gulzar Ahmad Nayik and Ab Raouf Malik

CRC Press
Taylor & Francis Group
Boca Raton London New York

CRC Press is an imprint of the
Taylor & Francis Group, an **informa** business

First edition published 2023
by CRC Press
6000 Broken Sound Parkway NW, Suite 300, Boca Raton, FL 33487–2742

and by CRC Press
4 Park Square, Milton Park, Abingdon, Oxon, OX14 4RN

CRC Press is an imprint of Taylor & Francis Group, LLC

© 2023 selection and editorial matter, Rafiya Mushtaq, Gulzar Ahmad Nayik, and Ab Raouf Malik; individual chapters, the contributors

ISBN: 978-1-032-14556-3 (hbk)
ISBN: 978-1-032-14557-0 (pbk)
ISBN: 978-1-003-23992-5 (ebk)

DOI: 10.1201/9781003239925

Typeset in Times
by Apex CoVantage, LLC

We dedicated this book to our beloved family

Contents

Preface

Apples are some of the finest fruits of the world. Apples are a commonly grown fruit in most of temperate regions of the world. For the last 20 years apple production has expanded more than 50 percent, and the biggest growth has been in China. India is the fifth largest producer of apples in world. Apples provide humans with a valuable source of phytonutrients and natural antioxidants, as well as important minerals.

We would like to thank and acknowledge all the contributors for their fruitful contributions and their dedication to editorial guidelines and timeline. We are fortunate to have the opportunity to collaborate with many international experts from Australia, Pakistan, China, Iran, Turkey, etc. We would like to thank colleagues from the production team of Taylor & Francis for their constant help during the editing and production process. Finally, we as editors acknowledge that this book may contain minor errors or gaps. Suggestions, criticism and comments are always welcome; so please do not hesitate to contact us for any relevant issue.

Foreword

Nazir Ahmad Ganai
Vice Chancellor
SKUAST-K, J&K-India

Apples are the premier fruit crop grown in temperate regions of the world. The crop holds immense economic importance in the global fruit industry and contributes to the GDP of many countries and states. In terms of fruit production, India ranks as the fifth largest producer of apples in world with an annual production of 86.14 million tons. This crop has an extraordinary diversity, and it has many health benefits as a valuable source of phytonutrients, natural antioxidants and important minerals. For these reasons, apples have been extensively studied from time to time and many interventions have been put forth such as the development of high density planting systems, breeding rootstocks of different characters like dwarfing, disease resistance and the development of varieties with diverse colors and yield potential. It is therefore important to understand, document and update preharvest and postharvest production methods implemented in recent times for harvesting the maximum possible benefits of the crop.

Therefore, the book *Apples: Preharvest and Postharvest Technology*, with 20 chapters covering different topics on preharvest and postharvest production, will be a comprehensive technology guide for all stakeholders including scientists, scholars, researchers and students. The interpretation of subject matter in each chapter of this book has been very carefully and systematically drawn and organized by eminent authors from all over the world. An attempt has been made by the authors to add new knowledge to the existing literature by including information particularly on high density plantations, advances in varietal improvement and rootstock breeding, modern irrigation and fertigation practices, innovative apple processing technologies and the bioactive compounds of apples. This book will bridge the gap between the old and modern methods of production and help many students, scholars and researchers understand the preharvest and postharvest handling of this important crop.

I hope and believe that the book will be an addition to the scientific knowledge of apple growing and will be of great value to scientists, students and scholars.

Prof. Nazir Ahmad Ganai
Vice-Chancellor
SKUAST-K, J&K-India

Editors' biographies

Rafiya Mushtaq has a PhD in Fruit Science from Sher-e-Kashmir University of Agricultural Sciences & Technology of Kashmir, Jammu and Kashmir, India. Her research focuses on high density apple plantations and has been awarded India's prestigious fellowship 'Prime Minister's Fellowship for Doctoral Research' for her PhD research. She is currently a senior project associate at Sher-e-Kashmir University of Agricultural Sciences & Technology of Kashmir. Dr. Rafiya has authored/co-authored 20+ research articles and ten book chapters. She has presented her research at many national and international conferences including the International Society for Horticultural Science (ISHS) conference and has won many best presentation awards. She has also been awarded the International Travel Support by Department of Science & Technology, Govt. of India for presenting her PhD work at the International Society for Horticultural Science Conference. Dr. Rafiya has received many awards including the best innovative award; the Excellence in research award; appreciation for innovation by Western Sydney University, Australia; and the National Initiative for Development and Harnessing Innovations (NIDHI) Prayas Grant award by the Department of Science and Technology, Government of India. Her scientific interests include fruit physiology, molecular breeding and metabolomic and transcriptomic studies in fruit crops.

Gulzar Ahmad Nayik completed his master's degree in Food Technology from Islamic University of Science & Technology, Awantipora, Jammu and Kashmir, India, and his PhD from Sant Longowal Institute of Engineering & Technology, Sangrur, Punjab, India. He has published over 75 peer-reviewed research and review papers and 35 book chapters, and he has edited eleven books with Springer, Elsevier and Taylor & Francis. Dr. Nayik has also published a textbook on food chemistry and nutrition and has delivered several presentations at various national and international conferences, seminars, workshops and webinars. Dr. Nayik was shortlisted twice for the prestigious Inspire-Faculty Award in 2017 and 2018 from the Indian National Science Academy, New Delhi, India. He was nominated for India's prestigious National Award (Indian National Science Academy Medal for Young Scientists 2019–2020). Dr. Nayik also fills the roles of editor, associate editor, assistant editor and reviewer for many food science and technology journals. He has received many awards, appreciations and recognitions and holds membership in various international societies and organizations. Dr. Nayik is currently editing several book projects with Elsevier, Taylor & Francis, Springer Nature, Royal Society of Chemistry, etc.

Ab Raouf Malik, PhD in Fruit Science, is currently an assistant professor in Division of Fruit Science, Faculty of Horticulture, Sher-e-Kashmir University of Agricultural Sciences & Technology of Kashmir, Jammu and Kashmir, India. He started his career as senior research fellow at the Central Institute of Temperate Horticulture and has also served as assistant floriculture officer in the Department of Floriculture, Govt. of Jammu and Kashmir. His research focuses on the development of new and improved cultivars in apples and other fruit crops through various approaches including mutation breeding. He has also worked on fruit nutrition and high density plantation and has published his research in many nationally and internationally reputed journals. He has also presented his work at many conferences. He is currently the coordinator and PI of many multi-institutional funded projects focusing on the creation of modern orcharding systems in apple, walnut, and other fruit crops. Dr. Malik has demonstrated many technologies like rejuvenation of apple orchards, high density apple plantations, etc. through his projects which have enhanced the livelihood of many farmers with more than 50 success stories. Dr. Malik has received many awards including the best innovator award.

Contributors

Muhammad Aamir
National Institute of Food Science & Technology
University of Agriculture
Faisalabad, Pakistan

School of Food Science
Washington State University
Pullman, USA

Muhammad Afzaal
Department of Food Sciences
Government College University
Faisalabad, Punjab, Pakistan

Asif Ahmad
Institute of Food and Nutritional Sciences
PMAS-Arid Agriculture University
Rawalpindi, Pakistan

M Feza Ahmad
Department of Fruit and Fruit Technology
Bihar Agricultural University
India

Suheel Ahmad
ICAR-Indian Grassland & Fodder Research
    Institute, Regional Research Station
Srinagar, Jammu and Kashmir, India

Aftab Ahmed
Department of Nutritional Sciences
Government College University
Faisalabad, Punjab Pakistan

Mohammed Tauseef Ali
Division of Fruit Science
Sher-e-Kashmir University of Agricultural
    Sciences & Technology of Kashmir
Shalimar, Srinagar, Jammu and Kashmir, India

Shinawar Waseem Ali
Department of Food Technology, Faculty of
    Agricultural Sciences
University of the Punjab
Lahore, Pakistan

Deelak Amin
Plant Virology and Molecular Pathology
    Laboratory (NAHEP-AI), Division of Plant
    Pathology
Sher-e-Kashmir University of Agricultural
    Sciences & Technology
Shalimar, Srinagar, Jammu and Kashmir, India

Faqir Muhammad Anjum
Ifanca Pakistan Halal Apex (Pvt.) Ltd.
Faisalabd, Pakistan

Iqra Ashfaq
National Institute of Food Science &
    Technology
University of Agriculture
Faisalabad, Pakistan

Huda Ateeq
Department of Food Sciences
Government College University
Faisalabad-Pakistan

Tabish Jehan Been
Division of Fruit Science
Sher-e-Kashmir University of Agricultural
    Sciences & Technology
Srinagar, Jammu and Kashmir, India

Chetak Bishnoi
PAU
Regional Research Station
Bathinda, Punjab, India

Adnan Bodhla
Department of Agricultural Engineering
Khwaja Fareed University of Engineering &
    Information Technology
Punjab, Pakistan

Basharat N. Dar
Department of Food Technology
Islamic University of Science & Technology
Awantipora, Jammu and Kashmir, India

Shaila Din
Division of Fruit Science
Sher-e-Kashmir University of Agricultural
    Sciences & Technology
Shalimar, Srinagar, Jammu and Kashmir, India

Suhail Fayaz
Department of Agronomy, School of Agriculture
Lovely Professional University
Punjab, India

Nazir Ahmad Ganai
Vice Chancellors' Secretariat
Sher-e-Kashmir University of Agricultural
    Sciences and Technology of Kashmir
Shalimar, Srinagar, Jammu and Kashmir, India

Mumtaz A. Ganie
Krishi Vigyan Kendra Shopian
Sher-e-Kashmir University of Agricultural
    Sciences & Technology
Shalimar, Srinagar, Jammu and Kashmir, India.

Gousia Gani
Division of Food Science and Technology
Sher-e-Kashmir University of Agricultural
    Sciences & Technology
Shalimar, Srinagar, Jammu and Kashmir, India

Irfan Gani
Division of Floriculture and Landscaping
Sher-e-Kashmir University of Agricultural
    Sciences & Technology
Shalimar, Srinagar, Jammu and Kashmir, India

Ali Gharaghani
Department of Horticultural Science, School of
    Agriculture
Shiraz University
Shiraz, Iran

Ali Akbar Ghasemi-Soloklui
Nuclear Agriculture Research School
Nuclear Science and Technology Research
    Institute (NSTRI)
Karaj, Iran

Mehdi Gararazhian
Peyvand Khavaran Ind. & Cultivation
    Company
South Khorasan, Iran

Archi Gupta
Department of Horticulture
Lovely Professional University
Phagwara, Punjab, India

Navjot Gupta
PAU, Regional Research Station
Bathinda, Punjab, India

Muzzamal Hussain
Department of Food Sciences
Government College University
Faisalabad-Pakistan

Syed Zameer Hussain
Division of Food Science and Technology
Sher-e-Kashmir University of Agricultural
    Sciences & Technology
Shalimar, Srinagar, Jammu and Kashmir, India

G I Hassan
Division of Fruit Science
Sher-e-Kashmir University of Agricultural
    Sciences & Technology
Shalimar, Srinagar, Jammu and Kashmir, India

Shazia Hassan
Division of Fruit Science
Sher-e-Kashmir University of Agricultural
    Sciences & Technology of Kashmir
Shalimar, Srinagar, Jammu and Kashmir, India

Ali Imran
Department of Food Sciences
Government College University
Faisalabad, Punjab, Pakistan

Muhammad Zafar Iqbal
Faculty of Plant Protection
Yunnan Agricultural University
Kunming, Yunnan, China

Nusrat Jan
Division of Food Science and Technology
Sher-e-Kashmir University of Agricultural
    Sciences & Technology
Shalimar, Srinagar, Jammu and Kashmir, India

Smruthi Jayarajan
Department of Horticulture
Lovely Professional University

Amardeep Kour
PAU
Regional Research Station
Bathinda, Punjab, India

A. Khalil
Division of Fruit Science
Sher-e-Kashmir University of Agricultural
Sciences & Technology of Kashmir
Shalimar, Srinagar, Jammu and Kashmir, India

Shumaila Khan
Department of Agricultural Engineering
Khwaja Fareed University of Engineering &
Information Technology
Punjab, Pakistan
Phagwara, Punjab, India

Ab Shakoor Khanday
Krishi Vigyan Kendra Kulgam
Sher-e-Kashmir University of Agricultural
Sciences & Technology of Kashmir
Shalimar, Srinagar, Jammu and Kashmir, India

Afiya Khursheed
Division of Vegetable Science
Sher-e-Kashmir University of Agricultural
Sciences & Technology
Shalimar, Srinagar, Jammu and Kashmir, India

Shoaib Kirmani
ICAR-Central Institute of Temperate Horticulture
Rangreth, Srinagar, Jammu and Kashmir,
India

Chamman Liaqat
Institute of Food and Nutritional Sciences
PMAS-Arid Agriculture University
Rawalpindi, Pakistan

Aabid H. Lone
Mountain Research Center for Field Crops
Khudwani
Sher-e-Kashmir University of Agricultural
Sciences & Technology of Kashmir
Srinagar, Jammu and Kashmir, India

Insha Majid
Division of Fruit Science
Sher-e-Kashmir University of Agricultural
Sciences & Technology
Shalimar, Srinagar, Jammu and Kashmir, India

Ab Raouf Malik
Division of Fruit Science
Sher-e Kashmir University of Agricultural
Sciences and Technology of Kashmir
Shalimar, Srinagar, Jammu and Kashmir, India

M. A. Mir
Division of Fruit Science
Sher-e-Kashmir University of Agricultural
Sciences & Technology
Shalimar, Srinagar, Jammu and Kashmir,
India

Mahrukh Mir
Division of Fruit Science
Sher-e-Kashmir University of Agricultural
Sciences & Technology
Shalimar, Srinagar, Jammu and Kashmir, India

Yasir Hanif Mir
Division of Soil Science & Agricultural
Chemistry, FOA
Sher-e-Kashmir University of Agricultural
Sciences & Technology of Kashmir
Srinagar, Jammu and Kashmir, India

Mojtaba Kordrostami
Nuclear Agriculture Research School
Nuclear Science and Technology Research
Institute (NSTRI)
Karaj, Iran

Khalid Mushtaq
Division of Fruit Science
Sher-e-Kashmir University of Agricultural
Sciences & Technology of Kashmir
Shalimar, Srinagar, Jammu and Kashmir, India

Rafiya Mushtaq
Division of Fruit Science
Sher-e-Kashmir University of Agricultural
Sciences & Technology
Shalimar, Srinagar, Jammu and Kashmir, India

Asha Nabi
Plant Virology and Molecular Pathology
Laboratory (NAHEP-AI), Division of Plant
Pathology
Sher-e-Kashmir University of Agricultural
Sciences & Technology
Shalimar, Srinagar, Jammu and Kashmir, India

Sajad-un-Nabi
ICAR-Central Institute of Temperate
  Horticulture
Rangreth, Srinagar, Jammu and Kashmir,
  India

Mohammadebrahim Nasrabadi
Agriculture and Livestock Binalood Company
Khayyam, St. Neyshabour, Iran

Farheen Naqash
Faculty of Horticulture
Sher-e-Kashmir University of Agricultural
  Sciences & Technology of Kashmir
Shalimar, Jammu and Kashmir, India

N. Nazir
Division of Fruit Science
Sher-e-Kashmir University of Agricultural
  Sciences & Technology of Kashmir
Shalimar, Srinagar, Jammu and Kashmir, India

Bushra Niaz
Department of Food Sciences
Government College University
Faisalabad-Pakistan

Bilal A. Padder
Plant Virology and Molecular Pathology
  Laboratory (NAHEP-AI), Division of Plant
  Pathology
Sher-e-Kashmir University of Agricultural
  Sciences & Technology
Shalimar, Srinagar, Jammu and Kashmir, India

Kuldeep Pandey
Division of Fruits and Horticultural Technology
ICAR-Indian Agricultural Research Institute
New Delhi, India

Jessica Pandohee
Centre for Crop and Disease Management,
  School of Molecular and Life Sciences
Curtin University
Bentley, WA, Australia

E. A. Parray
Division of Fruit Science
Sher-e-Kashmir University of Agricultural
  Sciences & Technology of Kashmir
Shalimar, Srinagar, Jammu and Kashmir, India

Shahid Qayoom
Division of Fruit Science
Sher-e-Kashmir University of Agricultural
  Sciences & Technology
Shalimar, Srinagar, Jammu and Kashmir, India

Moin Qureshi
Department of Field Crops, Faculty of Agriculture
Akdeniz University
Antalya, Turkey

Shafiya Rafiq
Amity Institute of Biotechnology
Amity University Rajasthan
Jaipur, India

Rajni Rajan
Department of Horticulture
Lovely Professional University
Phagwara, Punjab, India

Rovidha S. Rasool
Plant Virology and Molecular Pathology
  Laboratory (NAHEP-AI), Division of Plant
  Pathology
Sher-e-Kashmir University of Agricultural
  Sciences & Technology
Shalimar, Srinagar, Jammu and Kashmir, India

Farhan Saeed
Department of Food Sciences
Government College University, Faisalabad
Punjab, Pakistan

Aanisa Manzoor Shah
Division of Soil Science & Agricultural
  Chemistry, FOA
Sher-e-Kashmir University of Agricultural
  Sciences & Technology of Kashmir
Srinagar, Jammu and Kashmir, India

M. Abas Shah
ICAR-Central Institute of Temperate Horticulture
Rangreth, Srinagar, Jammu and Kashmir, India

Mehraj D. Shah
Plant Virology and Molecular Pathology
  Laboratory (NAHEP-AI), Division of Plant
  Pathology
Sher-e-Kashmir University of Agricultural
  Sciences & Technology
Shalimar, Srinagar, Jammu and Kashmir, India

M. K. Sharma
Division of Fruit Science
Sher-e-Kashmir University of Agricultural
  Sciences & Technology of Kashmir
Shalimar, Srinagar, Jammu and Kashmir, India

Haamiyah Sidiq
Division of Food Science and Technology
Sher-e-Kashmir University of Agricultural
  Sciences & Technology
Shalimar, Srinagar, Jammu and Kashmir, India

Shamim A. Simnani
Division of Fruit Science
Sher-e-Kashmir University of Agricultural
  Sciences & Technology
Shalimar, Srinagar, Jammu and Kashmir, India

Jatinder Singh
Division of Horticulture
Lovely Professional University
Phagwara, Punjab, India

Sumaira-Hamid
Department of Biosciences
Integral University
Lucknow, Uttar Pradesh, India

Sana B. Surma
Plant Virology and Molecular Pathology
  Laboratory (NAHEP-AI), Division of Plant
  Pathology
Sher-e-Kashmir University of Agricultural
  Sciences & Technology
Shalimar, Srinagar, Jammu and Kashmir, India

Ifrah Usman
Department of Food Sciences
Government College University
Faisalabad, Pakistan

Ab Waheed Wani
Department of Horticulture
Lovely Professional University
Phagwara, Punjab, India

Fehim J. Wani
Division of Agri. Economics & Statistics,
  Faculty of Agriculture
Sher-e-Kashmir University of Agricultural
  Sciences & Technology of Kashmir
Wadura-Sopore, Jammu and Kashmir, India

S. A. Wani
Faculty of Horticulture
Sher-e-Kashmir University of Agricultural
  Sciences & Technology of Kashmir
Shalimar, Jammu and Kashmir, India

Sajad Ahmad Wani
Department of Food Technology
Islamic University of Science & Technology
Awantipora, Jammu and Kashmir, India

Sajad Mohd Wani
Division of Food Science and Technology
Sher-e-Kashmir University of Agricultural
  Sciences & Technology
Shalimar, Srinagar, Jammu and Kashmir, India

Iftisam Yaseen
Division of Fruit Science
Sher-e-Kashmir University of Agricultural
  Sciences & Technology
Shalimar, Srinagar, Jammu and Kashmir,
  India

1 History, Distribution, Production and Taxonomic Classification

Shumaila Khan, Muhammad Zafar Iqbal,
Adnan Bodhla and Tabish Jehan Been

CONTENTS

1.1 INTRODUCTION

Apples are a very significant temperate fruit which are grown in Europe and Asia (Janick and Moore, 1996). Its varieties form the largest group of the temperate fruit varieties that has a major share in the fresh market. It has been domesticated to Europe, North America, and the Middle East. *M. asiatica* Nakai (soft apple) was the primarily apple grown (Zhang *et al.*, 2018) in China and nearby areas and is considered a hybrid composite species derived mainly from *M. sieversii* with *M. prunifolia* in southern and eastern Asia. People in the past used to collect the seeds of apple trees due to some specific favorable traits, and it led to the spread of the most desirable cultivars around the globe. In the early 3500 B.C., fresh and dried apple forms helped in the spread of apple varieties throughout the Mediterranean and Asian extended distance trade routes, and the spread eventually helped in the development of new areas for the cultivation of apples. Apples have been documented differently in different civilizations showcasing apple history. The range of cultivars varies within each country depending upon the consumer demand and its growing conditions. Theophrastus was perhaps the first scientist to bring back apples to Greece from the conquests of Alexander the Great in 320 B.C.; he also studied the crop and its modes of propagation.

1.2 HISTORY

It is believed that apples originated from the Tian Shan mountains. Even though the derivation of today's apples is not very strong, possibly apples evolved from Central Asia's wide-ranging apple woodlands, chiefly in the mountainous region of Kazakhstan (Cornille *et al.*, 2014). Archaeobotanists have stated that, initially, 38 species were involved in the domestication of apples (Bamforth *et al.*, 2014). Over the past decade there has been a lot of progress in revising the history of apples due to better availability of

DOI: 10.1201/9781003239925-1

phenotypic and genetic data as well as archaeobotanical data for both wildlife and planted cultivars of apples (Cornille *et al.*, 2014). Farming and cultivation of apples started in the beginning of the Neolithic period around the civilization of human beings (Morgan and Richards, 1993; Juniper *et al.*, 1996).

About 2000 years ago, *M. asiatica* Nakai (soft apple) was the primary apple grown (Zhang *et al.*, 2018) in China and nearby areas and considered a hybrid composite species derived mainly from *M. sieversii* with *M. prunifolia* in southern and eastern Asia. Later in the 19th and 20th centuries, *M.* × *domestica* became the major cultivated species, and it was introduced in North America, Europe, Australia, New Zealand, Russia, Japan and over the other parts of the world (Sakurai *et al.*, 2000). *M. microMalus* Makino, *M. baccata, Malus floribunda* Siebold ex Van Houtte, *Malus zumi* (Matsum.), *M. sargentii* and *Malus* × *atrosanguinea* (Spath) C.K. Schneid were among those species that were used to breed the most commercial apple cultivars (Janick, 2006).

Theophrastus was perhaps the first scientist to bring apples to Greece from the conquests of Alexander the Great in 320 B.C.; he also studied the crop and its modes of propagation. He defined grafting and other ways the apple trees should be cared for and the use of dwarfing rootstocks. Apples have been used both by Muslims and Christians in their religions, and this has played a contributory role in the proliferation of apples all over the world, especially in Africa and Europe. According to the Royal Horticultural Society of England, about 200 species of apples have been identified since 1826. The dominance of the Roman kingdom that spread north to western Europe led to apple crossing with the local *M. sylvestris* (crab apple), an otherwise restrained apple (Janick, 2006). By the 1st century A.D. various varieties were chronicled and had attained an important position in Roman cooking, drugs, and aesthetics. Lately, *Malus* species have been grown along borders of fields, as complementary gardens for commercial production or as trees of overstorey in the pasture (Cornille *et al.*, 2015).

In the last century, high density apple production was started by using the dwarfing rootstocks, and production was increased and strengthened in both the temperate and tropical region crops of the world at high altitudes. In order to increase the orchard efficiency, training systems were also introduced. Apples are a unique fruit crop with a variety of rootstocks that can create different tree forms, and the crop responds quite well to the new management techniques and training systems developed lately (Cornille *et al.*, 2015). As the research goes on and more shreds of evidence are collected, it is concluded that integrated production is required to lessen the use of pesticides or increase the organic production (Cornille *et al.*, 2014).

1.3 BOTANY

The genus *Malus*, according to most authorities, has 25–30 species and several subspecies of supposed crab apples. The cultivated apple is produced as a result of interspecific hybridization. The name *Malus* × *domestica* spp. is normally considered as scientifically suitable (Korban and Skirvin, 1984). The main ancestor of apples, *Malus sieversii*, is wildly growing in Tien Shan, the holy Mountains at the border among the Soviet Union formers (Russia), to the border edge of the Caspian Sea and Western China (Morgan and Richards, 1993; Ferree and Warrington, 2003).

Malus requires light to grow well; however it can be acclimatized to a wide range of altitudes ranging from 370 to 640 N. It can be grown in a vast range of soil conditions, and wide environments can be suitable for growth in the jungle territories, from conifers and oak to lime-like deciduous plants (Caminiti *et al.*, 2011). Given its high light demand and low competitive abilities, wild apples are naturally scattered in gaps within the forests and at the edge of the forests or in extremely dry or wet sites at which other plants are less competitive and most individuals are less than 10 m tall (Cornille *et al.*, 2015).

1.4 TAXONOMIC CLASSIFICATION

Apple fruit belongs to the *Malus* genus from *Rosaceae* family (deciduous tree or shrub) (Ferree and Warrington, 2003) and has various species, different colors, etc. Green and red are found in the Northern Hemisphere (temperate zone) with low temperatures (Bianzino *et al.*, 2011).

The taxonomy of *Malus* has been published (Phipps *et al.*, 1991). Family *Rosaceae* has some salient features that are found in all members like cooking apples, crab apples, wild apples, eating apples, etc. The subfamily of *Rosaceae*, which is known as *Maloideae*, also has other related fruits like pears, medlars, quinces and loquats divided based on DNA, chromosomes, morphological traits (foliage, reproductive parts, inflorescence), crossability or molecular polymorphism. In the *Malus* genus, three subgenera have been retained: (i) *Chloromeles*, (ii) *Malus* (iii) and Sorbo*Malus*. Several people have included the *Malus* species in other genera: *Eriolobus* includes M. Roem (*M. trilobata*) and *E. trilobata* (Poir.); *Docyniopsis* includes Decne. (*M. tschonoskii*), *D. tschonoskii* (Wall.) and presumably would include *M. formosana, M. doumeri and M. melliana*. They recommended that additional work may bear the addition of the genus *Docyniopsis* as part of the genus *Docynia* and the rise of subgenus *Chloromeles* (Campbell *et al.*, 1991).

The genus *Malus* is very diverse genetically (Wu and Chen, 2013). The discovery of 'McIntosh' (1796, USA), 'Rome Beauty' (1848, USA), 'Jonathan' (1826, USA), 'Granny Smith' (1868, Australia), 'Cox Orange' (1850, UK), 'Golden Delicious' (1890, USA) and 'Red Delicious' (1880, USA) were of historic importance. Varieties of cultivated apples have a rather slender genetic base; the two cultivars 'Golden Delicious' and 'Delicious' are most predominant. Commercially the world's most important growing apple cultivars belong to the species *Malus × domestica* Borkh.

Some other species also have some impact on commercial apple production, and almost all scab resistant cultivars that are commercially available have M. × *floribunda* Siebold ex Van Houte as their ancestor. Different genes for disease resistance have also been obtained from a wide scale of other *Malus* species such as *M. micromalus* Makino, *M.* × *atrosanguinea* (Spaeth) C. Schneider, *M. baccata* jackii Rehder and *M. sargentii* Rehder (Korban and Skirvin, 1984). *M. sieversii* has the same qualities as *M.* × *domestica* and also is very diverse as shown by the number of collection trips to Central Asia. The *Malus sieversii* apple variety is common in Kazakhstan (Harris *et al.*, 2002). In the chapter, "The story of the Apple", Juniper referred to it as *Malus pumila* for both the wild Central Asian apple and the domesticated one (Juniper and Mabberley, 2009). On the other hand, Coart and colleagues (2006) raised a question on this premise and declared *Malus × domestica* for the domestic apple and the *Malus sieversii* apple variety to be the right name for the wild apple of Central Asia. *M.× domestica* is believed to originate from *M. sylvestris* Mill (Mabberley *et al.*, 2001) as a hybrid, *M. dasyphyllus* Borkh/*M. praecox* Borkh/*M. sylvestris* and *M. pumila* var. praecox (Pall.) Ponomar (Korban and Skirvin, 1984). Presently, *M. sieversii* (Ledeb.) (Robinson *et al.*, 2001) is ubiquitous in Central Asia's highlands of advancements among 1200 m and 1800 m.

Genetic constitution and its modifications have a long-lasting effect on apple production, which in turn decides how the varieties will grow in orchards. Apple production was dominated by cultivars, such as 'McIntosh' (1800s), 'Jonathan' (1820s), 'Cox's Orange Pippin' (1830s), 'Granny Smith' (1860s), 'Fun' (1870s), 'Golden Delicious' (1890s) and 'Braeburn' (1940s), which formed the most preferred varieties around a 100 years ago. By this time, apples had reached all corners of the globe as immigrants from the old world brought them to their home countries. In Asia, these varieties often included local varieties selected from the development of species, i.e., *M. prunifolia* and its cultivated species *M. asiatica* (Morgan and Richards, 1993). The domesticated apple crops were developed and introduced in 1930s and 1940s; their orchards, formed in the 1960 and 1970s, included the 'Royal Gala' ('Kidd's Orange Red' × 'Sweet Gold'), Jonagold ('Sweet' × 'Jonathan'), 'Fuji' ('Ralls Janet' × 'Delicious') and 'Elstar' ('Ingrid Marie' × 'Golden Delicious'). They completely changed the existing crop varieties. China leads the growth and production of apples due to the introduction of the 'Fuji' variety of apples.

1.5 WORLDWIDE APPLE PRODUCTION

World production for 2021/22 is projected up 1.0 million tons to 81.6 million (www.nationalbeef-wire.com). In comparison with many other tree fruits, the apple tree has a more northern range due to relatively cold hardiness and late blooming. In recent years apple production worldwide has increased to some extent and is supposed to further increase (Table 1.1). However, experts

TABLE 1.1

Current Trends in the Production of Apples Worldwide (×1000 tons) (data from Faostat, 2020)

Area	2017	2018	2019	2020
Australia	313730	268355	265150	262966
Belgium	86236	231300	241860	168030
Brazil	1307642	1203007	1222949	983247
China	41391451	39235019	42426578	40501041
France	1695949	1740350	1753500	1619880
Germany	596666	1198520	991450	1023320
India	2265000	2327000	2316000	2734000
Italy	1921272	2466990	2303690	2462440
Japan	735200	756100	701600	720405
Mexico	714149	659692	761483	714203
Morocco	820547	696950	809762	778866
Netherlands	227000	269100	273000	220000
New Zealand	470000	480000	557680	562058
Turkey	3032164	3625960	3618752	4300486
USA	5240670	4644790	5028526	4650684

estimate a considerable increase in consumption and production in the coming years, mainly in Asian countries. For Europe and North America, the production of apples is predicted to remain stable. The main varieties in China are 'Red Delicious' (9.7%), 'Fuji' (60.4%), 'Golden Delicious' (6.2%), 'Gala' (2.8%), 'Jonagold' (3.0%) and others (1.9%) (Kellerhals, 2009). China's primary target markets are Russia and Southeast Asia as the economic developments in China and emerging economies surrounding China itself contribute to the high requirement of apple utilization in Asia. Also, it is supposed that a huge quantity of apples will not be exported from China to European and North American countries in the near future. Different sources show that 68.2% of all apples from China is exported to Southeast Asia and about 54.1% goes to countries in South Asia. China exports apples to different preferred countries including Hong Kong (30.3%), Malaysia (59.6%), India (19.8%), Indonesia (69.2%), Singapore (56.9%), Bangladesh (44.6%), Philippines (92.9%) and Thailand (63%).

1.6 PROPAGATION AND TRANSPORTATION

The propagation method adopted for the variety grafted and planted from root suckers might be different, except for *M. sieversii* which is sucker free. People also used the practices of cloning and shifting the desirable plant types to their desirable horticultural growing areas with the seasonally grazed animals. Open-pollinated progenies and trees were called elite varieties among all horticultural species that are seen in the forests nowadays. Intentionally or unintentionally, the ancestral species of apples took the final and longest trade route toward China to Europe and through Central Asia to the Middle East repeatedly. Some varieties like *M.* × *domestica* along with *Malus prunifolia* (Willd.) Borkh., *M. mandshurica* (Maxim.) Kom., *Malus baccata* (L.) Borkh and *Malus sieboldii* (Regel) Rehder complex may have ascended with hybridization with native species toward China. On the other hand, hybridization with other local cultivars of *Malus orientalis* Uglitzk and *M. sylvestris* into the west is still speculated (Nikiforova *et al.*, 2013; Volk *et al.*, 2013).

Along with the spread of various religions, apples continued to be prudently sustained, unfluctuating over battles and problematic epochs, in the religious foundation gardens all over Europe's and Iberia's groves. They also gained national importance in several countries and got a position in the early diet food of Scandinavians, Franks, Celts, Gauls and other individuals of northern areas of Europe in both dried or roasted forms. Fruit orchard preservation was invigorated as an uncomplicated ascetic ability, and various abbeys established hefty groves with numerous *M.* × *domestica* cultivars (Sakurai *et al.*, 2000). Similarly, in the Eastern Mediterranean and Iberia (Islamic world), pomology was well-regarded and skills of training, grafting and pruning under Koranic teachings were appreciated and became highly developed. In the 13th century, apples progressively became extensively planted in royal and local gardens all over Europe. Fresh apples were rarely consumed; it is more likely they were consumed with a more enhanced taste when cooked with other spices like sucrose and honey. At the end of the 17th century a minimum of 120 species of apples were introduced to Europe. The England Royal Horticultural Society approved at least 1200 different cultivars in the year 1826. The 18th and 19th centuries saw varieties of apples classified and recognized depending on their appropriateness for their uses (Korban and Skirvin, 1984). The late 19th century and early 20th century presented the highest number of varieties in apple farming in Europe with thousands of small orchards producing hundreds of locally well-liked cultivars. In the 20th century imported fruit from New Zealand, America, South Africa and Australia European country orchards increased in order to enhance the size of the fruit, on one hand, and to decrease the orchard area, on the other hand, growing only one variety at a time.

1.7 DISTRIBUTION OF APPLES

Apple varieties were categorized into three wide classes: cooking varieties, cider varieties and dessert varieties which were different from each other in many aspects like highlight color, odor, proportions, smoothness, flavor, freshness, and crispness. Varieties also showed variable contents like carbohydrates (sugar), acidic contents, tannin, vitamins A and C and dietary fibers (Omasheva *et al.*, 2015, 2017).

Originally apple species could be grown in temperate climate areas of the hemisphere. Approximately 2000 years ago, apples were only eaten by Europeans but the selection of varieties increased. Before the creation of the United States of America, a comparatively good number of introduced varieties were accepted in Europe (Xiang *et al.*, 2020). Different varieties of apples were grown around the United States for ornamental purposes by Native Americans by traveling workers. A prominent traveling worker was John Chapman, an expert nurseryman who planted apple trees widely in Ohio and Indiana.

The dispersal of *Malus* trees belonging to *Rosaceae* has seen continuous seed dispersal from devious mammals (humans and other animals like bears and deer). The Holocene can be directly proportioned to the dispersal of apples. Reduction in the population of individuals may lead to a reduction in the densities of dispersers (Wu and Chen, 2013). Lately, the rosaceous trees with fleshy fruits exist in fragmentary populations across European and Asian countries (Omasheva *et al.*, 2015). Botanists consider *Malus niedzwetzkyana* to have its species with a larger range of distribution in Central Asia to the Afghanistan border (Omasheva *et al.*, 2015, 2017).

There were larger ancestral inhabitants for today's domesticated apples, and it appears that the absence of diversification is a reaction to low-quality gene flow joined with the restricted time since the glacial retreat. The deficiency of the primary seed dispersers for the European wild apple prompted decreased capacity to take possession of new regions during the early and mid-Holocene. The analysis of archaeobotanical findings of apple seeds from Neolithic destinations in Europe implies that the reach of the wild apple was more limited before the three centuries (Richards *et al.*, 2009). Later, based on genetic investigations, outlines were made pointing to the developed apple that is domesticated across Europe and West Asia is by and large a half breed comprising of species circulated across the mainland (Cornille *et al.*, 2014). *Malus domestica* Borkh. is one of the most commercially and culturally natural and significant fruits on the planet today; and it is grown in all mild zones.

1.8   APPLE PRODUCTION AND ITS ASSOCIATED CONCERNS

Even though temperate climate zones are considered best for apple growth, commercial apple production is more often seen in those countries and those places that have comparatively better techniques for apple production and marketing. Because of the global climate change, there is a need to introduce heat tolerant varieties into cultivation such as 'Fuji' and 'Granny Smith' apart from introducing improved irrigation technologies that will help in the acceptable production of apples to grow successfully into climate regions that are comparatively warmer than others (O'Rourke, 2018). Since the Second World War there has been an increasing growth trend in the world apple production. Even though the growth rate slowed down in the 1980s, it leaped ahead in the 1990s due to just one factor: the extraordinary production system in China. In the start of the 1990s, Chinese apple production was about 4 million tons (Ferree and Warrington, 2003). It increased by more than fivefold by the end of the decade. Since the mid-1980s, China has seen a tremendous increase in apple cultivation, while the rest of the world has seen a considerable decline in apple land harvesting since the 1990s (Graham and O'Rourke, 2019). According to (Caminiti *et al.*, 2011), in the assortment of better quality phenotypes and their successive arbitrary breeds, the most efficient way of increasing the number of desirable alleles is mass selection because there is a huge preservative variance prevailing in the heritage of most traits. In the F1 progenies, most of the fruit breeders frequently choose superior genotypes in apples while the advanced selections are mostly made on the basis of market necessities and customer demands. These days the marketplaces are showing a great trend toward the growing of apples worldwide including varieties such as 'Delicious', 'Gala', 'Golden Delicious', 'Braeburn' and 'Fuji'. As a sustainable element of farming systems, apple varieties also have to meet requirements like satisfying different ecological and economical demands. According to different environmental concerns, there is a need for serious hard work and effort to produce high-quality apple varieties and disease-resistant cultivars particularly resistant to diseases like apple scab (*Venturia inaequalis*) and powdery mildew (*Podosphaera leucotricha*) which form the most significant and major diseases of fungus in different apple growing regions of the world.

The bacterial disease (*Erwina amylovora*) fire blight poses a significant hazard to the production of apples and is very difficult to manage. In temperate climates disease, various exotic cultivars permit a considerable decrease in input of pesticides in apple orchards where a larger amount of pesticide treatments are usually needed. In sustainable and adapted systems, the production of apples could build up the role of the apple crop as a vital and energetic foodstuff and play an imperative role in increasing the safety of food. However, none of the resistant cultivars have yet infiltrated the market in comparison to different apple cultivars such as 'Gala' and 'Golden Delicious'. On the other hand, the achievement of disease-resistant cultivars at the trade point is not associated with their disease conflicting attributes, as the flourishing of new varieties is basically disregarded by retailers and wholesalers who distinguish the quality of apples as the main customer desire and consequently drive the marketing policy.

1.9   CLIMATE, MANAGEMENT AND TECHNOLOGY

Specific data about the climate variable effects has been provided by horticultural scientists such as temperature effects on fruit yield from experimental trees plots. New techniques of management have also been created, such as new pruning and training systems and dwarf trees, that have also shown to affect the crop yield. At the more elemental level, recent developments in the research on physiological and hormonal associations between the apple trees has provided greater information on blossoms and the production of fruit. Practically though, to predict marketable crop levels yearly is very complicated. There are various essential climatic necessities which decide the regions where apples can be developed (Hogg, 1970). In Britain, these regions, for most parts of the country, fall in the hotter areas. The highest altitudes where apples are grown is normally as a rule between sea level (120 m) and receiving a precipitation of around 500–700 mm. While surrounded by these fundamental climatic prerequisites there is an enormous scope of climate factors that can be exploited.

According to some experienced crop producers, in absence of frost, hot months like May and June can generally provide great production for apples (Barlow and Cumming, 1975). The premise of endogenic behavior in the apple tree is dependent upon the hormonal extent and their circulation all through the plant organs. The complicated frameworks included are, as a rule, slowly disentangled. Hormones have explicit impacts at various locales in the framework and may show various impacts at different dimensions. The commencement of buds goes before blossom arrangement which thus goes before organic fruit production. The underlying bud is not entirely settled by the harmony flanked by cytokinin hormones from roots or leaves and gibberellins hormones from the seeds.

1.10 CONCLUSION

Apples are a very significant temperate fruit which is grown in Europe and Asia. Apples has a very unique agricultural farming narration and hereditary resources which leads to the production of a yield of superb varieties through a lot of propagation methods and different programs. Genetic constitution and its modifications have a long-lasting effect on apple production which in turn decides how the varieties will grow in orchards. Even though temperate climatic zones are considered to be best for apple growth, commercial apple production is more often seen in those countries and those places that have comparatively better techniques for apple production and marketing. These days, different challenges have necessitated the activation of similar processes in an intellectual way and through new traditions. The rising in the genomics paraphernalia in the species will speed up genomics-enabled proliferation techniques, promising regular delivery of apples that will ultimately donate to their uninterrupted cultivation as well as individual nourishment and health.

REFERENCES

Bamforth, C.W. and Ward, R.E., eds. 2014. *The Oxford handbook of food fermentations.* Oxford University Press.

Barlow, H.W.B. and Cumming, I.G. 1975. Effect of early summer weather on the yield of Lane's Price Albert apple. In: H.C. Pereira (ed.). *Climate and the orchard.* Commonwealth of the Agricultural Bureau, pp. 44–47.

Bianzino, A.P., Raju, A.K. and Rossi, D. 2011. Apples-to-apples: a framework analysis for energy-efficiency in networks. *ACM SIGMETRICS Performance Evaluation Review* 38(3): 81–85.

Caminiti, I.M., Noci, F., Muñoz, A., Whyte, P., Morgan, D.J., Cronin, D.A. and Lyng, J.G. 2011. Impact of selected combinations of non-thermal processing technologies on the quality of an apple and cranberry juice blend. *Food Chemistry* 124(4): 1387–1392.

Campbell, R.J., Fell, R.D. and Marini, R.P. 1991. Canopy position, defoliation, and girdling influence apple nectar production. *HortScience* 26(5): 531–532.

Coart, E.L.S., Van Glabeke, S., De Loose, M., Larsen, A.S. and Roldán-Ruiz, I. 2006. Chloroplast diversity in the genus Malus: new insights into the relationship between the European wild apple (*Malus sylvestris* (L.) Mill.) and the domesticated apple (*Malus domestica* Borkh.). *Molecular Ecology* 15(8): 2171–2182.

Cornille, A., Feurtey, A., Gelin, U., Ropars, J., Misvanderbrugge, K., Gladieux, P. and Giraud, T. 2015. Anthropogenic and natural drivers of gene flow in a temperate wild fruit tree: a basis for conservation and breeding programs in apples. *Evolutionary applications* 8(4): 373–384.

Cornille, A., Giraud, T., Smulders, M.J., Roldán-Ruiz, I. and Gladieux, P. 2014. The domestication and evolutionary ecology of apples. *Trends in Genetics* 30(2): 57–65.

Ferree, D.C. and Warrington, I.J., eds. 2003. *Apples: botany, production, and uses.* CABI.

Graham, C. and O'Rourke, B.K. 2019. Cooking a corporation tax controversy: apple, Ireland and the EU. *Critical Discourse Studies* 16(3): 298–311.

Harris, S.A., Robinson, J.P. and Juniper, B.E. 2002. Genetic clues to the origin of the apple. *Trends in Genetics* 18(8): 426–430.

Hogg, W.H. 1970. Climatic factors in apple growing. *Experimental Horticulture* 21: 67–74. https://www.nationalbeefwire.com/apples-world-markets-trade-world-production-for-2021-22-is-projected-up-1-0-million-tons-to-81-6-million-as-china-supplies-edge-upward-and-several-eu-member-states-recover-from-the-previous-year-s-damaging-weather

Janick, J. 2006. The PRI apple breeding program. *HortScience* 41(1): 8–10.

Janick, J. and Moore, J.N., eds. 1996. *Fruit breeding, tree and tropical fruits* (Vol. 1). John Wiley & Sons.

Juniper, B.E. and Mabberley, D.J. 2009. *The story of the Apple*. Timber Press.

Juniper, B.E., Watkins, R. and Harris, S.A. 1996, September. The origin of the apple. In: *Eucarpia Symposium on Fruit Breeding and Genetics* 484: 27–34.

Kellerhals, M. 2009. Introduction to apple (*Malus× domestica*). In: *Genetics and genomics of Rosaceae* (pp. 73–84). Springer.

Korban, S.S. and Skirvin, R.M. 1984. Nomenclature of the cultivated apple [*Malus x domestica* Borkh]. *HortScience* 19(2): 177–180.

Mabberley, D.J., Jarvis, C.E. and Juniper, B.E. 2001. The name of the apple. *Telopea* 9(2): 421–430.

Morgan, J. and Richards, A. 1993. The book of apples. *Nature* 366(6456): 641–641.

Nikiforova, S.V., Cavalieri, D., Velasco, R. and Goremykin, V. 2013. Phylogenetic analysis of 47 chloroplast genomes clarifies the contribution of wild species to the domesticated apple maternal line. *Molecular Biology and Evolution* 30(8): 1751–1760.

O'Rourke, A.D. 2018. *The world apple market*. Routledge.

Omasheva, M.E., Chekalin, S.V. and Galiakparov, N.N. 2015. Evaluation of molecular genetic diversity of wild apple *Malus sieversii* populations from Zailiysky Alatau by microsatellite markers. *Russian Journal of Genetics* 51(7): 647–652.

Omasheva, M.Y., Flachowsky, H., Ryabushkina, N.A., Pozharskiy, A.S., Galiakparov, N.N. and Hanke, M.V. 2017. To what extent do wild apples in Kazakhstan retain their genetic integrity? *Tree Genetics & Genomes* 13(3): 1–12.

Phipps, J.B., Robertson, K.R., Rohrer, J.R. and Smith, P.G. 1991. Origins and evolution of subfam. Maloideae (*Rosaceae*). *Systematic Botany* 1: 303–332.

Richards, C.M., Volk, G.M., Reilley, A.A., Henk, A.D., Lockwood, D.R., Reeves, P.A. and Forsline, P.L. 2009. Genetic diversity and population structure in *Malus sieversii*, a wild progenitor species of domesticated apple. *Tree Genetics & Genomes* 5(2): 339–347.

Robinson, J.P., Harris, S.A. and Juniper, B.E. 2001. Taxonomy of the genus *Malus* Mill. (*Rosaceae*) with emphasis on the cultivated apple, *Malus domestica* Borkh. *Plant Systematics and Evolution* 226(1): 35–58.

Sakurai, K., Brown, S.K. and Weeden, N. 2000. Self-incompatibility alleles of apple cultivars and advanced selections. *HortScience* 35(1): 116–119.

Volk, G.M., Henk, A.D., Richards, C.M., Forsline, P.L. and Chao, C.T. 2013. *Malus sieversii*: a diverse Central Asian apple species in the USDA-ARS national plant germplasm system. *HortScience* 48(12): 1440–1444.

Wu, S. and Chen, J. 2013. Using pullulan-based edible coatings to extend shelf-life of fresh-cut 'Fuji' apples. *International Journal of Biological Macromolecules* 55: 254–257.

Xiang, Q., Fan, L., Zhang, R., Ma, Y., Liu, S. and Bai, Y. 2020. Effect of UVC light-emitting diodes on apple juice: inactivation of *Zygosaccharomyces rouxii* and determination of quality. *Food Control* 111: 107082.

Zhang, Q., Zhou, B.B., Li, M.J., Wei, Q.P. and Han, Z.H. 2018. Multivariate analysis between meteorological factor and fruit quality of Fuji apple at different locations in China. *Journal of Integrative Agriculture* 17(6): 1338–1347.

2 Orchard Planning, Establishment and Soil Management

Suheel Ahmad, Ab Raouf Malik, Shoaib Kirmani, Sajad-un-Nabi, M. Abas Shah, Mumtaz A. Ganie and Rafiya Mushtaq

CONTENTS

2.1   INTRODUCTION

Horticulture and livestock husbandry have a major role in the economic development of temperate regions (Ahmad *et al*., 2018a; Singh *et al*., 2018). In fruit production in India, the Union Territory of Jammu and Kashmir accounts for 48% of the total temperate fruit production followed by Himachal Pradesh and Uttarakhand (Ahmad *et al*., 2017). Among these, Jammu and Kashmir has emerged as the leading apple producing region in the country, and the state has witnessed a 16 times increase in the area and 60 times increase in production from the past decade, which has led to an increase in the productivity of apples by about six times (Bhat *et al*., 2013). Although, the state production

DOI: 10.1201/9781003239925-2

in apples is the highest in the country (11.29 t ha^{-1}), however, as compared to the yields of advanced countries (55 to 60 t ha^{-1}) it is much less (Anonymous, 2009).

As per the Department of Horticulture of the erstwhile state of Jammu and Kashmir, there are more than 355000 ha under fruit orchards. There is substantial area available under these orchards (as interspaces) which can easily be utilized for growing fodder crops like forage grasses and cover crops for orchard floor management (Ahmad et al., 2021).

Apple orchards under traditional cultivation have no more than 280 plants per hectare. Monoculture of old and traditional cultivars, aging orchards and a lack of new cultivars have resulted in poor productivity and fruit quality. Increasing the number of plants per hectare by using dwarfing inducing clonal rootstocks and quality planting material (QPM) would raise productivity by 3–4 times while also improving yield quality. It will also shorten the time it takes for farmers to prepare for an early harvest.

Most of the interspaces in fruit orchards are not utilized for any productive use, hence they become exposed to weed infestations. Growing of such intercrops that are well-matched with the main crop, i.e. fruit trees, could reduce the weed infestations especially during early stages of orchard establishment and aid in increasing the yield as well as enhance soil fertility. Interculture of perennial forages in these orchards help in forage resource augmentation that could boost up the livestock sector in which a widening gap between demand and supply of fodder is the main cause of declining productivity. Such an integration of fruits and forages in the form of hortipastoral systems (fruit orchards + forage + livestock) where the alleyways between fruit tree rows are efficiently utilized for growing grass-legume mixtures. This approach has been very promising with regard to diversification and orchard floor or soil management. Hortipastoral systems demonstrate improved biomass, add organic matter and other important nutrients, offer beneficial soil micro-fauna and flora, reduce run-off and result in better infiltration (Ahmad et al., 2017; Ahmad et al., 2018b).

Orchard planning and land preparation is a prerequisite for a successful establishment of a fruit orchard. This is of paramount importance, but this aspect is often ignored by researchers and experienced orchardists alike. Establishing a fruit orchard is a long-term venture and any mistake or inattention at early stages can adversely affect the productivity and proceeds. Hence, knowledge and understanding regarding development and after-care becomes obligatory for sustainable fruit production. Fruit orchards are still being cultivated on traditional lines in India. Therefore, the aim of this chapter is to present a scientific course of action for the latest technologies for the establishment of orchards.

2.2 ORCHARD ESTABLISHMENT

Modern techniques should be considered by the orchardists for establishing orchards, especially high density or ultra high density apple orchards. One should have knowledge about the factors that should be kept in mind for selecting the site, land preparation and climatic requirements of plants. Careful planning for orchard establishment is of great importance to ensure steady high returns with the increasing age of plants. Since the production of fruits from plants is a long-term undertaking, poor initial decisions can be costly and impossible to correct later. Apple orchards under traditional cultivation have no more than 280 plants per hectare. Monoculture of old and traditional cultivars, aging orchards and a lack of new cultivars have resulted in poor productivity and fruit quality. Increasing the number of plants per hectare by using dwarf-inducing clonal rootstocks and quality planting material (QPM) would raise productivity by 3–4 times while also improving yield quality. It will also shorten the time it takes for farmers to prepare for an early harvest (Usha et al., 2015). There are many factors that should be carefully considered before establishing an orchard.

2.2.1 SITE SELECTION

One of the most important managerial decisions that an orchard manager has to make is selection of proper site which has considerable effect over the quality, productivity and life of the apple orchard. Fruit trees can be very productive if established at an appropriate site with a suitable climate. The selected site should preferably be free from frost pockets and exposure to strong winds and it should have ample sunlight and be well drained for sustainable fruit production. The basic inputs like planting material, irrigation facilities, plant protection agrochemicals and marketing channels should be accessible. Invasive species and other weeds have to be cleared, and the site should be plowed well and leveled before establishing an orchard. It is very important that the site should be free any wanted debris, etc. before the layout. Deep disk plowing followed by laser leveling should be done for virgin lands for apple tree plantations. Proper irrigation channels should be devised in the orchard. In hilly and sloping areas, contour terracing followed by leveling within the terraces should be done. For improvement of physical and chemical conditions of orchard soil before planting is taken up, it is recommended to raise a green manure crop, like vetch, pea, red or white clover, sainfoin, etc. and plow it in situ. On the boundaries of the orchard, it is essential to plant fast growing multipurpose trees and shrubs like *Salix* spp., *Morus* spp., *Ulmus* spp., *Populus* spp., etc. as wind breaks in areas where strong winds prevail. It is desirable to collect soil samples prior to planting to check the availability of essential nutrients. Soil samples should be collected from multiple locations for best results (Usha *et al.*, 2015). Preparation of the site should begin at least 6 months to 1 year before planting.

2.2.2 SELECTION OF VARIETIES/CULTIVARS AND ROOTSTOCK

The selection of proper cultivar (Table 2.1) is a vital decision and should never be taken lightly. It should not be left to the local nurserymen nor be based upon the nursery stocks that are available. It is better to place an order a year before so that the agency/propagating nursery will have sufficient time to propagate the desired cultivar on a desired rootstock and with desired feathering. Sufficient attention must be paid to the choice of rootstock (Table 2.2) because rootstocks modify

TABLE 2.1

Traditional and High Density Orcharding – A Comparison

Characteristics	Traditional orcharding	High density orcharding
Planting density	100–250 plants per hectare	400–8888 plants per hectare
Training systems	Modified central leader system, Central leader, and Open center	Tall spindle, Espalier, Vertical axis, Bi-axis, SolAxe, etc.
Rootstocks	Seedling (Crab apple, 'Maharaji', etc.)	Clonal rootstocks (M-9, M-7, MM-106, MM-111, Bud-9, etc.)
Varieties	Non-spur type ('Red Delicious', 'Golden Delicious', 'American Apirouge', etc.)	High yielding mainly spur type ('Oregon spur', 'Red Chief', 'Well Spur', 'Super Chief', 'Silver Spur'); Non-spur type ('Geromine', 'Cooper IV', etc.)
Precocity	Bearing starts after 6–10 years of plantation	Highly precocious (bearing starts from 2nd to 4th year of planting)
Productivity	Low (< 30 tons per hectare)	High (> 40 tons per hectare)
Fruit quality	Low; due to low light interception within the dense canopy	High
Input use efficiency	Low	High
Disease incidence	High; due to low air circulation within the dense canopy	Low
Mechanization	Difficult owing to spreading canopy	Easy

(Source: Mir *et al.*, 2017)

TABLE 2.2

Apple Rootstocks and Their Characteristics

Vigour	Rootstocks	Characteristics	Rootstocks
Ultra-dwarf	M-27, P-22, P-16, P-2, B-469, G-65	Cold hardy	B-491, B-490, B-9-2, P-18, P-22, K-14, Novole, Alnarp 2, Robusta 5
Dwarf	M-9, G-11, G-41, G-16, G-935, G-214	High temperature tolerant	M-7, MM-109
Semi-dwarf	M-7, MM-106, M-26, B-9, P-1, G-202, G-69, G-210, G-890	Drought tolerant	MM-111, KC-1, KC-1-48-41
Vigorous	MM-104, MM-109, MM-111	High moisture tolerant	MM-116, M-7, MM-104
Tolerant to soil pH	M-9, MM-106	Resistant to powdery mildew	P series (P-1, P-2, P-16, P-18)
Resistant to crown and root rot	B-9, B-491, MAC-9, O-3, P-2, Novole, G-30, G-65	Resistant to latent viruses	B-9, MAC-9, C-6, Novole

Source: Ahmed *et al.*, 2016; Mir *et al.*, 2017

tree morphology and fruit quality. Rootstocks also impart resistance to certain pests or diseases or abiotic stresses and also help the scion to withstand unfavorable soil and climatic conditions. As rootstocks influence tree size and vigor, hence, disease free and fresh should only be used. The rootstock should be vigorous, homogeneous and healthy. The orchardists should purchase the desired stock from certified nurseries.

2.2.3 PLANTING DESIGN AND DENSITY

Trees must be set at the optimum planting distance to obtain the most efficient and profitable use of land. In other words, the primary objective is to develop maximum bearing surface in minimum time and space. An ideal layout system is the one that accommodates optimum plant population with suitable crop geometry and where fruit trees get adequate space for their growth and development. Other factors governing this should be considerations of convenience for orchard cultural operations and aesthetic appearance. There are many orchard planting systems that are in vogue in various agro-climatic conditions. However, in India, especially in Kashmir, generally square and rectangular systems are used in commercial orcharding. The earlier practice of giving wider spacing between two trees was based on the concept that there should not be any significant interplant competition for vital resources like light, water and nutrients. With ever increasing land values, production costs and the need for early returns on invested capital, there is a worldwide tendency to use HDP. A north-south orientation of the rows is preferred over east-west orientation owing to better interception of light (Ferree and Warrington, 2003).

There are various planting designs in use.

2.2.3.1 Square System

A square system of planting is the most common system owing to easy layout. In this system, the plant to plant and row to row system is same and has the advantage of utilizing the central place between four adjacent trees to raise pollinizer cultivars for efficient pollination management. Such a layout allows intercropping in two directions.

2.2.3.2 Rectangular System

As the name suggests, each plant is plant at four corners to form a rectangle. There is unequal distribution of space per fruit tree because the spacing between rows is more than between trees.

However, the wider alleyways allow easy intercultural operations, and the interspaces could be used for growing intercrops without much effect on the fruit trees.

2.2.3.3 Hexagonal System

This system allows the fruit trees to be planted on the corners of an equilateral triangle; the six trees therefore make a hexagon with the seventh tree in the center. Although fruit trees are planted at equal distances, the layout of this system is little difficult and interculture is difficult compared to other systems of planting.

2.2.3.4 Quincunx System

Such a system is similar to the square method of planting with one extra plant in the middle of the square.

2.2.3.5 Triangular System

Fruit trees are planted in this system in the same way as the square system, with the exception that those in the even numbered rows are placed midway between those in the odd rows rather than opposite them. Each fruit tree occupies more area, but it accommodates less trees as compared to square system.

2.2.3.6 Contour System

It is commonly followed in hilly regions in which fruit trees are planted along the contour across the slope. The main principle of this system is to reduce soil erosion and help the conservation of soil moisture so as to make slopy areas suitable for raising fruits. In one of its manifestations, a terrace system is usually followed in which a flat strip of land is formed across the slope lying level along the contours. However, the spacing under the contour system is not always uniform.

2.3 ORCHARD FLOOR/SOIL MANAGEMENT IN FRUIT ORCHARDS

2.3.1 TILLAGE OR CLEAN CULTIVATION

In this method of management, regular tilling at different intervals is done in the interspaces, particularly for removal of weeds and bushes. However, this type of soil management, although commonly used, has several disadvantages. It destroys the structure of soil and removes soil organic matter due to frequent tillage (Ahmed *et al.*, 2016). It also causes injury to the feeder roots that are not too deep in the profile and fruit trees may have diminutive growth. Clean culture speeds up the process of mineralization of nutrients that later on may deplete some essential nutrients in the soil. Additionally, frequent tillage causes more soil and water loss (Merwin *et al.*, 1995; Merwin, 2003).

2.3.2 CLEAN CULTURE WITH COVER CROPS

In this system, a cover or a green manure crop is raised in the orchard interspaces. Use of leguminous cover crops like peas, cowpeas or French beans in the intercrops helps the management of apple orchards. In most places certain permanent forage cover crops like red clover, white clover, alfalfa or sainfoin are grown in the alleyways to control or avoid soil erosion and to produce fodder for livestock. Because they are leguminous crops, they establish in a short period and fix atmospheric nitrogen and conserve soil moisture. Among the various weed control measures in fruit orchards, growing perennial intercrops not only help in weed suppression but also add substantial organic matter to the soil (Merwin, 2003). Although the use of herbicides for controlling weeds is very effective, it has a negative consequence not only for animal consumption but for humans as well (Aktar *et al.*, 2009). It has been advocated that growing perennial forage crops act as cover crops and maintain continuous soil cover, diversify provisioning services (availability of fruit, fodder, fuelwood, milk, meat, etc.), reduce weed growth and improve overall soil quality (Glenn & Welker, 1989l; Granatstein and Sanchez, 2009; Ahmad *et al.*, 2018b; Ahmad *et al.*, 2021).

2.3.3 SWISS SANDWICH SYSTEM

The 'sandwich' system of orchard floor management first came from Switzerland as a new method of using crops as living mulches combined with a system of reduced/modified tillage. In this method annual or perennial crops are sown in a wide strip in the alleyway and the soil on the sides of the strip are frequently tilled (Weibel *et al.*, 2007; Tahir *et al.*, 2015). This method has been reported to provide weed control under limited costs without having any negative influence on tree growth and fruit yield (Stefanelli *et al.*, 2009). In Kashmir, a very innovative method of this sandwich system (horti-agripastoral system) is used where a three way combination is used (Figure 2.1).

In the tilled strip, garlic is grown following a proper spacing, whereas a perennial grass-legume mixture is raised in the alleyways to providing fodder to livestock (Ahmad *et al.*, 2018b). However, there are certain challenges these sandwich systems face such as using mulch mixture with negligible competition, rodent issues and managing proper vegetative growth without any negative influence on overall growth and quality of fruit (Tahir *et al.*, 2015).

2.3.4 SOD

In sloping lands a permanent cover of grass is raised in the orchard interspaces and no tillage is given. The main objectives of using sod is to reduce erosion and minimize dust, improve trafficability in the orchard, minimize compaction due to tillage machinery, encourage predator arthropod populations and suppress weeds (Hogue and Neilson, 1987). However, the main drawback of this system is that the grasses are exhaustive crops and compete with fruit trees, so a suitable grass-legume mix in appropriate proportions should be used (Ahmad *et al.*, 2021).

2.3.5 SOD MULCH

In this soil management practice the vegetation or sod is frequently cut and material is allowed to remain on the ground to act as a bio-mulch (Merwin, 2003). In both sod and sod mulch, more nitrogen application is required because grasses utilize more soil nitrogen for their growth.

FIGURE 2.1   Garlic in the tree row and grass-legume mix in the alleyway.

2.4   FORAGE COVER CROPS FOR SOIL MANAGEMENT IN ORCHARDS

The orchard floor makes up a large part of the total orchard agro-ecosystem, although not much has been reviewed with respect to the research and management of fruit tree horticulture and plant protection in terms of pest and disease management. However, there exist lot of scope and opportunities to improve overall orchard productivity through maneuvering of orchard floors the with best soil management practices, especially the use of perennial forage crops (Granatstein and Sanchez, 2009; Ahmad *et al.*, 2017; Ahmad *et al.*, 2018b; Ahmad *et al.*, 2021).

Orchard floor or soil management is, no doubt, imperative to the wellbeing and productivity of fruit orchards with management practices effecting the growth of fruit trees, yield of fruits and quality of fruits. However, out of so many soil management practices, the recommended method comprises maintaining a weed-free tree row (using either tillage, herbicides or both) and a perennial cover cropping in the alleyway (Parker *et al.*, 1993; Dabney *et al.*, 2001; Merwin, 2004; Ahmad *et al.*, 2018b). Clover (legume) and grass mixtures have been reported to increase shoot growth and yield in apple trees (Kuhn and Pedersen, 2009). It has been reported that interculture of annual crops in fruit orchards decreases soil fertility and increases the chance of soil erosion. On the other hand, perennial cover cropping maintains a continuous soil cover, increases water penetration and reduces runoff and soil erosion. Hence, cultivation of perennial forage crops in the alleyways improves overall soil quality and also helps in diversification of produce (Figure 2.2, Figure 2.3 and Figure 2.4).

FIGURE 2.2   An apple-based hortipasture system at Central Institute of Temperate Horticulture, Srinagar experimental farm.

FIGURE 2.3   Forage harvesting operations in an apple orchard.

FIGURE 2.4 *Dactylis* + red clover in an apple orchard.

Perennial forage crops are generally grown in mixtures as grass-legume mixes in optimum ratios owing to their capability to enhance forage yield and to provide fodder with balanced nutrition (improved forage quality) for livestock (Barnett and Posler, 1983; Giambalvo *et al.*, 2011; Geirus *et al.*, 2012). Orchard floor or soil management methods involving perennial grasses and legumes (hortipastoral systems, i.e., integrated fruit-fodder systems) are important not only in fruit production but for maintaining soil structure and fertility, decreasing weed invasions, harboring beneficial arthropods and reducing erosion of soil (Atucha *et al.*, 2011). However, the productivity and quality of forage crops in such fruit-based agroforestry systems could be enhanced by growing shade-loving grass and legume mixtures in optimum ratios. The selection of appropriate grass and legume species, as well as maintaining the optimal balance between the two species in grass-legume stands, is a major problem for these integrated systems (Kyriazopoulos *et al.*, 2013).

2.4.1 Fruit Orchards and Forage Production

Farmers find it challenging to designate appropriate areas for fodder and forage crops due to the restricted acreage available (Anonymous, 2012; Wani *et al.*, 2014). In the Himalayan region there is a big imbalance between fodder production and demand (Ahmad *et al.*, 2016). Around 64 lakh MT of green fodder and 35 lakh MT of dry fodder were produced in the former state of Jammu and Kashmir. However, the demand for green fodder was 139.13 lakh MT while the demand for dry fodder was 58.53 lakh MT (Anonymous, 2013). As a result, higher fodder production is required to meet the nutritional needs of cattle and to improve declining productivity. Furthermore, the agro-climatic conditions in the majority of the Himalayan region need the cultivation of even more fodder for storage as silage for lean periods (prolonged winter).

Livestock, as well as horticulture, play a predominant role in the economic development of the Himalayan region, and both sectors form an essential part of agriculture. More than 70 percent of the population depends on agriculture as their main occupation. According to the state's 18th live-stock census (Wani and Wani, 2007), the state had 3.45 million cattle, 1.05 million buffalo, 3.68 million sheep and 2.07 million goats. In the last decade the total cattle population has remained almost stagnant but the annual growth rate in crossbred cattle was 4.49 percent per annum. Buffalo has registered a growth rate of 2.94 percent per annum during 1997–2007, whereas sheep and goats had a growth rate of 1.5 and 1.05 percent per annum respectively. The fodder scenario of the erstwhile state of Jammu and Kashmir with such a burgeoning population of livestock has been presented in Table 2.3.

TABLE 2.3

Fodder Balance Sheet on Dry Matter Basis ('000 tons)

Region	Availability	Requirement	Deficit
Jammu	5545.58	8188.00	2642.42 (32.27)
Kashmir	1866.49	3635.00	1768.51 (48.65)
Ladakh	108.57	740.00	631.43 (85.32)
Jammu and Kashmir	7420	12563	5142.36 (40.93)

Note: Figures in parenthesis represent percentage
Source: Wani *et al.*, 2014

There is a smaller scope of increasing forage availability by bringing more area under fodder cultivation. Hence, it becomes more important that efforts are made to enhance fodder resources through development of alternate land use systems. Horticulture has a special importance for the economy of people in hilly regions. In the fruit map of India, the state of Jammu and Kashmir accounts for 48 percent of total temperate fruit production followed by Himachal Pradesh and Uttarakhand. As per the department of Horticulture of the erstwhile state of Jammu and Kashmir, there are more than 355000 Ha under fruit orchards. There is substantial area available under these orchards (as interspaces) which can easily be utilized for the growing fodder crops like forage grasses and cover crops that can lead to orchard floor management (Wani *et al.*, 2014). Such orchards can serve as niche areas for augmentation of forage resources that could improve the declining livestock productivity by providing nutritious fodder which is a main constraint hampering the development of livestock sector in the Himalayan region. It is planned to establish perennial grasses and legumes in phases to utilize available orchard land for many benefits under this fruit and forage integration program (Table 2.4).

The region has a great potential for production of various temperate horticultural crops. It has created a unique distinction in production of quality apples, pears and other dry fruits like almonds and walnuts. Agro-climatic conditions in the region range from sub-tropical to sub-temperate to temperate to cold dry. Each agro-climatic zone has its own capacity for growing specific fruits which allows growing a wide range of fruits over the majority of the year. (Table 2.5).

Green fodder demand is increasing, especially during the winter, due to the deterioration of grazing areas and the lack of considerable expansion of other forage resources. A horticulture system incorporating the integration of fruit trees with pasture (grass and/or legume) could be created utilizing a proper processes to supplement fodder resource availability (Sharma, 2004; Kumar and Chaubey, 2008; Khan and Kumar, 2009). Intervention of pasture components in the interspaces of fruit plants (Table 2.6) is useful not only for proper orchard floor management but also for supplying forage to livestock.

Misri (1988) conducted a study on the green forage yield of various combinations of grasses and legumes in an apple orchard and reported that rye grass + red clover and *Dactylis* + red clover combinations recorded green forage yield of 48 and 42 tons per hectare, respectively.

Singh (1995) has reported that orchard and rye grasses have been found to be the best grass species and clovers and alfalfa the ideal perennial forage legumes for growing in the interspaces of apple trees. Makaya and Gangoo (1995) reported that the average green fodder yield of orchard grass, red clover and white clover was 22.03, 24.96 and 24.58 tons per hecatare, respectively, as compared to the natural vegetation (14.64 t ha^{-1}) while working on the yield of perennial grasses and legumes in almond orchards.

Mughal and Bhattacharya (2002) identified several agroforestry systems in the North Western Himalayan region that included plantations of trees on the boundaries, agri-silviculture, horti-silviculture, horti-pasture, horti-silvi-agriculture and kitchen gardens. Ram *et al.* (2005), while working on the performance of *Ziziphus mauritiana* based hortipasture systems, found that growing Guinea

TABLE 2.4
Area Under Fruits in Jammu and Kashmir (area in 000 hectares)

Year	Apple	Pear	Apricot	Cherry	Other fresh	Walnut	Almond	Other dry	Total fruits
2004–05	107.93	10.54	4.93	2.55	41.62	74.89	15.43	0.42	258.31
2005–06	111.88	11.00	5.16	2.59	43.60	77.22	15.55	0.41	267.41
2006–07	119.04	11.25	5.43	2.75	46.24	81.39	16.37	0.62	283.09
2007–08	127.80	12.10	4.78	3.14	48.32	82.05	16.40	0.55	295.14
2008–09	133.10	12.35	4.92	3.30	49.65	84.56	17.18	0.56	305.62
2009–10	138.19	12.55	5.00	3.41	50.57	87.28	17.54	0.60	315.14
2010–11	141.71	12.53	5.85	3.46	53.50	89.78	17.65	0.58	325.06
2011–12	154.72	13.21	6.05	3.48	54.11	83.61	16.41	11.19	342.78

Source: Anonymous, 2012–13; Malik and Choure, 2014

TABLE 2.5
Suitable Fruit Crops for Different Zones

Zone	Districts	Fruit crop
Temperate zone	Anantnag, Pulwama, Shopian, Budgam, Baramulla, Srinagar, Ganderbal, Kulgam, Kupwara and Budgam	Apple, Almond, Pear, Cherry, Walnut and Apricot
Intermediate zone	Doda, Rajouri, Poonch and parts of Udhampur	Peach, Plum, Apricot, Olive and pomegranate
Sub-tropical zone	Jammu, Kathua, Samba and parts of Udhampur	Mango, Citrus, Ber and Aonla
Cold arid zone	Leh and Kargil	Apricot and Apple

Source: Ahmad and Verma, 2011

TABLE 2.6
Suitable Grasses and Legumes for Apple Orchards

Zone	Grasses	Legumes
Temperate zone	*Dactylis glomerata, Festuca arundinacea, Lolium perenne, Phleum pratense, Bromus unioloides, Phalaris spp., Poa pratensis, Lolium multiflorum, Agrostis spp. Avena sativa*	*Trifolium pratense, T. repens, Onobrychis viciifolia, Medicago sativa, Trifolium alexandrinum*
Intermediate zone	*Dactylis glomerata, Festuca arundinacea, Lolium perenne, Dicanthium annulatum, Chloris gayana, Chrysopogon fulvus, Heteropogon contortus, Setaria spp., Avena sativa*	*Trifolium alexandrinum, Stylosanthus hamata, Macroptelium atropupreum*
Sub-tropical zone	*Dicanthium annulatum, Chloris gayana, Chrysopogon fulvus, Heteropogon contortus, Cenchrus ciliaris, C. setigerus, Paspalum notatum, Avena sativa*	*Trifolium alexandrinum, Stylosanthus hamata, Stylosanthus scabra*
Cold arid zone	*Festuca arundinacea, Avena sativa, Phalaris spp., Dactylis glomerata*	*Medicago sativa, Medicago falcate, Lotus corniculatus, Astragalus spp., Caragana spp., Melilotus officinalis, Cicer microphyllum*

Source: Ahmad *et al.*, 2016

grass in combination with Carribean stylo resulted in significantly produced forage with higher crude protein yields compared to Guinea grass + Carribean stylo + Dinanath grass + Carribean stylo and natural pasture. Ram *et al.* (2006) conducted a study on Annona-based hortipasture systems and found that the total crude protein yield was significantly increased in intercropping of *Stylosanthus hamata* + buffel grass as compared to *Stylosanthus scabra.*

Ram and Parihar (2008) reported that growing *Chrysopogon fulvus* with *Stylosanthus hamata* in 1:1 row proportion resulted in increased forage yield as compared to monocropping of either grass or legume species.

Grass-legume intercropping, in addition to increasing the biomass production, provides better quality forage rich in protein and carbohydrates which would be helpful to improve milk production and animal health (Ram, 2008; Ram and Parihar, 2008).

Forage legumes, especially white clover (*Trifolium repens* L.) and red clover (*Trifolium pratense* L.) have been utilized for improving the forage production and soil nitrogen status throughout the world. White clover is a stoloniferous perennial forage legume with a well-developed root system (Baker and Williams, 1987; Frame *et al.*, 1998). On the other hand, red clover is also a perennial legume with an erect growth habit having fusiform roots and is drought tolerant owing to a deep taproot system. It provides high quality forage having high suitability and nutritional value (Thomas, 2003).

Tall fescue (*Festuca arundinaceae*) is a very productive perennial bunchgrass with deep roots (Stephenson and Posler, 1988) and is adapted to a wide range of climatic conditions (Charlton and Stewart, 2006). Its chief characteristics are drought and heat tolerance, good growth during summers (Reed, 1996) and compatibility with legumes (Lowe and Bowdler, 1995). Orchard grass (*Dactylis glomerata* L.) is also a widespread perennial bunchgrass that is well-suited to shaded conditions (Lin *et al.*, 1999; Devkota *et al.*, 2009). It is considered one of the most suitable grasses in temperate agroforestry systems (Peri *et al.*, 2007; Koukoura and Kyriazopoulos, 2007; Kyriazopoulos *et al.*, 2013).

Cover cropping in almond orchards has been reported to improve soil quality in comparison to tillage by enhancing organic matter, aggregate stability and microbial activity (Ramos *et al.*, 2010). Hoagland *et al.*, (2008) have reported that growing mixed legumes in an apple orchard enhanced organic carbon and total nitrogen in soil.

Several studies have shown that the roots of fruit trees grew deeper and had better growth under grass covers than pre-herbicide treatments (Yao *et al.*, 2009).

Other Paybacks of Grass-Legume Mixtures in Fruit Orchards
- It improves soil fertility and improves soil tilth, structure and water-retention capacity.
- It promotes the growth and multiplication of beneficial soil organisms
- It aids in capturing and recycling soil nutrients.
- It helps reduce soil erosion
- It helps reduce the runoff thereby improving water quality.
- It helps with weed suppression.
- It improves orchard access during wet weather.

2.4.2 COVER CROPS IN DISEASE MANAGEMENT

The use of cover crops has been studied in great detail in annual crops for management of pathogens, but little has been done to identify the effect of cover cropping in perennial systems. Generally, such crops are grown between seasons during a small window period and mixed into the soil in situ as green manure. This approach has been quite useful in annual cropping systems, but in perennial systems the crops remain in the field for several years. The yield of perennial crops often reduces with time due to amassing soil-borne pathogens in the root zone of plants (Urbez-Torres *et al.*, 2014). Mazzola and Manici, 2012 have reported decreased crop yield and replant issues in perennial monocultures. The detrimental soil microbial community restrains the growth of neighboring crop plants (Civitello *et al.*, 2015). In less diverse systems, a parasite of the host finds it easier to find a suitable host. Increased diversity in production systems, like agroforestry systems, makes hosts more difficult to locate and disease outbreaks become less frequent which leads to the 'dilution effect' linked with more biodiversity (Keesing *et al.*, 2011). Apart from abiotic filters contributing to the formation of soil microbial communities (Lauber *et al.*, 2008), it has been reported that the ability of plants to "train" their associated microbial communities is of paramount importance (Fanin *et al.*, 2016). However, there are limitations to control the diversity of crop plants in cropping systems; cover crop species characteristics and diversity could be an efficient way to enrich soil microbial diversity and suppress soil-borne diseases that cause production losses (Garbeva *et al.*, 2004). Cover cropping can be done before planting or during production in the interspaces (between plants). The perennial forage cover crops can be cut at any stage if they are found to affect the main crop. The addition of organic matter to the soil by cover crops has been found to improve the competitiveness of beneficial, non-pathogenic microbes (Finney *et al.*, 2017). These beneficial microbes could aid in out-competing soil-borne pathogens, thereby helping in protecting the main crop from diseases (Chen *et al.*, 2011).

2.4.3 COVER CROPS AND PEST MANAGEMENT

The most popular method of orchard floor management is the development of ground vegetation alleys between tree rows with vegetation-free strips in the tree rows (Tahir *et al.*, 2015; Saunders and Luck, 2018). A variety of cover crops, mostly grasses, are planted and maintained in the alley between the tree rows. Ground cover provides additional sources of nectar, pollen and shelter for arthropods over time which in general leads to an increase in diversity and an abundance of insects and other arthropods, both beneficial and harmful. Ground cover flora may provide optional/additional food sources to natural enemies, act as their breeding site, provide shelter for over-seasoning and act as refuge from potentially harmful crop management practices (Mailloux *et al.*, 2010; Simon *et al.*, 2010; Saunders *et al.*, 2013). Therefore, ground covers usually support a higher number and kind of natural enemy and pollinator populations. Orchard ground covers have been extensively studied for their agronomic aspects in terms of their effect on water and nutrient management in orchards, soil properties and suitability of various grasses/cover crops for cultivation in orchards. Habitat manipulation studies focused on the effect of understory vegetation on beneficial and harmful arthropods in orchard systems have been reviewed by Prokopy (1994); Bugg and Waddington (1994); Landis *et al.* (2000); Simon *et al.* (2010, 2011) and Herz *et al.* (2019). In general, most of the studies are based on the manipulation of understory plants or plant assemblages or the naturally occurring ground cover grasses and other weeds. Different assemblages are reported to help control a variety of insect and mite pests through enrichment of natural enemy diversity and abundance. For example, grassy and flower strips are believed to reduce the incidence of rosy apple aphid, *Dysaphisplantaginea* (Wyss *et al.*, 1995; Cahenzli *et al.*, 2019); intercropping with aromatic plants such as ageratum, French marigold and basil are reported to help reduce the incidence of spiraea aphid, *Aphis spiraea* in apple orchards (Song *et al.*, 2013) and ground covers with buckwheat, phacelia or alyssum are reported to help control *Tortricidae* (Irvin *et al.*, 2006). Living mulches within almond orchards have been found to provide habitat and resources for potential wild pollinators, especially native bees (Saunders *et al.*, 2013). Such insects provide precious ecosystem services, especially to pollinator-dependent crops. Therefore, evidence suggests that effectively designed and properly chosen wildflowers (legumes like sainfoin, birdsfoot trefoil, red and white clover) in the alley ways of apple orchards can contribute to the enrichment of wild bees and other pollinators. Therefore, orchard management can be optimized to enable the ground cover vegetation for effective pest regulation and pollination services on case specific basis.

2.5 CONCLUSION

There is great scope for expansion of livestock-oriented activities in the Himalayan region because significant quantities of the demand of livestock produce is met through import from neighboring states of Haryana, Punjab, Delhi and Rajasthan. Due to a lack of fodder and the inability to increase the area under fodder cultivation due to demographic pressures, the growing of perennial grasses and legumes in the interspaces of fruit orchards offer a unique opportunity to mitigate the fodder shortages and, at the same time, help manage the orchard floor in a sustainable manner. Lot of information is available on cultivation of green fodder in irrigated areas, but more emphasis should be on integration of fruits and forages. Scanty data is available on fruit tree and forage crop interactions in temperate agroforestry systems. Hence, we believe that more research is needed in exploring new species introductions with appropriate mixtures to optimize production and sustainability of these systems. The trade-offs and synergies of fruit tree-forage associations need to be studied in detail covering a wide range of aspects like the effects on weed suppression, soil quality and buildup of soil fertility, and the interactions should be exploited by evaluating different forage species under given soil and climatic conditions (Jose *et al.*, 2004). The promotion of fodder crop output through improved agronomic practices and the use of modified seed is critical for the development of horti-pastoral models. There should be focused and coordinated programs on establishment, development and management of hortipastures. To encourage more farmers/orchardists and educated youth in the animal husbandry sector, hortipastoral models need to be developed along with introduction of fodder conservation techniques. Such an approach would augment the supply of nutritious fodder thereby ensuring sustainable livestock production in the region.

REFERENCES

Ahmad, M.F. and Verma, M.K. 2011. Temperate fruit scenario in Jammu and Kashmir: Status and strategies for enhancing productivity. *Indian Horticulture Journal* 1(1): 1–9.

Ahmad, S., Bhat, S.S. and Mir, N.H. 2021. Intercropping in almond orchards with grasses/legumes enhanced soil fertility and weed suppression in a temperate region. *Range Management and Agroforestry* 42(1): 30–37.

Ahmad, S., Khan, P.A., Mughal, A.H., Qaiser, K.N., Zaffar, S.N., Mir, N.H. and Bhat, S.S. 2018a. Evaluation of apple based hortipastoral systems in Kashmir Himalaya. *Multilogic in Science* 25: 308–310.

Ahmad, S., Khan, P.A., Verma, D.K., Mir, N.H., Sharma, A. and Wani, S.A. 2018b. Forage production and orchard floor management through grass/legume intercropping in apple based agroforestry systems. *International Journal of Chemical Studies* 6: 953–958.

Ahmad, S., Khan, P.A., Verma, D.K., Mir, N.H., Singh, J.P., Dev, I. and Roshetko, J. 2017. Scope and potential of hortipastoral systems for enhancing livestock productivity in Jammu and Kashmir. *Indian Journal of Agroforestry* 19: 48–56.

Ahmad, S., Singh, J.P., Khan, P.A. and Ali, A. 2016. Pastoralism and strategies for strengthening rangeland resources of Jammu and Kashmir. *Annals of Agri-Bio Research* 21(1): 49–54.

Ahmed, N., Singh, D.B., Srivastava, K.K., Mir, J.I., Sharma, O.C., Sharma, A. and Raja, W.H. 2016. Scientific apple cultivation. *CITH Technical Bulletin* 01/2016. Srinagar, India: ICAR-CITH, 48p.

Aktar, W., Sengupta, D. and Chowdhury, A. 2009. Impact of pesticides use in agriculture: Their benefits and hazards. *Interdisciplinary Toxicology* 2(1): 1–12.

Anonymous. 2009. *Production and Area Statements for 2008–2009*. Department of Horticulture, Government of Jammu & Kashmir, pp. 139–154.

Anonymous. 2012. *Digest of Statistics*. Directorate of Economics and Statistics, Planning and Development Department, Government of Jammu & Kashmir.

Anonymous. 2013. JK targets 20 lakh mt milk production this fiscal. *The Greater Kashmir*, 23rd March 2013. http://jammu.greaterkashmir.com/news/2013/Mar/23/jk-targets-20-lakh-mt-milk-production-this-fiscal-50.asp

Atucha, A., Merwin, I.A. and Brown, M.G. 2011. Long-term effects of four groundcover management systems in an apple orchard. *HortScience* 46(8): 1176–1183.

Baker, M.J. and Williams, W.M. 1987. *White Clover*. Oxon, UK: CAB International, 534p.

Barnett, F.L. and Posler, G.L. 1983. Performance of cool-season perennial grasses in pure stands and in mixture with legumes. *Agronomy Journal* 75: 582–586.

Bhat, R., Wani, W.M., Banday, F.A. and Sharma, M.K. 2013. Effect of intercrops on growth, productivity, quality and relative economic yield of apple cv. Red Delicious. *SKUAST Journal of Research* 15(1): 35–40.

Bugg, R.L. and Waddington, C. 1994 Using cover crops to manage arthropod pests of orchards: a review. *Agriculture, Ecosystems & Environment* 50: 11–28.

Cahenzli, F., Sigsgaard, L., Daniel, C., Herz, A., Jamar, L., Kelderer, M., Jacobsen, S.K., Kruczyńska, D., Matray, S., Porcel, M. and Sekrecka, M. 2019. Perennial flower strips for pest control in organic apple orchards: A pan-European study. *Agriculture, Ecosystems & Environment* 278: 43–53.

Charlton, D. and Stewart, A. 2006. *Pasture and Forage Plants for New Zealand*. Dunedin, New Zealand: Grassland Association and New Zealand Grassland Trust, 96p.

Chen, L.H., Yang, X.M. Raza, W. Luo, J. Zhang, F.G. Shen, Q.R. 2011. Solid-state fermentation of agro-industrial wastes to produce bioorganic fertilizer for the biocontrol of Fusarium wilt of cucumber in continuously cropped soil. *Bioresource Technology* 102: 3900–3910.

Civitello, D.J., Cohen, J., Fatima, H., Halstead, N.T., Liriano, J., McMahon, T.A., Ortega, C.N., Sauer, E.L., Sehgal, T., Young, S., Rohr, J.R. 2015. Biodiversity inhibits parasites: Broad evidence for the dilution effect. *Proceedings of National Academy of Sciences USA* 112: 8667–8671. doi:10.1073/pnas.1506279112

Dabney, S.M., J.A. Delgado, and D.W Reeves. 2001. Using winter cover crops to improve soil and water quality. *Communications in Soil Science and Plant Analysis* 32: 1221–1250.

Devkota, N.R., Kemp, P.D., Hodgson, J., Valentine, I. and Jaya, I.K.D. 2009. Relationship between tree canopy height and the production of pasture species in a silvopastoral system based on alder trees. *Agroforestry Systems* 76(2): 363–374.

Fanin, N., Fromin, N. and Bertrand, I. 2016. Functional breadth and home-field advantage generate functional differences among soil microbial decomposers. *Ecology* 97: 1023–1037. doi:10.1890/15-1263.1

Ferree, D.C. and Warrington, I.J. 2003. *Apples: Botany, Production, and Uses*. Oxon, UK: CABI Publishing, 660p.

Finney, D.M., Buyer, J.S. and Kaye, J.P. 2017. Living cover crops have immediate impacts on soil microbial community structure and function. *Journal of Soil Water Conservation* 72: 361–373.

Frame, J., Charlton, J.F.L. and Laidlaw, A.S. 1998. *Temperate Forage Legumes*. Oxon, UK: CAB International, 327p.

Garbeva, P., van-Veen, J.A. and van Elsas, J.D. 2004. Microbial diversity in soil: Selection of microbial populations by plant and soil type and implications for disease suppressiveness. *Annual Review of Phytopathology* 42: 243–270.

Geirus, M., Kleen, J., Logus, R. and Taube, F. 2012. Forage legume species determine the nutritional quality of binary mixtures with perennial ryegrass in the first production year. *Animal Feed Science and Technology* 172(2012): 150–161.

Giambalvo, D., Ruisi, P., Di Miceli, G., Frenda, A.S. and Amato, G. 2011. Forage production, N uptake, N2 fixation, and N recovery of berseem clover grown in pure stand and in mixture with annual ryegrass under different managements. *Plant Soil* 342: 379–391.

Glenn, D.M. and Welker, W.V. 1989. Sod proximity influences the growth and yield of young peach trees. *Journal of the American Society for Horticultural Science* 114(6): 856–859.

Granatstein, D. and Sanchez, E. 2009. Research knowledge and needs for orchard floor management in organic tree fruit systems. *International Journal of Fruit Science* 9(3): 257–281.

Herz, A., Cahenzli, F., Penvern, S., Pfiffner, L., Tasin, M. and Sigsgaard, L. 2019. Managing floral resources in apple orchards for pest control: Ideas, experiences and future directions. *Insects* 10(8): 247. doi:10.3390/insects10080247

Hoagland, L., Carpenter-Boggs, L., Granatstein, D., Mazzola, M., Smith, J., Peryea, F. and Reganold, J.P. 2008. Orchard floor management effects on nitrogen fertility and soil biological activity in a newly established organic apple orchard. *Biology and Fertility of Soils* 45: 11–18.

Hogue, E.J. and Neilson, G.H. 1987. Orchard floor vegetation management. *Horticultural Reviews* 9: 377–430.

Irvin, N.A., Scarratt, S.L., Wratten, S.D., Frampton, C.M., Chapman, R.B. and Tylianakis, J.M. 2006. The effects of floral under storeys on parasitism of leafrollers (Lepidoptera: Tortricidae) on apples in New Zealand. *Agricultural and Forest Entomology* 8: 25–34.

Jose, S., Gillespie, A.R. and Pallardy, S.G. 2004. Interspecific interactions in temperate agroforestry. *Agroforestry Systems* 61(1): 237–255.

Keesing, F., Oberoi, P., Vaicekonyte, R., Gowen, K., Henry, L., Mount, S., Johns, P., Ostfeld, R.S. 2011 Effects of garlic mustard (*Alliaria petiolata*) on entomopathogenic fungi. *Ecoscience* 18: 164–168. doi:10.2980/18-2-3385

Khan, T.K. and Kumar, S., Eds. 2009. *Physical Hortipastoral Models, their Management and Evaluation*. Jhansi, India: IGFRI.

Koukoura, Z. and Kyriazopoulos, A.P. 2007. Adaptation of herbaceous plant species in the understorey of *Pinus brutia. Agroforestry Syst*ems 70(1): 11–16.

Kuhn, B.F. and Pedersen, H.L. 2009. Cover crop and mulching effects on yield and fruit quality in un-sprayed organic apple production. *European Journal Horticulture Sciences* 74: 247–253.

Kumar, S. and Chaubey, B.K. 2008. Performance of aonla (*Emblica officinalis*)-based hortipastoral system in semi-arid region under rainfed situation. *Indian Journal of Agricultural Sciences* 78(9): 748–751.

Kyriazopoulos, A.P., Abraham, E.M., Parissi, Z.M., Koukoura, Z. and Nastis, A.S. 2013. Forage production and nutritive value of *Dactylis glomerata* and *Trifolium subterraneum* mixtures under different shading treatments. *Grass and Forage Science* 68(1): 72–82.

Landis D.A., Wratten, S.D. and Gurr, G.M. 2000. Habitat management to conserve natural enemies of arthropod pests in agriculture. *Annual Review of Entomology* 45(1): 175–201.

Lauber, C.L., Strickland, M.S., Bradford, M.A. and Fierer, N. 2008. The influence of soil properties on the structure of bacterial and fungal communities across land-use types. *Soil Biology and Biochemistry* 40: 2407–2415. doi:10.1016/j.soilbio.2008.05.021

Lin, C.H., McGraw, R.L., George, M.F. and Garrett, H.E. 1999. Shade effects on forage crops with potential in temperate agroforestry practices. *Agroforestry Systems* 44(2): 109–119.

Lowe, K.F. and Bowdler, T.M. 1995. Growth, persistence, and rust sensitivity of irrigated perennial temperate grasses in the Queensland subtropics. *Australian Journal of Experimental Agriculture* 35(5): 571–578.

Mailloux, J., Le Bellec, F., Kreiter, S., Tixier, M.S. and Dubois, P. 2010. Influence of ground cover management on diversity and density of phytoseiid mites (Acari: Phytoseiidae) in Guadeloupean citrus orchards. *Experimental and Applied Acarology* 52(3): 275–290.

Makaya, A.S. and Gangoo, S.A. 1995. Forage yield of pasture grasses and legumes in Kashmir valley. *Forage Research* 21(3): 152–154.

Malik, Z.A. and Choure, T. 2014. Horticulture growth trajectory evidences in Jammu and Kashmir (A lesson for apple industry in India). *Journal of Business Management & Social Sciences Research* 3(5): 45–49.

Mazzola, M. and Manici, L.M. 2012. Apple replant disease: role of microbial ecology in cause and control. *Annual Review of Phytopathology* 50(50): 45–65. doi:10.1146/annurev-phyto-081211-173005

Merwin, I.A. 2003. Orchard floor management systems. In: D.C. Ferree and I.J. Warrington (Eds.), *Apples: Botany, Production and Uses* (pp. 303–315). Warrington, UK: CABI.

Merwin, I.A. 2004. *Groundcover Management Effects on Orchard Production, Nutrition, Soil and Water Quality* (pp. 25–29). Ithaca, NY: New York Fruit Quarterly.

Merwin, I.A., Rosenberger, D.A., Engle, C.A., Rist, D.L. and Fargione, M. 1995. Comparing mulches, herbicides and cultivation as orchard ground cover management systems. *HortTechnology* 5(2): 151–158.

Mir, J.I., Singh, D.B., Ahmed, N., Sharma, O.C., Sharma, A., Srivastava, K.K., Lal, S., Kumawat, K.L., Raja, W.H., Jan, A. and Kirmani, S.N. 2017. *High Density Plantation in Apple*. Srinagar, India: ICAR-CITH, 17p.

Misri, B. 1988. Forage production in alpine and sub-alpine region of North West Himalaya. In: P. Singh (Ed.), *Pasture and Forage Crops Research* (pp. 43–55). Jhansi: RMSI, IGFRI.

Mughal, A.H. and Bhattacharya, P.K. 2002. Agroforestry systems practiced in Kashmir Valley of Jammu and Kashmir. *Indian Forester* 128(8): 846–852.

Parker, M.L., Hull, J. and Perry, R.L. 1993. Orchard floor management affects peach rooting. *Horticulture Science* 118: 714–718.

Peri, P.L., Lucas, R.J. and Moot, D.J. 2007. Dry matter production, morphology and nutritive value of *Dactylis glomerata* growing under different light regimes. *Agroforestry Systems* 70(1): 63–79.

Prokopy, R.J. 1994. Integration in orchard pest and habitat management. *Agriculture Ecosystems Environment* 50: 1–10.

Ram, S.N. 2008. Productivity and quality of pasture as influenced by planting pattern and harvest intervals under semiarid conditions. *Indian Journal of Agricultural Research* 42(2): 128–131.

Ram, S.N., Kumar, S., Roy, M.M. 2005. Performance of jujube (*Ziziphus mauritiana*)-based hortipasture system in relation to pruning intensities and grass-legume associations under rainfed conditions. *Indian Journal of Agronomy* 50(3): 181–183.

Ram, S.N., Kumar, S., Roy, M.M. and Baig, M.J. 2006. Effect of legumes and fertility levels on buffel grass *(Cenchrus ciliaris)* and Annona *(Annona squamosa)* grown under horti-pasture system. *Indian Journal of Agronomy* 51(4): 278–282.

Ram, S.N. and Parihar, S.S. 2008. Growth, yield and quality of mixed pasture as influenced by potash levels. *Indian Journal of Agricultural Research* 42(3): 228–231.

Ramos, M.E., Benitez, E., Garcia, P.A. and Robles, A.B. 2010. Cover crops under different managements vs. frequent tillage in almond orchards in semiarid conditions: Effects on soil quality. *Applied Soil Ecology* 44(1): 6–14.

Reed, K.F.M. 1996. Improving the adaptation of perennial pasture grasses. *New Zealand Journal of Agricultural Research* 39: 457–464.

Saunders, M.E. and Luck, G.W. 2018. Interaction effects between local flower richness and distance to natural woodland on pest and beneficial insects in apple orchards. *Agricultural and Forest Entomology* 20: 279–287.

Saunders, M.E., Luck, G.W. and Mayfield, M.M. 2013. Almond orchards with living ground cover host more wild insect pollinators. *Journal of Insect Conservation* 17(5): 1011–1025.

Sharma, S.K. 2004. Horti-pastoral based land use systems for enhancing productivity of degraded lands under rainfed and partially irrigated conditions. *Uganda Journal of Agricultural Sciences* 9: 320–325.

Simon, S., Bouvier, J.C., Debras, J.F. and Sauphanor, B. 2010. Biodiversity and pest management in orchard systems: A review. *Agronomy for Sustainable Development* 30: 139–152.

Simon, S., Bouvier, J.C., Debras, J.F. and Sauphanor, B. 2011. Biodiversity and pest management in orchard systems. In: E. Lichtfouse, M. Hamelin, M. Navarrete and P. Debaeke (Eds.), *Sustainable Agriculture*, Vol. 2. Dordrecht: Springer. doi:10.1007/978-94-007-0394-0_30

Singh, J.P., Ahmad, S., Radotra, S., Dev, I., Mir, N.H., Deb, D. and Chaurasia, R.S. 2018. Extent, mapping and utilization of grassland resources of Jammu and Kashmir in Western Himalaya: a case study. *Range Management and Agroforestry* 39: 138–146.

Singh, V. 1995. Technology for forage production in Hills of Kumaon. In: C.R. Hazra and Bimal Misri (Eds.), *New Vistas in Forage Production* (pp. 197–202). Jhansi: RMSI, IGFRI.

Song, B., Tang, G., Sang, X., Zhang, J., Yao, Y. and Wiggins, N. 2013. Intercropping with aromatic plants hindered the occurrence of Aphis citricola in an apple orchard system by shifting predator – prey abundances. *Biocontrol Science and Technology* 23: 381–395.

Stefanelli, D., Zoppolo, R., Perry, R. and Weibel, F. 2009. Organic orchard floor management systems for apple effect on rootstock performance in the midwestern United States. *HortScience* 44: 263–267.

Stephenson, R.J. and Posler, G.L. 1988. The influence of tall fescue on the germination, seedling growth and yield of bird's foot trefoil. *Grass and Forage Science* 43(3): 273–278.

Tahir, I.I., Svensson, S.E. and Hansson, D. 2015. Floor management systems in an organic apple orchard affect fruit quality and storage life. *HortScience* 50(3): 434–441.

Thomas, R.G. 2003. Comparative growth forms of dryland forage legumes. In: D.J. Moot (Ed.), *Legumes for Dryland Pastures* (pp. 19–26). Grassland Research and Practice Series No. 11. Wellington, New Zealand: New Zealand Grassland Association.

Urbez-Torres, J.R., Haag, P., Bowen, P. and O'Gorman, D.T. 2014. Grapevine trunk diseases in British Columbia: incidence and characterization of the fungal pathogens associated with black foot disease of grapevine. *Plant Disease* 98: 456–468. doi:10.1094/pdis-05-13-0524-re

Usha, K., Thakre, M., Goswami, A.K. and Nayan, D.G. 2015. *Fundamental of Fruit Production*. New Delhi: Division of Fruits and Horticultural Technology, Indian Agricultural Research Institute, 245p.

Wani, S.A., Shaheen, F.A., Wani, M.H. and Saraf, S.A. 2014. Fodder budgeting in Jammu and Kashmir: Status, issues and policy implications. *Indian Journal of Animal Sciences* 84(1): 54–59.

Wani, S.A. and Wani, M.H. 2007. Livestock *Crop Production System Analysis for Sustainable Production System in various Agro-Climatic Zones of J & K*. Lead paper presented in 90th Annual Conference of Indian Economic Association, October 2007.

Weibel, F., Tamm, L., Wyss, E., Daniel, C., Haseli, A. and Suter, F. 2007. Organic fruit production in Europe: Successes in production and marketing in the last decade, perspectives and challenges for the future development. *Acta Horticulture* 737: 163–172.

Wyss, E., Niggli, U. and Nentwig, W. 1995. The impact of spiders on aphid populations in a strip-managed apple orchard. *Journal of Applied Entomology* 119: 473–478.

Yao, S.R., Merwin, I.A. and Brown, M.G. 2009. Apple root growth, turnover, and distribution under different orchard groundcover management systems. *HortScience* 44(1): 168–176.

3 Recent Advances in Varietal Improvement and Rootstock Breeding

*Tabish Jehan Been, M. A. Mir, Ab Raouf Malik,
Iftisam Yaseen, Mahrukh Mir, and Rafiya Mushtaq*

CONTENTS

DOI: 10.1201/9781003239925-3

3.1   INTRODUCTION

Apples, with their most commonly accepted ancestor as *Malus sieversii* Lebed., are believed to have originated in Central Asia from where they disseminated to China and Europe through the 'Silk Route' (Juniper and Mabberley, 2006). They came as a result of domestication by interspecific hybridization (Harris *et al.*, 2002). With the exception of a few triploid and tetraploid cultivars of little mercantile importance, most of the cultivars share the same chromosome number of 2n=34 (Juniper and Mabberley, 2006). Propagation of apples was done by seeds mostly because of their abundant availability and ease of propagation. But due to their heterozygous nature, apple cuttings (to some extent) and rootstocks (largely) have been used for apple propagation for around 2000 years (Roach, 1985). Apples extended from their diversity center, roughly corresponding to present-day southwest China, Kirgizia and Kazakhstan as well as the Himalayan foothills on the other side, via Central Asia to the old Silk Road in Near East and then through the Romans and Greeks to Europe. *Malus sieversii*, a wild species found throughout Central Asia, is now considered as ancestral form. This species is diversified with wild trees having a wide range of forms, colors and tastes found in Kazakhstan and other autonomous Central Asian republics founded as a result of Soviet Union's disintegration, particularly in the Alma Ata region. Recent collection expeditions to Central Asia have confirmed that *M. sieversii* is a very diversified species that possesses all of the characteristics of *M. domestica* (Forsline *et al.*, 1994; Forsline, 1995). Vavilov (1930), in his expeditions, discovered wild Trony apples in Turkestan and Caucasus woods producing fruits of various sizes, and some were of good quality. Cross-fertilization occurred naturally with *M. orientalis*, discovered in Caucasus, on the way to the Europe. It spread eastward through southern China and eastern Asia where cross-fertilization occurred most likely with *M. prunifolia* and *M. baccata*. Dissemination occurred in a northerly route via Transcaucasus to southern Russia. Prior to Christ in the seventh century to the fifth century, there were already plantation-like fruit cultivations in the Persian Empire. The spread of apples, which are known for their enormous genetic variability, has always been coupled by suitable selection resulting in the proliferation of delicious and large-fruited forms that can be viewed as archetypes of today's cultivars. One prerequisite was that one learned how to vegetatively proliferate fruit coppices, i.e. to refine them, so as to maintain and reproduce the best apples in this way over and over again. Around 2500–3000 years ago, Babylonian gardeners may have deduced this. Apples were first domesticated in this way.

Rootstocks also hold a very important position in the plant system of temperate fruits, including apples. Rootstocks are the machinery that work continuously to provide anchorage, conduct water and minerals and, most importantly, protect itself and the above ground portion against various biotic and abiotic stresses. A large number of factors in the scion are determined by the stock including the scion growth and habit, the flowering phenology, copious flowering, the tendency of fruit set and yield potential (Nimbolkar *et al.*, 2016). Rootstocks in apples can be either of seedling origin (raised from the seeds) or can be clonally propagated (multiplied through layering, cutting or micro propagation).

3.2 APPLE VARIETAL DOMESTICATION: PAST AND PRESENT

The Romans and Greeks are thought to have practiced apple cultivation, which they extended throughout Asia and Europe as a result of their invasions and travels. Later on, cultivation was also centered on medieval monasteries. Cultivars were chosen and propagated from the beginning as grafting had been known for around 2000 years ago. Many named varieties were recognized by end of thirteenth century, and the names 'Costard' and 'Pearmain' were attained from this time. The Romans also identified the names of various apple varieties and spread them throughout Europe, particularly in the west and north (Hurt, 1916). Stock containers were discovered from drowned Viking ships (1000–500 A.D.) that included apples as a preserved food ingredient among several other commodities. Several old varieties that are still recognized today are descended from seeds of these varieties from the Middle Ages and later. The first types of apples to arrive in America were undoubtedly brought by colonists from Europe in the late Middle Ages.

There are around 7000 apple cultivars listed, although only a few are important in global production. 'McIntosh' (1800s), 'Jonathan' (1820s), 'Cox's Orange Pippin' (1830s), 'Granny Smith' (1860s), 'Delicious' (1870s) and 'Golden Delicious' (1890s) are examples of nineteenth-century cultivars (Janick *et al.*, 1996; Noiton and Alspach, 1996). The majority of them are clones selected from chance seedlings or from controlled hybridizations.

Most apple varieties were chance seedlings chosen by the fruit producers until the latter half of twentieth century. Although there are over 10000 cultivars known, only a few dozen are commercially cultivated around the world (Way *et al.*, 1990). In 1983, the world's best-known varieties were mostly chance seedlings explored in the eighteenth or nineteenth centuries, many of which were created in North America: 'Delicious' (6.3 million t, origin: United States), 'Golden Delicious' (6.3 million t, origin: United States), 'Cox's Orange Pippin' (1.7 million t, origin: England), 'Belle de Boskoop' (0.7 million t, origin: The Netherlands), 'Rome Beauty' (0.8 million t; origin: United States), 'Jonathan' (0.5 million t, origin: United States) and 'Granny Smith' (0.6 million t, origin: Australia). Apples were introduced to the United States as seeds, and seeds received from cider mills were sown resulting in millions of seedlings being cultivated and evaluated by the fruit growers. In the eighteenth century, nurseryman Jonathan Chapman of Leominster, Massachusetts, is acknowledged with disseminating apple seeds from cider mills of Western Pennsylvania into the wildernesses of Ohio, Pennsylvania and Illinois (Morgan and Richards, 1993). In the nineteenth century and the early twentieth century, the assessment of such seedlings was the most significant achievement of American pomology. The 'Delicious' apple (discovered in the Iowa in 1872) continues to dominate the US production, while the 'Golden Delicious' apple (discovered in the West Virginia in 1905) continues to dominate the Europe. Other regional cultivars include 'McIntosh' in the northeastern United States, 'Jonathan' in the Midwest, 'Granny Smith' in Australia and 'York Imperial' in Shenandoah Valley.

Thomas Andrew Knight (1759–1838), who developed first cultivars of the known parentage, is credited with introducing controlled breeding of apples. This strategy is still used in all present apple breeding studies, but it was notoriously unsuccessful in comparison to other fruits until recently. In retrospect, the reason was the poor parental choice. Regardless of fact that hundreds of seedlings of apples were named, the only major US cultivars developed from controlled breeding in first half of twentieth century were 'Idared' and 'Cortland'. Seedlings developed from controlled hybridization started to establish a position in world production of apples in last 25 years. Nine (36%) of the 25 cultivars obtained from the crosses that are developed in Japan are the seedlings of the 'Golden Delicious' (Bessho *et al.*, 1993). 'Braeburn', the latest sensation from the New Zealand, is, ironically, a chance seedling.

3.3 POPULAR APPLE VARIETIES

The past 2000 years saw thousands of varieties named and multiplied. In the third century B.C. varietal information was limited to just seven varieties which grew to about 36 varieties. Initial fruit breeders developed most of the varieties for obtaining cold hardiness, resistance to diseases and pesticide, late blooming, etc. Apple breeding programs for varietal improvement started in the American state of Iowa followed by Czechoslovakia, Germany, England and Sweden. The parents used mostly included 'Northern Spy', 'Cortland', 'Rome Beauty', 'Winesap', 'Wolf River', etc. By 1970s, the important varieties were all grown from 'Delicious', 'Golden Delicious', 'McIntosh', 'Rome Beauty', etc. It is believed that the first variety grown was 'Empire'. At that time, it was known that more than 25 years are needed until you obtain the progeny of a cross that could be utilized in commercial planting. 'Empire' was the first variety grown after a breeding program on a large scale. Similarly, another variety, 'Magness', through its evaluation process, proved that it took almost 25 years to commercialize a variety after a breeding program. The US apple market was affected by another variety that gained quite popularity, i.e., 'Granny Smith' that was discovered in Australia in the nineteenth century. With the turn of the century, people became attracted toward some other types including 'Gala', 'Braeburn' and 'Fuji'. 'Fuji' turned out to be very profitable for the growers, while 'Braeburn' couldn't survive in the market for long as it suffered from bitter pit problems.

The new era of varietal development is such where patenting of the newly developed varieties is going to play a major role in determining the final marketing ability of the new ones. The varieties ruling the supermarkets are very difficult to replace because of their high quality. The challenge for researchers is to develop varieties that are not only good quality but are disease resistant, can be stored for a long time, bears regularly and blooms late to avoid frost conditions. Adoption of molecular techniques in this regard can prove to be very helpful, but multi-location testing of newly developed cultivars will definitely take years before their final release.

3.3.1 RED DELICIOUS SPORTS

3.3.1.1 Red Delicious 'Redkan'

'Redkan' is a mutation of 'Top Red' discovered in 1998 in South Tyrol. The vigor is typical with the spur character and is a mid-early variety. Pollinizers for this variety include 'Golden Delicious', 'Idared', 'Granny Smith' and 'Gala'. Fruit color is a dark red that covers the entire surface. It's a medium-sized, long and ribbed shaped fruit with a high total sugar content. The skin is smooth and thick. Flesh is crispy and firm and greenish white in color. It has moderate resistance to sunburn.

3.3.1.2 Red Delicious Superchief 'Sandidge'

It is a mutation of 'Red Chief' and was discovered by Charles R. Sandidge (Washington state). It is a typical spur variety with a medium vigor. It is a mid-early variety with 'Granny Smith', 'Idared', 'Golden Delicious', 'Fuji' and 'Gala' as pollinizers. Fruit is flushed and striped with a deep red color. The flavor is quite sweet with a low acidity level. The size is medium-large, with distinct edges. The skin is thick and smooth with numerous white lenticels. Flesh is juicy, crispy and firm with a greenish white color. High and consistent yields, as well as early and uniform fruit coloration across the tree, is the main characteristics of this variety. Fruit shape is consistent with five visible cusps all around the calyx. Spacing within rows is minimum.

3.3.1.3 Red Delicious Scarlet Spur 'Evasni'

It is a mutation of 'Oregon Spur(r) Trumdor' from the Van Well nursery (Washington state). A spur type and low to medium vigor is characteristic of this variety. Pollinators recommended for this variety include 'Gala', 'Fuji' and 'Idared'. It is a mid-early variety tinged and bright red in color. It is medium-large and relatively flat in shape. The skin is thick and smooth with numerous lenticels which are white in color. The flesh is crispy, firm and white in color. It is a precious and high yielding variety and is harvested earlier than other clones of 'Red Delicious'. The coloration is excellent even under unfavorable conditions.

3.3.1.4   Red Delicious 'RedVelox® Stark Gugger'

Red Delicious 'RedVelox® Stark Gugger' is a mutation of the standard 'Red Delicious'. The tree has a medium vigor and blooms a few days earlier than the 'Golden Delicious'. Recommended pollinizers include 'Granny Smith', 'Idared', 'Golden Delicious' and 'Gala' apples. It gives dark red fruit with a stripped and blushed appearance. The flavor is sweet and low in acidity with a conical shape like typical 'Red Delicious'. The skin is very soft, and the flesh is crispy, firm and white in color. The entire tree has a uniform and early coloration and matures approximately 2–3 weeks before other clones. It is harvested one week before standard 'Red Delicious'. It is not susceptible to alternate bearing.

3.3.2   McIntosh

These apples have red and green color and are a perfect blend between sweet and acidic; its flavor is called vinous. They are the national fruit of Canada. Its flesh is white, and it ripens in late September. It can be stored for quite a while, but the flavor goes away very fast. It is a very popular variety of Ontario, Canada.

3.3.3   Golden Delicious

The flesh of the apple is sweet, and the flesh is full of juice, but it almost gives a taste similar to 'Red Delicious'. It is a very popular apple variety in the United States. It is very easy to propagate and has a good storage ability after harvesting. The problem it deals with is that it's normally harvested when green and kept in storage for several months before it is sold in the market, which doesn't reveal its original flavor. Otherwise, if it is harvested at its proper time, it tastes just like sugarcane.

3.3.4   Granny Smith

'Granny Smith' form the most likable variety for those consumers who love tart flavor. They are the most eaten variety of green apples. It's crisp when eaten and remains in the market year round. It was discovered in Australia in around the 1860s. The exceptional keeping quality makes it easy to transport throughout the world without being affected. When stored, the flavor tends to become sweet. But its market is also seeing a decline because of the preference of bicolored varieties in the market nowadays.

3.3.5   Braeburn

With the parents as 'Lady Hamilton' and 'Granny Smith', these apples were first discovered in New Zealand. They are a perfect blend of a sugary and tart flavor. These apples present red/orange colored streaks on a green/yellow background. They have a good storage quality and satisfies all the requirements of the international supply chains. They are crisp and juicy and are considered as a standard for other apples.

3.3.6   Pink Lady

The most distinctive property of these apples is that they have a good sugar to acid ratio. The apples that aren't properly qualified for the 'Pink Lady' are marketed as 'Cripps' apples. The parents of this variety are 'Golden Delicious' and 'Lady Williams'. They taste a bit sour initially but end up sweet. They are used in pies and eaten raw.

3.3.7 GALA SPORTS

3.3.7.1 'Premier Star'
This Royal Gala sport was introduced from New Zealand and was initially introduced with a view to get commercialized in the Asian markets. Its bright color and high sugar content make it ideal for that market.

3.3.7.2 'Galmac' ('Camelot'®)
This hybrid of 'Gala' and 'Jerseymac', developed in Switzerland, has good quality but a short harvest period for peak quality. Apples ripen in late July and are ready for sale on August 1st, Switzerland's National Day.

3.3.7.3 'Genesis'™
'Genesis'™ a new apple from the Yummy Fruit Company in New Zealand, is a red, sweet, firm and crisp hybrid of 'Braeburn' and 'Royal Gala'.

3.3.7.4 'Gradiyel'
This hybrid of 'Christmas Rose' and 'Gradigold' was developed in France and has a semi-upright habit, free flowering and fruiting and a long storage life. It was granted USPP #22,974 by the US Patent and Trademark office.

3.3.8 HONEYCRISP SPORTS

3.3.8.1 'Royal Red Honeycrisp'™ (LJ-1000)
In November 2011, this sport was discovered in Washington state and granted USPP #22,244 by the US Patent and Trademark office. LJ-1000 is said to color faster and contain more sugar than regular 'Honeycrisp'. Willow Drive Nursery is the only place to buy it, and sales began in 2013 (Lehnert, 2012).

3.3.9 JONAGOLD SPORT

3.3.9.1 'Jonastar'
'Jonastar,' which received USPP #20,590 by the US Patent and Trademark office, has a lovely red stripe on a red background. Its earlier coloring ensures that the fruits are fully colored when harvested.

3.3.10 JONATHAN SPORTS

3.3.10.1 'Chrisolyn' (Cambell Jonathan)
'Chrisolyn' is a sport of 'Robinson' and is distinguished by alternating dark and bright red stripes as opposed to the fainter stripes of the standard 'Jonathan.' USPP # 21,300 was granted in 2010 by the US Patent and Trademark office.

3.3.10.2 'Junami'® ('Milwa' or 'Diwa')
Inova Fruit manages the varieties 'Junami', 'Ruebens' and 'Wellant' varieties in the Netherlands, and it has excellent web pages for marketing and information for growers of their varieties. Rainier

Fruit sells this Swiss-developed variety in the United States where it was granted USPP# 19,615 in January 2009 by the US Patent and Trademark office. 'Junami' was created by crossing a hybrid of 'Idared' x 'Maigold' with 'Elstar,' and is marketed to young consumers for its juiciness and crunch (Smolka et al., 2010).

3.4   STRATEGIES FOR VARIETAL IMPROVEMENT

Because the basic requirements for developing new cultivars are constantly changing and becoming more difficult, for example, better performance than existing cultivars, large fruit size, early maturity, firm skin, and without resetting, better preserving capacity, high content of sugar and so on, it is necessary to implement various strategies for their improvement, some of which are listed in the next section:

3.4.1   INTRODUCTION AND ASSESSMENT OF LOW CHILLING VARIETIES

In Himachal Pradesh, climate change has resulted in a shift of apple belts, and the impacts of heat waves have resulted in the following consequences:

- Early flowering occurs.
- Due to erratic rainfall during the second fortnight of April followed by a drop in temperature, fruit set is poor.
- Optimum fruit set occurs at 24°C while in the region 26°C has been experienced for 17 days.

According to Singh *et al.* (2016), the introduction of low chilling cultivars could be a viable option for minimizing the effects of climatic change on the production of apple.

3.4.2   INTRODUCTION AND DEVELOPMENT OF REGULAR BEARING AND PRECOCIOUS VARIETY

Through selection, Das *et al.* (2012) proposed the Central Institute of Temperate Horticulture (CITH) Lodh apple, a clonally selected 'Red Delicious' cultivar that develops earlier than 'Red Delicious' and produces higher yields with better coloration in fruit in marginal chilling zones.
The following are the features of this variety.

- The variety was developed by the CITH in Srinagar in 2012. It is the clonal selection from an old plantation of 'Red Delicious'.
- It is a very early bearing, regular bearing, mid-season blooming variety. Trees have a spreading growth habit and fruits that are large-to-medium in size and oblong or conical in shape.
- The calyx end is conspicuous and bluntly lobed. The skin is yellow on the ground and dark shiny red on top with medium lenticels. The fruit stalk is short-medium in length, with a firmness of 8.5 kg/cm^2 and a total soluble solids content of 13.6 0B.
- With a long shelf life, it is resistant to major pests and diseases. The variety is ideal for cultivation in the mid to high hills of the northwestern Himalayan agro-climatic zones where low fruit set on traditional high chilling apple cultivars has been observed due to climate change.

Brown and Maloney (2015) developed a hybrid variety called 'SnapDragon', which gets its juicy crispness from its 'Honeycrisp' parent and has a spicy-sweet flavor that was a big hit with taste testers. It was introduced as "New York 1" in 2010 and is similar to 'Honeycrisp' but lacks leaf disorders and a bitter pit. It is a firm, early-ripening, high-quality variety with a high sugar level.

Kotoda *et al.* (2010) identified, isolated and characterized two apple flowering locus (FT)-like genes, MdFT1 and MdFT2, and mapped them on distinct linkage groups (LG), LG 12 and LG 4, with partial homology. MdFT1 and MdFT2 expression patterns differed in that the former was found to be primarily expressed in apical buds of fruit-bearing shoots during the adult phase, whereas the latter was found to be primarily expressed in reproductive organs. Because transgenic Arabidopsis expressing MdFT1 or MdFT2 flowered earlier than wild-type plants, both genes had the potential to induce early flowering. Furthermore, overexpression of MdFT1 altered the expression of other endogenous genes and conferred early flowering in apples indicating that MdFT1 may function by promoting flowering through gene expression changes and that other genes may play an important role in apple flowering regulation. The long juvenile period of fruit trees prevents early cropping and efficient breeding, and these findings will be useful in uncovering the molecular mechanism of flowering as well as developing new methods to shorten the juvenile period in various fruit crops.

3.4.3 DEVELOPMENT OF VARIETIES HAVING HIGH YIELD AND EXCELLENT QUALITY

Through hybridization, Brown and Maloney (2015) created the variety 'RubyFrost' (NY2), whose parents are 'Braeburn' and 'Autumn Crisp'. It was released as 'New York 2' in 2010. The fruits are high in vitamin C and have a long storage and shelf life. The apples are best eaten fresh.

3.4.4 DEVELOPMENT OF FIRM APPLE STRAINS

Atkinson *et al.* (2012) investigated the effect of suppressing the PG1 expression in the apple cultivar 'Royal Gala' (it has high levels of PG1 and typically softens during fruit ripening). They discovered that 'Royal Gala' apples harvested at different times of the year when MdPG1 is suppressed are firmer than controls. They discovered increased intercellular adhesion, and cell wall analysis revealed a higher molecular weight distribution of CDTA-soluble pectin as well as a change in yield and composition of pectin. In contrast, structural analysis revealed that ruptured cells with free juice in pulled apart sections indicated improved intercellular connection integrity and subsequent cell rupture. They also observed reduced expansion in the apple hypodermis in PG1-deficient lines resulting in densely packed cells in this layer, and this change appeared to be associated with reduced transpiration loss in fruit.

The arctic apples are non-browning apples. Using the RNAi approach, a company in Canada (Okanagan) has genetically modified 'Granny Smith' and 'Golden Delicious' by inhibiting four MdPPO genes. When compared to the control, the resulting transgenic lines ('Arctic Granny Smith' and 'Arctic Golden Delicious') showed a significant reduction in browning (Xu, 2015).

Varieties with greater adaptability and resistance to biotic and abiotic stress are being developed. In New Zealand, a new variety called 'Jazz' has been developed which is a cross between 'Braeburn' and 'Royal Gala' and is adaptable to a variety of climates. At harvest the fruit is extremely firm with a rich flavor and moderate acidity. Laurens *et al.* (2005) reported that 'Ariana', a hybrid of 'Florina' × 'Prima' developed in France, has good resistance to powdery mildew and fire blight. It possesses the V_f-gene, which confers resistance to *Venturia inequalis*. It has a bright blush colored skin with excellent flavor, eating quality and storage life.

3.5 SEEDLING AND CLONAL ROOTSTOCKS

Rootstocks raised from the seeds have the advantage of having a deep tap root system that can penetrate up to 20 feet (6 meters) in the ground if provided with proper conditions of soil and moisture for growth (Stone and Kalisz, 1991). These tap roots provide a good anchorage to support a huge tree and act as a source of protection during the times of less water availability or drought. A tree on a full sized rootstock can live up to 100 years. However, the major problem associated with the seedling rootstocks is the long juvenile pre-bearing period apart from the highly heterozygous nature of the tree formed with variability in both scion characteristics (fruit size and color) as well as the stock characteristics (root architecture, edaphic factors and tolerance to various pests and diseases). Also, the inability to plant these rootstocks in intensive orchard management systems and the difficulty in harvesting and management adds to their disadvantages. Most commonly used seedlings for raising such rootstocks include those of 'Golden Delicious', 'Yellow Newton', 'Wealthy', 'McIntosh', 'Granny Smith', 'Maharaji' and 'Red Gold' (all diploid) relative to the triploid ones (Kanwar, 1987).

On the other hand, the progenies raised through clonal/dwarfing rootstocks are true to type and produce uniform, comparatively smaller, productive trees that offer resistance to a large number of diseases including fire blight (*Erwinia amylovora*), wooly aphid (*Eriosoma lanigerum*) and root rot (*Phytophthora* sp.) viral infections. Many are able to tolerate adverse abiotic stresses as well (drought, extreme edaphic conditions, etc.). These superiorities make them more acceptable in the commercial apple fruit industry. But the flaws associated with them include shallow adventitious roots (about 1 meter deep) which are not able to provide a good anchorage needed to support a heavy fruit load on the tree. The most common problems of many of the clonal rootstocks are a weak graft union, propagation difficulty and liability to be influenced by various biotic and abiotic stresses (Marini and Fazio, 2018). In spite of their disadvantages, clonal rootstocks hold a promising future, and improvement work is happening continuously through various breeding programs. Some of these breeding programs have been terminated and many are still working on various breeding objectives.

3.6   APPLE ROOTSTOCK BREEDING OBJECTIVES

Improvement/breeding of clonal apple rootstocks (either through conventional breeding methods or through the biotechnological interventions), which is the current need of the fruit industry, is mostly done for the following objectives for rootstocks:

- Freestanding and free from suckering
- Precocious and widely adaptable
- Good anchorage to reduce the cost of a support system
- Easy to propagate
- Dwarfing
- Fully graft compatible
- Strong drought tolerance and resistance to weed competition for soil and water minerals
- Resistant to woolly apple aphid
- High resistance to severe winter cold
- Resistant to fire blight and collar rot
- Resistant to replant problem.

3.7   HISTORY OF APPLE ROOTSTOCK BREEDING

Even though apples had been domesticated for thousands of years, history had not seen the initiation of a proper breeding program for the rootstocks until the East Malling Research Program which started in Kent, England, initially by the director of East Malling Research Station, R. Wellington (1912) and carried on by R. G. Hatton (1914). Their program was based on the selection of already available material that had been collected from over 71 nurseries from England, Holland, France and Germany (Bunyard, 1920). This selection was very important because the already available material were called 'Paradise' apples which were further divided into two categories: 'French Paradise' and 'Broad-leaf Paradise'. These apples were traded across Europe without any information about their actual characteristics (Bunyard, 1920). Hence, EMR program was the first program associated with the selection and characterization of the rootstocks. It was in 1917 that the Malling series was produced from Type I to Type IX, and its name was gradually changed from the EM series to M series while the Roman numerals were changed to Arabic numerals (Marini and Fazio,

2018) (Table 3.1 and Table 3.2). Other rootstocks produced through crossings were released in the later years until the 1970s.

3.7.1 Subclones of M-9

It was the propagation problem in the Malling series rootstocks in the stool beds that caused horticulturists to opt for subcloning within the current series. The subclones formed as a result only differed in their growth characteristics from 10–15% compared to M-9. Their origin is a deciding factor. The most versatile of these clones is M9T337 which is used as a benchmark for M9. Flueren 56 is the most dwarfing clone tested so far (Table 3.3).

3.7.2 Malling Merton Series

After the release of the rootstocks for the Malling series, the attention of the scientists was drawn toward woolly apple aphids (WAA) which were very prevalent in most of the United States of America. As a result, a joint program was started by M. B. Crane of John Innes Horticultural Institute and H. M. Tydeman at Merton (England) and East Malling respectively (Preston, 1956).

TABLE 3.1

The First Nine Paradise Malling Apple Rootstock Selections and Their Renaming

Type	Common Name/Original Name	Malling Clone Name	Vigor
1	'Broad Leaved English Paradise'	M.1	Vigorous
2	'Doucin'	M.2	Vigorous
3	No name	M.3	Semi-dwarf
4	No name	M.4	
5	'Doucin Ameliore' (Improved Doucin)	M.5	Vigorous
6	'Rivers' Nonsuch Paradise'	M.6	Very vigorous
7	No name	M.7	Semi-dwarf
8	'French Paradise' (Clarke Dwarf)	M.8	Dwarf
9	'Jaune de Metz'	M.9	Dwarf

Source: Data from Ahmed *et al.*, 2016

TABLE 3.2

Characteristic Features of the Malling Series

Name of the Rootstock	Characteristic Features
M-7a	60–70% the size of a seedling tree, suckers heavily; prone to burr knots; tolerant to fire blight
M-9	Tree size approximately 25–30% of a seedling tree; very early bearing; very productive; needs support; promotes good fruit size; suckers heavily; somewhat prone to burrknots; resistant to fire blight; very susceptible to crown rot
M-26 (released in 1959) (M-16 × M-9)	Semi-dwarfing; most cold hardy of all the Malling series rootstocks; 40% of standard size; larger and sturdier than M-9
M-27 (released in 1971) (M-13 × M-9)	Very dwarfing rootstock (20% the size of standard); shows resistance to diseases like crown rot but not to fire blight; free from suckering problem and burrknots

Source: Data from Robinson *et al.*, 2004

TABLE 3.3
Most Important M9 Clone Rootstocks

Rootstock	Origin	Growth	Remarks
T337	The Netherlands-NAKB[a]	M9	Standard rootstock in Europe
Flueren 56	The Netherlands-Flueren Nursery	< M9T337	Has good productivity; shows increase in fruit color
M9 EMLA	UK	> M9T337	First virus free M-9 rootstock
Pajam1 (Lancep)	France, CFITL[b]	Like M9T337	High productivity; virgin soils
Pajam2 (Cepiland)	France, CFITL	> M9T337	Has more vigor than M9; can be used in replant problems
Nicolai clone (e.g., Nic-8, 13, 19, 29)	Belgium	> M9T337	Very productive; shows early bearing; some runners
Burgmer clone (B-719, 751, 984, 756)	Germany, Burgmer Nursery	< or > M9T337	Good production; free from virus

Source: Data from Lind *et al.*, 2003

[a] NAKB: General Netherlands Inspection Service of Woody Nursery Stock

[b] CTIFL: Centre Technique interprofessional des Fruits et Legumes

TABLE 3.4
Characteristic Features of the Malling Merton Series

Name of the Rootstock	Characteristic Features
MM-106- ('Northern Spy' × M-1)	70–80% the size of the seedling tree; shows early bearing; productivity is good; performs well in most soil types; cold hardy in late winter but susceptible in early winter; best performance if the soils are either loamy or sandy; poorly drained soils should be avoided; susceptible to crown rot and fire blight; small problem of suckering; prone to burrknots
MM-111- ('Northern Spy' × MM-793)	80–90% the size of a seedling tree; bears early; has good anchorage; moderate in cold hardiness and productivity; adapted to most soils; drought tolerant, but can't tolerate poorly drained soils; tolerant to crown rot on well drained soils; tolerant to fire blight; tree form is more upright; little suckering; susceptible to burrknots
Merton-793 ('Northern Spy' × M-2)	Vigorous; propagates readily with the help of stools and layers; resistance to wooly apple aphid and collar rot; suitable for the replant areas and for spur type varieties

Source: Data from Robinson *et al.*, 2004

This was a first of its kind program for breeding a series of rootstocks against an insect. It resulted in the selection of 15 seedlings out of 3500 which were named MM.101–115. 'Northern spy' was used as the parent for transferring the trait of resistance to WAA (Le Pelley, 1927). These years also saw the breeding of a few hybrids of the Malling series including M-25, M-27 and M-26. Meanwhile, the Merton Immune (MI) series was selected in the 1930s out of 800 seedlings (i.e., 778–793). Merton-793 became popularized, and it still in use as M-793.

The breeding program at East Malling Research (EMR) has restarted after a break and is continuing with the evaluation of the crosses made back in the 1900s. Current work going on at the EMR is the use of marker assisted selection in apple rootstock breeding, genome sequencing of apple rootstocks, transcriptome and metabolome analysis, genetic fingerprinting and the incorporation of metagenomics for studying the causes of 'replant' disease (Felicidad Fernandez, 2016) (Table 3.4).

3.7.3 EMLA SERIES

During the 1950s to 1970s, a joint breeding program between East Malling and Long Ashton research stations in England led to the release of the first virus certified series which was free from all prevalent latent viruses known at that time. It came to be known as EMLA series. The viruses were eliminated by heat treatment. By 1969, more than 80000 stool shoots followed by around 400000 stool shoots were distributed to nurseries. The EMLA clones included M.9 EMLA, M.7 EMLA, M.26 EMLA, M.27 EMLA, M.106 EMLA and M.111 EMLA. When grown on the yield trials, some showed more yield than the non-EMLA ones (M.9 EMLA) while some proved to be vigorous (MM.106) (Sastry and Zitter, 2014).

3.7.4 Polish Series

The inability of the M and MM series to survive in the extremely cold conditions of Poland prompted the breeders there to initiate a separate rootstock breeding program. It was started at the Research Institute of Pomology and Floriculture in 1954. In 20 years, this program led to the release of five important rootstocks including P1, P2, P14, P16 and P22. The cold hardy cultivar used as the parent was 'Antonovka'. In 1968, another series of 40 crosses were made for improving orchard and propagation characteristics of the rootstocks, which resulted in the selection of 6420 seedlings which were further reduced to 60 clones. Another rootstock, P60, was released in 1994 (Jakubowski and Zagaja, 2000). The Polish rootstocks have shown the fire blight susceptibility similar to M9. However, the clones P66 and P67 have shown some promise for replant disease and productivity (Zurawicz et al., 2008, 2013). Another series of rootstocks have also been developed through mutagenesis which include PM.1, PM.2, PM.3 and PM.4 (Table 3.5). These are the mutants of M26 and A2, and their relative vigor is more or less comparable to M26 (Jakubowski and Zagaja, 2000).

3.7.5 Budagovsky Series

A Russian breeding program started in 1938 led to the release of famous rootstocks like B.9, B.146, B. 491, etc. It was started in the Minchurin College of Horticulture. A characteristic feature of this series is its red colored leaves. It produces a tree size approximately 25–30% smaller than M.9 for

---

TABLE 3.5
Characteristic Features of the Polish Series

Name of the Rootstock	Characteristic Features
Polish 2 (P2) (M.9 × 'Antonovka')	30–40% the size of a seedling tree; bears early; has good productivity; requires support; behaves as hardy until mid-winter but can become susceptible in the end; gives maximum performance if the soils are well drained with very little suckering; burrknots are not formed
Polish 18 (P18)	100% the size of a seedling tree; slow bearing; moderate productivity; has good anchorage; hardy in early winter but maybe susceptible in late winter; resistant to collar rot; moderately resistant to fire blight; very little suckering; very few burrknots
Polish 16 (P16) ('Longfield' × M11)	20–30% the size of a seedling tree; very early bearing; very productive; needs support; appears very hardy but needs further testing; well adapted on most soils; resistant to collar rot; susceptible to fire blight; shows suckering problem; does not show much of a burrknot problem
Polish 22 (P22)	10–20% the size of a seedling tree; bears early; requires support; hardy in mid-winter very early bearing; productive; very hardy mid-winter but appears susceptible in late winter; has good adaptivity in most soils; soils resistant to collar rot; moderately susceptible to fire blight; very little suckering or burrknots

Source: Data from Robinson et al., 2004

TABLE 3.6

Characteristic Features of the Budagovsky Series

Name of the Rootstock	Characteristic Features
Bud.9 (B.9) (M-8 × 'Red Standard')	30–40% the size of a seedling tree; bears early; has good productivity; requires support; has good performance on well drained soils; can't withstand poorly drained soils; very resistant to collar rot; more tolerant to fire blight; moderate suckering; very few burrknots; susceptible to drought
Bud.10 (B.10) (M-27 × 'Robusta 5')	Tree size is 15% smaller as compared to tree size on M9T337; cold hardy; yield efficient; tolerant to fire blight; good root anchorage; tolerant to stress
Bud.146 (B.146)	20–30% the size of a seedling tree; bears early; has good productivity; requires support as the roots are brittle; susceptible to fire blight; several strains have been identified with variability existing between strains; suitable for vigorous cultivars on fertile sites; moderately prone to suckering and burrknots
Bud.491 (B.491)	20–30% the size of a seedling tree; bears early; has good productivity; requires support; grows well on well drained soils; susceptible to both collar rot and fire blight; suitable for vigorous cultivars on fertile sites; produces few suckers and burrknots

Source: Data from Robinson *et al.*, 2004

the same scion variety. It had an ontogenic resistance to fire blight (Russo *et al.*, 2008), however it is not tolerant to replant disease (Auvil *et al.*, 2011). It is an active breeding program; work on another rootstock of the Budagovsky series over a period of 10 years has resulted in the release of B.10, which is tolerant to fire blight with good root anchorage and stress tolerance (Table 3.6).

3.7.6   CORNELL-GENEVA SERIES

The Cornell-Geneva program became famous as a collaborative work initiated by Dr. Cummins and Dr. Aldwinkle in 1968 from the Cornell University Geneva campus and then joined by the United States Dept. of Agriculture (USDA) in 1998. Terence Robinson represented the Cornell University while Gennaro Fazio represented the USDA. 'Robusta 5' was used as the source of resistance to both fire blight and the woolly apple aphid, and the crosses were being made with the rootstocks of the Malling series. The rootstocks that were commercialized include G.11 and G.41. This program is also operational and makes crosses for resistance to drought and various biotic and abiotic stresses (Marini and Fazio, 2018). Almost all of the rootstocks are resistant to fire blight while most of them are also tolerant to replant disease. Another breeding program, which was started in the United States in 1959 at Michigan State University, saw the release of an important rootstock known as MAC-9 which became popularized as 'Mark'. It has been patented and licensed for propagation. It shows very poor performance in hot dry soils and is sensitive to drought and crown gall (Rom and Carlson, 1987). It is a dwarfing rootstock with good ability to bear heavy crop loads. It is quite susceptible to woolly apple aphid, fire blight and scab but has a good yield efficiency. Some of the other famous selections of this series include MAC-1, MAC-9 and MAC-11 (Table 3.7).

3.8   OTHER APPLE ROOTSTOCK BREEDING PROGRAMS

Apart from the major rootstocks breeding programs mentioned in this chapter, various other programs were initiated in a large number of other countries as well. Many of these programs were either terminated, but some are continuing to carry out further research (Table 3.8).

TABLE 3.7

Characteristic Features of Geneva® Apple Rootstocks

Traits	G.11	G.16	G.41(a)	G.213	G.214	G.814	G.935	G.222
Arranged in order by size (smallest to largest)	M.9 T337	M.9 T337	M.9T337	M.9T337	M.9/M.26	M.9/M.26	M.26	M.26
WAA resistance	No	No	High	High	High	No	No	High
Fire blight resistance	Resistant	Resistant	Very Resistant	Very Resistant	Very Resistant	Very Resistant	Very Resistant	Very Resistant
Replant disease complex resistance	Partial	Partial	Tolerant	Tolerant	Tolerant	Tolerant	Tolerant	No
Crown and root rots (Phytophthora)	Tolerant	Tolerant	Tolerant	Tolerant	Tolerant	Tolerant	Tolerant	Tolerant
Cold hardiness	Yes	Good mid-winter; bad early cold	Yes	TBD	Yes	Yes	Yes	Yes
Efficiency as good or better than M.9	Yes	Yes	Yes	Yes	Yes	Yes	Yes	Yes
Susceptibility to latent viruses	No	Yes	No	No	No	Yes	Yes	No

Source: Data from Jessica Lyga.

TABLE 3.8

Other Apple Rootstock Breeding Programs

Breeding program	Institute/University involved	Series	Objectives	Rootstocks
Canada	Agriculture and Agri-Food Canada (1959)	Ottawa series; Kentville series; Vineland series; SJM series	Winter-hardy rootstocks; resistant to crown rot; adapted to local environments	Ottawa 3, V.2, V.4, SJM-15, SJM-150, SJM-189, SJM-5198 and SJM-5128 (Khanizadeh *et al.*, 2005)
Germany	Dresden-Pillnitz program	Pillinitzer Supporter series (Fischer and Mildenberger, 1999)	Good propagation properties; nursery tree growth; yield and resistance to fire blight, apple scab and powdery mildew	Pillinitzer Supporter 1, Pillinitzer Supporter 2, Pillinitzer Supporter 3, Pillinitzer Supporter 4
Sweden	The Alnarp Fruit Tree Station breeding program	BM Series	Cold hardiness and propagation ease	Alnarp 2, BM342
The Czech Republic	Techobuzice (Prague)	JTE series	Winter hardiness; dwarfing and propagation ease (Cummins and Aldwinckle, 1983)	JTE-E, JTE-F, JTE-H, JTE-B, JTE-D, JTE-G (Wertheim, 1998)
Japan	Fruit Tree Research Station, Morioka, Japan	JM series	Disease resistance to local problems (Bessho and Soejima, 1992; Moriya *et al.*, 2008)	JM7, JM2, JM1, JM7, JM5 Marubakaido × M.9 clones (1972)

3.9 ADVANCED BIOTECHNOLOGICAL APPROACHES FOR ROOTSTOCK BREEDING IN APPLES

Due to many disadvantages of the conventional methods of rootstock breeding that were previously used in various rootstock breeding programs, including linkage drag and long time period requirement for the evaluation of the crosses, advanced biotechnological approaches offer a better means of obtaining the desired rootstock at a much faster rate and avoiding the problem of linkage drag to a greater extent. Some of the methods that can be employed in future for breeding new rootstocks are detailed in the following subsections.

3.9.1 TRANSGRAFTING

Transgrafting explains how the original method of grafting can be performed along with the genetic improvement of the fruit trees. In the practice of grafting, the scions with buds are grafted onto the rootstock bearing the roots (belonging to either the same species or different species). In transgrafting the genetic changes can affect the characteristics of both, and it can be brought about in three different ways which are as follows:

1. Non-GM scions grafted onto GM rootstocks
2. GM scions grafted onto non-GM rootstocks
3. GM scions grafted onto GM rootstocks.

The actual focus of the technique is to modify the rootstocks by recombinant technologies. Various examples of grafting of a scion onto a genetically transformed rootstock have been reported (Schaart and Visser, 2009). The aim was to modify the performance of the above parts of the fruit trees, including the fruits, without affecting their genetics.

In Smolka *et al.* (2010) study, three rolB transgenic dwarfing rootstocks from M26 and M9 were taken, and they were grafted upon by five non-transgenic scion cultivars. The results showed that the reduced vegetative growth was seen in all the transformed rootstocks irrespective of the scion cultivars when compared with the non-transgenic ones. There was a reduction in the flowering and fruiting as well, but the quality of the fruit did not seem to be affected in the transformed rootstocks. RT-PCR also confirmed that the rolB gene was expressed under field conditions. At the same time, the results confirmed that the rolB gene was not detectable in the scion cultivars, which was indicative of the fact that the gene had not been transferred to the scion. The research, therefore, suggests that rolB-modified rootstocks should be used along with vigorous scion cultivars so as to obtain good vegetative growth as well as yield.

3.9.2 CISGENESIS/INTRAGENESIS

A cisgenic plant is one which shows modification with one or more genes. It contains introns and flanking regions in a sense orientation. These genes have been taken from a donor plant which can be crossed with the available plant (Schouten *et al.*, 2006). On the other hand, intragenesis is similar to cisgenesis, but it's different in that it allows the formation of unique combinations of the DNA fragments. Here different new genes can be formed *in vitro* by joining different genetic elements including the promoters, coding region and the sequences of the terminator. This newly formed gene can be introduced into the available varieties. In order to breed a cisgenic apple line in apples, one insert of T-DNA was found in the chromosome 16 of 'Gala Galaxy'. It was found that the transformed line C44.4.146 with the transcript FB_ MR5 was found similar to the traditionally bred cultivars of Mr5. Also, three separate experiments, in which a Mr5 avirulent strain of *Erwinia amylovora* were performed using scissors or syringe. It was found that significantly less symptoms were seen in the shoots of the cisgenic lines in comparison to that of 'Gala Galaxy' apples that were not transformed (Kost *et al.*, 2015).

3.9.3 RNA INTERFERENCE

RNAi is an evolutionarily conserved mechanism. It is basically a post transcriptional gene silencing (PTGS). During this process, mRNA sequences are degraded at specific points by double stranded RNA. It has been widely used as a knockdown technology and to analyze gene functions in various organisms. It results in inhibition of translation or transcriptional repression. From the time it was discovered, it has shown tremendous potential in opening new pathways for fruit and other crop improvements. This technology is very effective and has many advantages over the antisense technology. It has been used with great success to improve various quality traits in plants genetically.

A study was conducted in order to develop dwarf apple lines by RNAi silencing of the MdGA20-ox gene. As a result, it was discovered that the gibberellin content of the leaves was decreased in the transgenic lines as compared to the non-transgenic controls, and the correlation between the gibberlic acid (GA) content and the internodal length was found to be positive to the order of 0.93 (Zhao *et al.*, 2016).

3.9.4 GENOME EDITING

The field of fruit biotechnology is all set to see the revolution of the effective combined ability of whole genome sequences and the genetic tools by introducing specific changes in the genome with great accuracy in order to discover developing phenotypes apart from introducing new varieties.

The genome editing tools like zinc finger nuclease (ZFN), tale nuclease (TALENs), (CRISPR) and CRISPR-Associated Protein (Cas9) and Oligonucleotide-directed mutagenesis (ODM) are used for precise genome editing of the organism (Limera *et al.*, 2017). The genome editing tools are explored in woody tree species. The genome editing techniques have a potential to develop the novel rootstock while precisely editing the gene of interest. Even though this method is very efficient, at the same time, it also has some disadvantages, which include random integration of different plasmid sequences into the host genome. Not only that, but these varieties released as a result may not be accepted for commercial production because of the genetically modified organisms (GMO)-regulations associated with them.

3.10   FUTURE THRUSTS

1. Introduction of new rootstocks with multiple resistance traits
2. Long term and large scale multi-locational testing of new hybrid/clonal rootstocks
3. Rootstock breeding needs to be initiated in prioritized crops and along different aspects
4. Development of complex hybrids through interspecific, intraspecific and intergeneric hybridization to develop more versatile rootstocks
5. Development and use of efficient interstocks
6. Survey, selection and evaluation of a large number of indigenous fruit species
7. Development of virus-free material for commercial rootstocks
8. Use of biotechnological approaches and molecular biological tools
9. Funding by the government to support rootstock breeding programs

3.11   CONCLUSION

Rootstocks and interstocks are an integral part of commercial fruit cultivation and need equal focus. In India, the development of rootstocks is not adequate. The existing new rootstocks and interstocks are mostly obtained from exotic sources and adopted after testing. Hence, more emphasis should be placed on collection of local or indigenous wild germplasm in different tree and vine fruit crops. Local rootstocks for apples need to be standardized. The potential to produce new rootstocks by either conventional means or in combination with biotechnological approaches is substantial and demonstrated.

REFERENCES

Ahmed, N., Mir, J.I. and Wasim, H.R. 2016. Advances in varietal and rootstock improvement in temperate fruits and nuts: Future strategies. In: *Temperate Fruits & Nuts* (pp. 167–199). New Delhi: The Horticultural Society of India.

Atkinson, R.G., Sutherland, P.W., Johnston, S.L., Gunaseelan, K., Hallett, I.C., Mitra, D., . . . and Schaffer, R.J. 2012. Down-regulation of POLYGALACTURONASE1 alters firmness, tensile strength and water loss in apple (Malus x domestica) fruit. *BMC Plant Biology* 12(1): 1–13.

Auvil, T.D., Schmidt, T.R., Hanrahan, I., Castillo, F., McFerson, J.R. and Fazio, G. 2011. Evaluation of dwarfing rootstocks in Washington apple replant sites. In: *IX International Symposium on Integrating Canopy, Rootstock and Environmental Physiology in Orchard Systems* 903: 265–271.

Bessho, H. and Soejima, J. 1992. Apple rootstock breeding for disease resistance. *Compact Fruit Tree* 25: 65–72.

Bessho, H., Soejima, J., Ito, Y. and Komori, S. 1993. Breeding and genetic analysis of apple in Japan. In: T. Hayashi *et al.* (Eds.), *Techniques on Gene Diagnosis and Breeding in Fruit Trees*. Japan: FTRS.

Brown, S. and Maloney, K. 2015. *Recent Advances in Apple Breeding, Genetics and New Cultivar*. Geneva, NY: Department of Horticultural Sciences, New York State Agricultural Experiment Station, Cornell University.

Bunyard, E.A. 1920. The history of the Paradise stocks. *Journal of Pomology and Horticultural Science* 1(3): 166–176.

Cummins, J.N. and Aldwinckle, H.S. 1983. Breeding apple rootstocks. In: *Plant Breeding Reviews* (pp. 294–394). Boston, MA: Springer.

Das, B., Ahmed, N., Ranjan, J.K., Attri, B.L., Sofi, A.A., Verma, R.K., Verma, M.K. and Tripathi, P. 2012. New apple variety released. *Indian Journal of Horticulture* 69(2): 1–4.

Felicidad Fernandez. 2016. *East Malling Rootstock Club.* www.emr.ac.uk/projects/crop-breeding-genetics/rootstock-breeding-club/ (accessed March 25, 2022).

Fischer, M. and Mildenberger, G. 1999. New Naumburg/Pillnitz pear breeding results. In: *Eucarpia symposium on Fruit Breeding and Genetics* 538: 735–739.

Forsline, P.L. 1995. Adding diversity to the national apple germplasm collection: Collecting wild apple in Kazakhstan. *New York Fruit Quart* 3(3): 3–6.

Forsline, P.L., Dickson, E.E. and Djangalieu, A.D. 1994. Collection of wild Malus, Vitus and other fruit species genetic resources in Kazakhstan and neighboring republics. *HortScience* 29: 433.

Harris, S.A., Robinson, J.P. and Juniper, B.E. 2002. Genetic clues to the origin of the apple. *Trends in Genetics* 18(8): 426–430.

Hurt, A. 1916. *Theophrastus Enquiry into Plants.* London: William Heinemann.

Jakubowski, T. and Zagaja, S.W. 2000. 45 years of apple rootstocks breeding in Poland. *Acta Horticulturae* 538: 723–727. https://doi.org/10.17660/ActaHortic.2000.538.131

Janick, J., Cummins, J.N., Brown, S.K. and Hemmat, M. 1996. Apples In: J. Janick and J.N. Moore (Eds.), *Fruit Breeding* (pp. 1–76). Dordrecht, The Netherlands: Springer, Kluwer Academic Publishers.

Jessica Lyga. Geneva® apple rootstocks comparison chart v.4. https://ctl.cornell.edu/wp-content/uploads/plants/GENEVA-Apple-Rootstocks-Comparison-Chart.pdf (accessed March 25, 2022).

Juniper, B.E. and Mabberley, D.J. 2006. *The Story of the Apple.* Portland, OR: Timber Press, 219pp.

Kanwar, S.M., 1987. *Apples. Production Technology and Economics.* New Delhi: Tata McGraw Publishing Company Limited, pp. 88–98.

Khanizadeh, S., Groleau, Y., Levasseur, A., Granger, R., Rousselle, G.L. and Davidson, C., 2005. Development and evaluation of St Jean-Morden apple rootstock series. *Horticulture Science* 40(3): 521–522.

Kost, T.D., Gessler, C., Jänsch, M., Flachowsky, H., Patocchi, A. and Broggini, G.A. 2015. Development of the first cisgenic apple with increased resistance to fire blight. *PLoS One* 10(12): 0143980.

Kotoda, N., Hayashi, H., Suzuki, M., Igarashi, M., Hatsuyama, Y., Kidou, S.I., Igasaki, T., Nishiguchi, M., Yano, K., Shimizu, T. and Takahashi, S. 2010. Molecular characterization of FLOWERING LOCUS T-like genes of apple (Malus× domestica Borkh.). *Plant and Cell Physiology* 51(4): 561–575.

Laurens, F., Lespinasse, Y. and Fouillet, A.A. 2005. New scabresistant Apple: Ariane. *HortScience* 40: 484–485.

Lehnert, R. 2012. DS-22 planned for this fall. New apple from a private breeder will be managed as a club variety. *Good Fruit Grower.*

Le Pelley, R. 1927. Studies on the resistance of apple to the woolly aphis (*Eriosoma lanigerum* Hausm.). *Journal of Pomology and Horticultural Science* 6(3): 209–241.

Limera, C., Sabbadini, S., Sweet, J.B. and Mezzetti, B. 2017. New biotechnological tools for the genetic improvement of major woody fruit species. *Frontiers in Plant Science* 8: 1418.

Lind, K., Lafer, G., Schloffer, K., Innerhofer, G. and Meister, H. 2003. Principles of organic fruit growing. In: *Organic Fruit Growing.* Wallingford, UK: CABI Publishing, p. 41.

Marini, R.P. and Fazio, G. 2018. Apple rootstocks: History, physiology, management, and breeding. In: *Horticultural Reviews,* vol. 45. Hoboken, NJ: John Wiley and Sons Inc., pp. 197–312. https://doi.org/10.1002/9781119431077.ch6

Morgan, J. and Richards, A. 1993. *The Book of Apples.* London: Ebury Press.

Moriya, S., Iwanami, H., Takahashi, S., Kotoda, N., Suzaki, K. and Abe, K. 2008. Evaluation and inheritance of crown gall resistance in apple rootstocks. *Journal of the Japanese Society for Horticultural Science* 77(3): 236–241.

Nimbolkar, P.K., Awachare, C., Reddy, Y.T.N., Chander, S. and Hussain, F. 2016. *Journal of Agricultural Engineering and Food Technology* 3(3): 183–188.

Noiton, D.A.M. and Alspach, P.A. 1996. Founding clones, inbreeding, coancestry, and status number of modern apple cultivars. *Journal of the American Society for Horticultural Science* 121: 773–782.

Preston, A.P. 1956. The control of fruit-tree behaviour by the use of rootstocks. *Annals of Applied Biology* 44(3): 511–517.

Roach, F.A. 1985. *Cultivated Fruits of Britain: Their Origin and History.* Oxford, UK: Basil Blackwell Publisher Ltd.

Robinson, T.L., Anderson, L., Azarenko, A., Barritt, B.H., Brown, G., Cline, J., Crassweller, R., Domoto, P., Embree, C., Fennell, A., Ferree, D., Garcia, E., Gaus, A., Greene, G., Hampson, C., Hirst, P., Hoover, E., Johnson, S., Kushad, M. and Marini, R.E. 2004. Performance of Cornell-Geneva rootstocks across North America in multi-location Nc-140 rootstock trials. *Acta Horticulturae* 658: 241–245.

Rom, R.C. and Carlson, R.F. 1987. *Rootstocks for Fruit Crops.* New York: Wiley.

Russo, N.L., Robinson, T.L., Fazio, G. and Aldwinckle, H.S. 2008. Fire blight resistance of Budagovsky 9 apple rootstock. *Plant Disease* 92(3): 385–391.

Sastry, K.S. and Zitter, T.A. 2014. *Plant Virus and Viroid Diseases in the Tropics: Volume 2: Epidemiology and Management.* Dordrecht, Heidelberg, New York, London: Springer Science & Business Media.

Schaart, J.G. and Visser, R.G.F. 2009. *Novel Plant Breeding Techniques. Consequences of New Genetic Modification-based Plant Breeding Techniques in Comparison to Conventional Plant Breeding.* Bilthoven: Cogem.

Schouten, H.J., Krens, F.A. and Jacobsen, E. 2006. Cisgenic plants are similar to traditionally bred plants: International regulations for genetically modified organisms should be altered to exempt cisgenesis. *EMBO Reports* 7(8): 750–753.

Singh, N., Sharma, D.P. and Chand, H. 2016. Impact of climate change on apple production in India: A review. *Current World Environment* 11: 1–9.

Smolka, A., Li, X.Y., Heikelt, C., Welander, M. and Zhu, L.H. 2010. Effects of transgenic rootstocks on growth and development of non-transgenic scion cultivars in apple. *Transgenic Research* 19(6): 933–948. doi: 10.1007/s11248-010-9370-0

Stone, E.L. and Kalisz, P.J. 1991. On the maximum extent of tree roots. *Forest Ecology and Management* 46(1–2): 59–102.

Vavilov, N.I. 1930. Wild progenitors of the fruit trees of Turkistan and the Caucasusband the problems of the origin of fruit trees In *Report of 9th International Horticultural Congress, Royal Horticultural Society,* London, pp. 271–286.

Way, R.D., Aldwinckle, H.S., Lamb, R.C., Rejman, A., Sansavini, S., Shen, T., Watkins, R., Westwood, M.M. and Y. Yoshida. 1990. Apples (*Malus*). In: J.N. Moore and J.R. Ballington Jr. (eds.). *Genetic Resources of Temperate Fruit and Nut.* Wageningen: International Society for Horticultural Science (Acta Hort. 290), pp. 1–62.

Wertheim, S.J. 1998. *Rootstock Guide: Apple, Pear, Cherry, European Plum.* Proef station voor de Fruitteelt (Fruit Research Station).

Xu, K. 2015. *Arctic Apples: A Look Back and Forward.* Geneva: Horticulture Section, School of Integrative Plant Science NYSAES, pp. 1–24.

Zhao, K., Zhang, F., Yang, Y., Ma, Y., Liu, Y., Li, H., Dai, H. and Zhang, Z. 2016. Modification of plant height via RNAi suppression of MdGA20-ox gene expression in apple. *Journal of the American Society for Horticultural Science* 141(3): 242–248.

Zurawicz, E., Bielicki, P., Czynczyk, A., Bartosiewicz, B., Buczek, M. and Lewandowski, M. 2008, August. Breeding of apple rootstocks in Poland-the latest results. *Acta Horticulturae* 903: 143–150. https://doi.org/10.17660/ActaHortic.2011.903.13

Zurawicz, E., Pruski, K., Lewandowski, M. and Szymajda, M. 2013. Effect of replant disease on growth of *Malus x domestica* 'Ligol' cultivated on P-series apple rootstocks. *Journal of Agricultural Science* 5(10): 28.

4 Advances in Propagation and Nursery Management: Methods and Techniques

*Ab Waheed Wani, Iftisam Yaseen, Suhail Fayaz,
Afiya Khursheed, Archi Gupta, G I Hassan, Rajni Rajan,
Shahid Qayoom, Irfan Gani and Smruthi Jayarajan*

CONTENTS

4.1 INTRODUCTION

Temperate fruit crops have been grown in the northwestern Himalayan states for generations with the majority of the plantations growing seedlings of origin with low production. Temperate fruits are currently cultivated on 4.45 lakh hectares with a yield of 25.8 lakh tons and an average productivity of 6.8 tons per hectare compared to a demand of 44 lakh tons per year. Apples are an important temperate fruit crop that accounts for around 52 percent of the total fruit acreage and 79 percent of the temperate fruit production. Other fruits such as pears, peaches, plums, walnuts, almonds, apricots, and others make up the rest of the area. The apple tree (*Malus x domestica* Borkh.) is among the most extensively cultivated fruit of the temperate climate in the world with a significant economic impact on many countries' fruit production. Plant propagation is important in horticulture because it allows the number of plants to be rapidly multiplied while retaining the desirable characteristics of the mother plants and reducing the plant's bearing age. Depending on the species, various techniques can be used to optimize a nursery production system or even to solve a specific propagation problem. Plant propagation is a fundamental human occupation. Civilization

DOI: 10.1201/9781003239925-4

may have begun when ancient man learned to plant and grow different types of plants to meet his and his animals' nutritional needs. Our cultivated plants are primarily the result of three general processes. Some plants were selected directly from wild species, but as a result of man's selective hand they evolved into types that differ greatly from their wild forms. Other types of plants, such as pears, strawberries, and prunes, arose as hybrids between species accompanied by changes in chromosome number. Another type of plant occurs naturally as a rare monstrosity. They may be useful to man despite being unadapted to their native environment (Srivastava, 1966). Fruit plants do not germinate when raised from seed, so they are generally propagated using vegetable methods either on their own roots by cutting or layering techniques or on the roots of other plants (rootstocks) by budding and grafting techniques. Because growing fruit trees on rootstocks has many advantages over growing fruit trees on their own roots, the majority of fruit plants are raised on rootstocks rather than on their own roots (Sharma, 2002).

4.2 PROPAGATION

Plant propagation is the controlled perpetuation or multiplication of plants with the goals of (i) increasing plant number and (ii) preserving essential plant characteristics. It is essentially two types.

4.2.1 SEXUAL PROPAGATION

Sexual reproduction is the multiplication of the plants through the formation of seeds from union of the male and the female gametes. Meiosis division occurs during fusion, and the chromosome numbers, as in the parents, are reduced to half, which returns to normal after fertilization. It has many benefits as well as drawbacks. Sexual propagation necessitates careful management of germination conditions and facilities as well as knowledge of the needs of the various types of seeds.

4.2.2 ASEXUAL PROPAGATION

Asexual propagation, also known as vegetative propagation, is the process of inducing a plant's vegetative parts, for example, its stems, roots, and/or leaves – to rejuvenate it into newer plant or, in some situations, a number of plants. The offspring is genetically similar to parent plant, with very few exceptions. Cuttings, layering, budding, grafting, and micropropagation are the main asexual propagation methods of propagation. Each method of propagation has its own advantages and disadvantages. The method of asexual propagation is easier and faster method of propagation (Hartmann *et al.*, 2009).

People's dietary needs have changed as a result of changes in living standards, and fruits now play a significant role in the diet. Thus, the area under the fruits has rapidly increased in recent decades, but the production of standard planting materials of various temperate fruits is primarily confined to a small number of registered nurseries producing quality plants. Healthy, high-quality fruit plants are the foundation of profitable fruit production. As a result of the increased demand for quality plants, the need to modernize our fruit nurseries in temperate growing regions has become more pressing. Furthermore, plant propagation has evolved into a highly specialized industry requiring a significant investment in certified and highly developed skill and technology.

Dormancy is a problem directly related to germination and storage of seeds. Seed dormancy might be classified as endogenous (caused by factors related to embryos) and exogenous (due to factors related to endosperm, tegument, or other barriers imposed by the fruit). It can also be primary or natural (trait inherent to the seed) and secondary or induced (caused by unfavorable conditions to germination) (Vivian *et al.*, 2008). Stratification to break the dormant phase of seeds is used frequently in treating seeds and preparing them for germination. Seed stratification is performed in cold storage facilities at 2–4°C in rare situations. The seeds can be planted in peat, sand, or without a substrate, but they must not be allowed to dry out (Wertheim and Webster, 2003). During stratification many changes occur in the seeds, including an increase in the permeability of the seed coat to water, an increase in its softness, an increase in gas exchange, and an increase in the activity of enzymes (Jauron, 2000). Efficiency of stratification in apples could be improved by application of certain compounds (potassium nitrate, ethephon, carbon monoxide, hydrogen peroxide), phyto-hormones (gibberellins, 6 benzylaminopurine, jasmonic acid), and physical methods (heat shock and pulsed radiofrequency) (Grzesik *et al.*, 2017). Although there has been a significant rise in area, production, and productivity, apple productivity in India is still quite low when compared to that of industrialized countries. This is primarily due to a lack of quality planting material and seeds, the fact that seedling trees have a long juvenile period and begin bearing after 7–8 years, the presence of aged unproductive seedling orchards, land holdings that are small with a lack of irrigation, the use of poor genetic stocks/varieties/hybrids, inefficient supervision and non-adoption of packages of practices, a high pest and disease incidence, massive preharvest and postharvest losses, and a scarcity of qualified human resources. Thus, in industrialized countries, micropropagation is a well-established way of creating millions of identical disease-free plants, but it has failed to catch on in India. Apple micropropagation has aided in the creation of infection-free, healthy plants as well as the quick production of desirable scions and rootstocks. From a practical and economic standpoint, several reliable procedures for both rootstocks and scions have been established in the apple industry during the previous few decades. Various internal and external conditions, compris-ing ex vitro (e.g., genotype and physiological state) and in vitro settings, influence successful apple micropropagation that employs pre-existing meristems (culture of apical buds or nodal segments; e.g., media constituents and light).

4.3 CLONAL PROPAGATION

4.3.1 CUTTINGS

In the summer or winter, cutting procedures are used to detach immature plant portions from the mother plant, which are subsequently stimulated to establish the roots. In the first scenario, leafy tips of shoot are taken in early summer or spring and rooted in high humidity glasshouses; they are known as the softwood cuttings. One-year-old parts of leafless shoots are gathered, and the roots are induced in an appropriate medium with bottom heat in the event of winter cuttings. Semi-hardwood cuttings collected in the late summer after shoot development has stopped have been utilized for propagation on occasion. Cutting stands out among the numerous strategies of vegetative propagation because it is a method of propagation in which detached portions of a plant, under controlled condi-tions, are propagated. In proper circumstances, it will establish roots and start a new life as a plant with identical traits to the one that gave origin to it (Pasqual *et al.*, 2001). A suitable rooting medium ensures the provision of appropriate moisture and ventilation to the bases. A good supply of oxygen to the base of the cuttings ensures the oxidation of the nutrition in the cuttings to produce the inter-nal energy required for cell division and for the formation of the root tips (Hammoudi and Dayoub, 1986). A quick dip of 2500 mg L^{-1} IBA and the use of coco peat + perlite media resulted in a higher percentage of rooted cuttings of MM-111 apple clonal rootstock (Dvin *et al.*, 2011). The sclerenchyma tissue between the bark and the crust of the cutting is an obstacle to the appearance and growth of the roots. Therefore, following the process of wounding, the callus is formed and when the cutting lixiviates, the root sprout develops and the wounded tissues are stimulated to cell division (Hartmann *et al.*, 2002). This happens according to a natural cumulative of auxins and carbohydrates. In addi-tion, the wound-damage tissues are also stimulated to produce ethylene which plays a big role in the cell division and accelerates the formation of adventitious roots (Abdel Wahab and Al-Dujaili, 2001). The rooting response of MM-111 apple rootstock to wounding shows that making a split wound at the base resulted in better roots and shoot parameters of cuttings than it does with no wounding (Upadhayay *et al.*, 2020). Dormant cuttings from plants that are 2 years old should be acquired for this system. Cuttings cut from the 1-year-old rootstock tops or plants may barely survive 10 percent to 20 percent of the time. During the winter, cuttings are clipped to a length of about 20 cm and to minimum a diameter of about 5 mm (July/August). Rooting of apple cuttings is affected by number of factors like the plant's physiological state and environmental circumstances. Rooting is influenced by the mother plant's nutritional state. Because of the nutritional balance, roots are encouraged. Zinc (Zn) is a critical element in the rooting of apple cuttings because it activates tryptophan, which is precursor of auxin and is necessary for the root growth. Because younger plants root more easily, the age of mother plant has an impact on a rootstock's capacity to reproduce. Genetic potential, which varies widely among rootstocks, is one of the most critical factors determining the rooting of apple cuttings. Less robust rootstocks (dwarfing) have a lower rooting rate in general. Balance of cytokinins, auxins, and gibberellins is required in terms of hormones. Exogenous auxin boosters, such as naphthalene or indolebutyric acid, can be used to increase auxin concentration and, as a result, rooting rate. In terms of environmental conditions, substrate temperature is one of the most critical aspects in determining rooting potential, with temperatures between 21°C and 26°C resulting in more root production. Extremely high temperatures should be avoided when working with herbaceous and semi-woody cuttings because they enhance tissue transpiration which can induce wilting and also encourage sprouting before roots. Water loss is one of the most common reasons for cuttings not to root, particularly in vegetative cuttings. The use of nebulization allows for a decrease which encourages roots. The rooting rate and root growth are influenced by the substrate which should offer appropriate water retention and porosity for this purpose.

4.3.2 BUDDING AND GRAFTING

Temperate fruit trees are most commonly grown in commercial nurseries through late-summer budding of the scion cultivars onto the rootstocks in field or through late-winter bench grafting onto the bare-rooted rootstocks. The most frequent form of fruit tree propagation is budding (Kumar, 2011), with chip budding and T-budding being the most popular. Budding uses propagation material of the scion more efficiently than grafting and thus is more popular than grafting. Although there are a variety of methods for grafting apple trees economically, the most common are cleft grafting and budding grafting.

4.3.2.1 Cleft Grafting

If rootstocks have been transferred, cleft grafting may be done at the nursery's site or the tabletop. It entails the removal of rootstocks and can also be performed in the sheds; it is not impacted by weather. Forks with three buds and a length of 5 to 8 cm should be clipped slightly above the last bud. The procedure of double cleft grafting is used. Cleft grafting is done during dormancy (Goldschmidt, 2014).

4.3.3 LAYERING SYSTEM

Layering is the most extensively used method for multiplying apple tree rootstocks. Young plants are not separated from the mother plant until the growth of adventitious roots in this arrangement. Adaptability to automation, good quality of the vegetative material, easy maintenance, cheap cost, and a high rate of rooting and quantity of the tillers are all characteristics of this system (Table 4.1).

Stool layering and Chinese layering are two ways utilized for layering. When compared to the Chinese technique, stool stacking necessitates a significantly larger number of plants for formation of the mother plants. But in the Chinese system, however, greater attention is paid to the phytosanitation quality of the plants in order to minimize the empty spaces in between the plants that are caused by their loss. The first year of both the systems is typically utilized only for the development of plant. There is no need to take any action until the initial shoots are trimmed in the winter, thus leaving branch tips at about 5 cm from surface of soil. However, some nurseries use the substrate

TABLE 4.1

Per Mother Plant, Production of the Rootstocks that are Rooted in the Layering Method

Rootstock	2nd year	3rd year	4th year	5th year	6th year	7th year	8th year
MM-106	2	4	4	9	11	10	13
MM-111	3	5	7	11	10	8	12
MM-25	2	5	8	12	12	9	11

TABLE 4.2

Propagation Material with Different Series

Root stock series	Clones	Native place
M	M-9, M-26, M-25, M-27, M-7	England
MM	MM-111, MM-106	England
MAC	Mark-24, Mark-9	United States
P	P-18, P-22, P-16, P-1, P-2	Poland
Budagovski	Bud-491, Bud-29, Bud-9, Bud-490	Russia
CG	CG-24, CG-44, CG-60, Novole, CG-10, G-80	United States
Ottawa	Ottawa 3, Ottava 8	Canada
Jork	Jork 9	
JM	JM-7, JM-8	Japan
MI	MI-793	England
Clones M-9	M-9 EMLA	Unknown
Marubakaido	Maruba	Japan
Others	Nicolai 29 (NIC 29)	France
	Bemali	Switzerland

technique to encourage roots in mother plants during the first year of planting (Modgil and Thakur, 2017). The gathered tillers can also be utilized for starting a new nursery even if the percentage of rooting is low in the year of the planting. Local dirt or sawdust might be utilized as a substrate. Sawdust must be chemically neutral and originate from trees that do not release any chemicals that might hinder rooting. Stool stacking entails planting a rootstock with a minimum diameter of 1 cm that is planted at a depth of roughly 30 cm. Two buds are left above ground level while planting the rootstock. Planting density varies per nursery and depending on the cultural gear available. Planting mother plants in two parallel rows at a spacing of 10 cm between the rows and plants is standard practice. A substrate protection is formed when the buds attain a length of 10 cm, leaving 2–3 cm of the end of the buds out of the soil. Two or three more protections should be applied throughout the vegetative cycle to achieve a rooting area of about 10 to 15 cm. Depending on the spacing, 15 to 25 thousand mother plants, i.e., rootstocks, are required for one hectare. From the fourth year on, each mother plant can yield 5–8 high-quality, well-rooted rootstocks. This number rises until the eighth year when it begins to level out (Table 4.2). Depending on their phytosanitary conditions, the same mother plants might be utilized for 15 or more years.

4.4 INTERSTOCKING

The application of interlayering with dwarfing rootstock between canopy cultivars and robust rootstocks can help to reduce the ultimate plant size. This is a method for reducing plant vigor, increasing efficiency of production, and improving the quality of fruit (Marcon Filho *et al.*, 2009).

Interstocking is a method that involves planting cuttings of dwarfing rootstock between canopy cultivars and rootstocks to diminish the vigor of robust rootstocks. This strategy is used to make use of several intriguing rootstock properties, including pest and disease resistance, anticipating fruiting, improving fruit quality, and making plant maintenance easier. It works well with strong rootstocks like Marubacaido, but it may also be utilized with semi-vigorous rootstocks like M-7 and MM-111. The M-9 interstocking is employed. It is cited as a possible usage for M-26, although it has the drawbacks of forming numerous burrknots, having greater vigor than M-9, and demonstrating greater variability in the growth of plant. The vigor loss is proportional to the length of interstocking, with the longer interstocking having a bigger impact on plant size reduction. It has produced excellent results when Marubacaido rootstock was used. Interstocking plants more than 20 cm apart has the drawback of promoting production of the sprouts which makes plants more susceptible to breaking.

4.5 USE OF PGRS FOR QUALITY PLANT MATERIAL: NEW TECHNIQUE IN PROPAGATION

One of the most essential variables in the growth of the fruit sector is the availability of high-quality planting material for propagation. As per the report of Planning Commission's Working Group on Horticulture, Plantation Crops, and Organic Farming for the XI 5-year Plan (2007–12), there were around 6330 registered nurseries in the public and private sectors. The total number of nurseries in Jammu and Kashmir is 77 in the public sector and 348 in the private sector. In order to fulfill the current need for more planting material, the manufacture of high-quality planting material is a must (Krishnan et al., 2014). Most of the apple cultivars that are produced in the nurseries don't possess lateral shoots after 1 year of the growth which is highly undesirable as far as the quality of planting material is concerned. Such plants also fetch lesser price and have little demand. Therefore, the maximum number of lateral shoots is an important consideration for the producers of apple trees. The plant quality parameters include height and diameter of the trees and the length and number of the lateral shoots. The formation of lateral shoots in young trees is the important aspect for inducing fruit production early and for quick economic returns in the orchard. If young trees are not managed properly then the leaders of the young trees hardly produce any lateral shoots (Jacyna and Puchala, 2004). The inability to create a suitable tree crown due to an inadequate number of feathers (sylleptic shoots) is restricting the formation of high-quality maiden trees. Some nursery trees are difficult to branch at the correct height to create crowns (Jacyna, 2001). Fruit trees and their varieties differ in their capacity to generate laterals. Because of their great apical dominance, many commercially significant cultivars develop few lateral branches in the nursery (Cline, 2000). According to Elfving and Visser (2005), the production of lateral branches is controlled by a balance of cytokinin and auxin levels, and diffusible auxin at high concentrations, generated mostly in the apical bud, suppresses the growth of lateral shoots. Traditional approaches for promoting branching (pinching leaves, shoot tips, or trimming shoot tips) may not always produce good results, hence bioregulators may be required (Csiszar and Buban, 2004). A number of techniques for causing lateral shoots have been devised. Plant bioregulators, in addition to nurserymen methods (pinching shoot tips or young leaves), have been shown to be effective in secondary shoot production. Tipping (pruning) can disrupt the apical dominance mechanism, encouraging buds that would otherwise be dormant (Elfving and Visser, 2007). All of these mechanical approaches are regarded to be ineffective compared to chemical applications. Promalin, the most widely applied branching agent, is a combination of benzyladenine (BA) and gibberellins (GA4+7). The fundamental method employed in the creation of premade seedlings as propagating material in apples has been the employment of phytorregulators. Both gibberellins and cytokinins can produce branching in apple tree seedlings because of growth stimulation of the spur (Greene et al., 2016). When applied in combination with the gibberellins (GA3 and GA 4+7) or applied alone in multiple applications, BA, the cytokinin-based regulator which is commonly applied as a chemical treatment of apple and pear trees enhances the formation of the lateral branches (Robinson and Miranda, 2014). The feathered nursery (3F, 5F, and

7F) apple plants' development through this technology has an important scope in the horticulture industry. Different branching agents produce a significant effect on the number of feathers of apple nursery plants. A maximum number of feathers (6.57) was recorded under BA+GA3 at 450 ppm (two times) with pinching, while the number of feathers (0.64) was recorded under untreated plants. It was noticed that the treatments in which BA was used in combination with GA3 resulted in a significantly higher number of feathers compared to the application of BA only (shahid *et al.*, 2017). This is because BA is responsible for overcoming apical dominance (Müller and Leyser, 2011) and enhancing feathering in apple nursery trees (Sazo and Robinson, 2011) while the main role of GA is to elongate feathers. GA_3 is known to increase the efficacy of BA by increasing the nutrient metabolites due to the nutrient sink created by GA_3 which is required for bud break to promote feathering. The most often utilized chemical for lifting the dormancy of summer buds is BA (Magyar and Hrotko, 2005). Synthetic cytokinin-active substances, such as BA, can also be employed efficiently for the same purpose. BA is most commonly used to train nursery plant canopies and young trees in the orchards that are planted new (Theron *et al.*, 2000). On the global market, there are several products that contain BA alone or with a combination of different plant hormones (primarily GA4+7) (Buban, 2000). Repeated BA treatments have a favorable effect and stimulate branching in the 1-year-old nursery trees of Idared/M.26, according to Hrotkó *et al.* (1996). The present findings are in agreement with the earlier reports of Palmer *et al.* (2011) who reported that in pears, when applied on its own, GA_3 enhanced feathering.

4.6 MICROPROPAGATION

Micropropagation is defined as a technology that permits for quick and large-scale multiplication of canopy material or rootstocks (Dobranszki and Silva, 2010) as well as generation of material free from viruses. This approach sparked a lot of excitement, but it was only utilized commercially for the rootstock multiplication for production of mother plants and then followed by conventional cultivation practices. Furthermore, because the phytosanitation conditions of the propagules stored under the aseptic conditions considerably shorten or may even eliminate the quarantine period, micropropagation improves the import/export of plant material. Plant propagation is independent of the season/time of the year as micropropagation procedures are performed in a controlled environmental setting (Bhatti and Jha, 2010). As a result, it can be done numerous times throughout the year. It's a highly sophisticated procedure that differs from variety to variety. However, irrespective of cultivar, tissue culture involves four stages of development: (i) in vitro establishment (ii) the multiplication (iii) the rooting and (iv) the acclimatization.

Totipotency, which means that every cell has ability to divide as well as to generate a whole organism's differentiated cells (Hartman *et al.*, 1997), is the key of micropropagation. The explant isolation in an appropriate and infection-free environment for the production of plantlets is essentially what the establishment phase entails. Explant types should be selected based on their capacity to adhere to in vitro conditions, with those containing the most meristematic tissue being favored. Explants like cuttings, seeds, meristems, buds, nodal segments, and stem apices may be utilized in in vitro establishment, and the degree of tissue differentiation used as well as the objective of micropropagation should be considered while selecting them. Because of their genetic stability, shoot tips or meristems can also be employed as explant (Bhattacharya and Singh 2018). Keeping in mind the goal of the in vitro culture, any of these explants could be used in apples (Kaushal *et al.*, 2005). For the establishment to succeed, explants must be disinfected with sodium hypochlorite or calcium before being placed in a culture medium. The inorganic salts (phosphorus, nitrogen, potassium, magnesium, calcium, boron, copper, cobalt, manganese, iron, iodine, and zinc), complex ingredients, organic chemicals, and passive substrates (vitamins, carbohydrates, and hormones) make up the culture media. In the establishment, medium dosages of cytokinin (0.5–1 mg L^{-1}) are employed and have a primal role to play. In most cases, the establishment period lasts 4–6 weeks. The multiplication process is the second step in which plantlets are multiplied in scale in the course of repeated subcultures using suitable multiplication methods, which results in parts that are either partitioned in small parts or are differentiated to produce new explants. Cytokinin levels should be increased during the generation of new buds/explants (Ashrafzadeh, 2020).

The material, after sufficient multiplication, is shifted to medium which lacks cytokinin for adventitious root development. The kind of auxin most commonly employed in this phase is AIB (indole-3-butyric acid). Root development is frequently hampered by light and cold conditions. Seedlings that have been kept in the dark (etiolated) have a better chance of rooting than those that have been cultivated directly in the light (Hartman *et al.*, 1997). The method of etiolation results in a number of morphological and the physiological changes, including the loss of phloem fibers, the stimulation of the elongation of cell wall, and the rise of undifferentiated parenchymal tissues. As far as apple trees are concerned, in vivo rooting is becoming increasingly popular. Explants are supplied with the auxin, i.e., rooting hormones, shortly after multiplication and consequently shifted to trays consisting of substrates like peat, sand, pine bark, etc. and are handled in same manner as when reproduced by the vegetative cuttings. Auxins like indoleacetic acid (IAA), indolebutyric acid (IBA), and naphthalene acetic acid are the most commonly utilized auxins for rooting cuttings mentioned in Figure 4.1. This reaction, according to Soni *et al.* (2011), can be explained by the propagating material's varied age, which may cause differential responsiveness to growth regulators. Following in vitro rooting, plants are acclimatized, which involves shifting plants to the substrate so as to give it water stress tolerance, disease resistance, as well as to transition the plant from the heterotrophic to an autotrophic stage (Soni *et al.*, 2011). Seedling acclimatization is the final and most challenging micropropagation step because of the low survival rate. Seedlings are moved into a greenhouse with a fogging system to prevent drying out. Because rootstock seedlings vary in size, it is essential that they be classified in order to achieve uniformity in mother plants (AbdAlhady, 2018).

FIGURE 4.1   Micro propagation of clonal apple rootstock MI-793. (a) the bud break in the axillary or terminal buds on the Murshage and Shookage medium having 1.0 mg/L BA, 0.1 mg/L IBA; (b) shoot elongation after a 2 month period; (c) production of shoots on the MS medium having 0.5 mg/L BA, 0.05 mg/L IBA; (d–f) root formation and callus development in shoots on media having 0.1 mg/L NAA, 0.7 mg/L IBA, and 0.8 mg/L IAA + 0.2 mg/l IBA; (g) rooted shoots in the two-step technique; (h) rooted shoots on supplemented media containing activated charcoal; (i) hardened plants that are cultivated in the paper cups.

4.7 OTHER BIOTECHNOLOGY BREAKTHROUGHS AND APPLE USES

Plants generated from apple in vitro cultures may have genetic differences (Dobranszki and Silva, 2010). Utilization of molecular markers to check genetic stability of micropropagated plants is an effective idea (Kacar et al., 2006). Rootstocks Merton 793 and MM-111 (Pathak and Dhawan, 2010, 2012b) and 'Gala' have substantially comparable DNA amplification profiles of the molecular markers like intersimple sequence repeats (ISSR) (Li et al., 2014, 2015). When the 'Gala' plantlets grown from the leaf fragments (Montecelli et al., 2000) or from the axillary buds of the rootstock EMLA 111 were evaluated using random amplified polymorphic DNA, no genetic differences were found (Gupta et al., 2009). Using random amplified polymorphic DNA (RAPD), Virscek-Marn et al. (1998) discovered little variability in plantlets derived from the leaves of 'Golden Delicious Bovey' and 'Goldspur'. Caboni et al. (2000) discovered polymorphism in plants obtained from leaves of 'Golden Delicious Bovey' and 'Goldspur' as opposed to no polymorphism in rootstock 'Jork 9' apical meristem derived plants. Modgil and Thakur (2017) used RAPD to examine if M-7 rootstock plantlets developed in vitro differed genetically from mother plants with no polymorphism. On the other hand, Noor Mohammadi et al. (2015) employed inter simple sequence repeat (ISSR) to discover substantial differences among in vitro generated plantlets and the mother plants, demonstrating up to 53 percent or 46 percent polymorphism in the rootstocks of M-7 and M-9. Pathak and Dhawan (2010), unlike Noor Mohammadi et al. (2015), discovered no genetic difference between tissue cultured clones of the rootstock MM-111 or in the rootstock Merton 793 (Pathak and Dhawan, 2012b) when examined by ISSR markers. These findings suggest that the number of subcultures, type of explant, method of shoot multiplication, as well as genotype, can also affect genetic stability of an offspring (Dobranszki and Silva, 2010). However, appropriate molecular method selection is also critical in discovering genetic polymorphism if any exists.

4.7.1 VIRUS ERADICATION

Meristem culture has become a powerful and successful tool for virus elimination from infected plants. In vitro thermotherapy-based methods, including combining thermotherapy with shoot tip culture, chemotherapy, micrografting or shoot tip cryotherapy, have been successfully established for efficient eradication of various viruses from almost all of the most economically important crops. Apple viruses can be successfully eradicated via thermotherapy (Paprstein et al., 2008; Hu et al., 2015, 2017; Vivek and Modgil, 2018; Wang et al., 2018a, 2018b), chemotherapy (Paprstein et al., 2013; Hu et al., 2015), combining the thermotherapy and chemotherapy (Hu et al., 2015), shoot tip culturing of adventitious buds (Wang et al., 2016), (Zhao et al., 2018).

4.7.2 INDUCTION OF POLYPLOIDIZATION IN VITRO

Induction of polyploidization could be brought about by two methods: mitotic polyploidization, which involves doubling of the somatic tissues (Ramsey and Schemske, 1998), or polyploidization employed directly in breeding experiments, hence, shortening the procedure by one generation. Because in vitro cultures provide a rather more regulated and controlled condition than greenhouse conditions, in vitro plant propagation therefore offers a significant potential to increase chromosomal doubling efficiency. Podwyszyńska et al. (2016) investigated the ploidy of the 70 genotypes of apple, and it was discovered that polyploids exhibit greater vegetative and reproductive features, as well as fruits, implying that producing polyploids in vitro has practical relevance. In vitro cultures of apples could also be employed to assess tolerance to biotic and abiotic stress. In vitro apple callus cultures are extremely useful for evaluating abiotic stress as well as examining various molecular mechanisms behind this process via the knowledge of gene expression and function.

4.7.3 BIOASSAYS AND BIOFERTILIZERS

The 'Royal Delicious' apple cultivar was utilized for novel bioassay for assessing polyphenol oxidase and browning activity in the response to organic volatile compounds produced by bacteria (Gopinath *et al.*, 2015). *A-proteobacterium* (Luteibacter rhizovicinus MIMR1), derived from the tainted in vitro 'Golden Delicious' cultures (Piagnani *et al.*, 2007), resulted in higher levels of the IAA (127 mg/l) and caused rooting in the tissue cultures of barley (*Hordeum vulgare* L.) implying that this apple-derived bacterium that promotes growth of plant can also function as a bio stimulant.

4.8 CONCLUSION

The goal of good propagation strategies, it is decided, is to provide the finest possible quality materials for new development regions. Quality apple material developed using advanced propagation techniques such as micro propagation and biotechnology approaches is beneficial for increased yield and production during a 25-year or longer life cycle. Advance propagation technologies are vital for the development of apple plants because they give a low-cost, high-quality material. As a result, choosing suitable planting materials is the first and most critical stage in the propagation process.

REFERENCES

AbdAlhady, M. R. A. 2018. *In vitro* propagation for peach rootstock (Nemaguard). *Egyptian Journal of Desert Research* 68(1): 1–13.

Abdel Wahab, S. A. S. and Al-Dujaili, A. H. 2001. The effect of pens taking, deformation and treatment dates on indole butyric acid on rooting apple apples. *Al-Rafidain Agriculture Journal* 32(3): 71–78.

Ashrafzadeh, S. 2020. *In vitro* grafting: Twenty-first century's technique for fruit tree propagation. *Acta Agriculturae Scandinavica, Section B – Soil & Plant Science* 70: 404–405.

Bhattacharya, A. and Singh, R. 2018. Optimization of micropropagation from nodal segments of apple (*Malus × domestica* Borkh.) cultivars golden delicious and red Fuji. *Current Journal of Applied Science and Technology* 31: 31–39.

Bhatti, S. and Jha, G. 2010. Current trends and future prospects of biotechnological interventions through tissue culture in apple. *Plant Cell Reports* 29: 1215–1225. https://doi.org/10.1007/s00299-010-0907-8.

Buban, T. 2000. The use of benzyladenine in orchard fruit growing: A mini review. *Plant Growth Regulation* 32: 381–390.

Caboni, E., Lauri, P. and D'angeli, S. 2000. *In vitro* plant regeneration from callus of shoot apices in apple shoot culture. *Plant Cell Reports* 19(8): 755–760.

Cline, M. G. 2000. Execution of the auxin replacement apical dominance experiment in temperate woody species. *American Journal of Botany* 87: 182–190.

Csiszar, L. and Buban, T. 2004. Improving the feathering of young apple trees in environment friendly way by modified benzyladenine application. *Journal of Fruit and Ornamental Plant Research* 351: 26–32.

Dobranszki, J. and Silva, J. A. T. 2010. Micropropagation of apple: A review. *Biotechnology Advances* 28: 462–488.

Dvin, S. R., Moghadam, E. G. and Kiani M. 2011. Rooting response of hardwood cuttings of MM111 Apple clonal rootstock to Indole butyric acid and rooting media. *Asian Journal of Applied Sciences* 4: 453–458.

Elfving, C. D. and Visser, B. D. 2005. Cyclanilide induces lateral branching in apple trees. *HortScience* 40(1): 119–122.

Elfving, C. D. and Visser, B. D. 2007. Improving the efficacy of cytokinin applications for stimulation of lateral branch development in young sweet cherry trees in the orchard. *HortScience* 42(2): 251–256.

Goldschmidt, E. E. 2014. Plant grafting: New mechanisms, evolutionary implications. *Frontiers in Plant Science* 5: 727. doi:10.3389/fpls.2014.00727

Gopinath, S., Kumaran, K. S. and Sundararaman, M. 2015. A new initiative in micropropagation: Airborne bacterial volatiles modulate organogenesis and antioxidant activity in tobacco (*Nicotiana tabacum* L.) callus. *In Vitro Cellular and Developmental Biology – Plant* 51: 514–523.

Greene, D. W., Crovetti, A. J. and Pienaar, J. 2016. Development of 6-benzyladenine as an apple thinner. *HortScience* 51(12): 1448–1451.

Grzesik, M., Górnik, K., Janas, R., Lewandowki, M., Romanowska-Duda, Z. and van Duijn, B. 2017. High-efficiency stratification of apple cultivar Ligol seed dormancy by phytohormones, heat shock and pulsed radio frequency. *Journal of Plant Physiology* 219: 81–90.

Gupta, R., Modgil, M. and Chakrabarti, S. K. 2009. Assessment of genetic fidelity of micropropagated apple rootstock plants, AMLA 111, using RAPD markers. *Indian Journal of Experimental Biology* 47: 925–928.

Hammoudi, M. R. and Dayoub, A. H. 1986. *Fundamentals of Vegetables and Fruits.* Aleppo University Publications, Ibn Khaldun Press, 468 p.

Hartman, H. T., Kester, D. E., Davies, F. T. and Geneve, R. G. 1997. *Plant Propagation: Principles and Practices.* 6th ed. Upper Saddle River, NJ: Prentice Hall.

Hartmann, H. T., Kester, D. E., Davies, F. T. and Geneve, R. L. 2002. *Hartmann and Kester·s Plant Propagation Principles and Practices.* 7th ed. Upper Saddle River, NJ: Prentice-Hall.

Hartmann, H. T., Kester, D. E., Davies, F. T. and Geneve, R. G. 2009. *Plant Propagation: Principles and Practices.* 7th ed. Upper Saddle River, NJ: Prentice Hall.

Hrotkó, K., Magyar, L. and Bubán, T. 1996. Improved feathering on one-year-old 'Idared' apple trees in the nursery. *HortScience* 8(1–2): 29–34.

Hu, G. J., Dong, Y. F., Zhang, Z. P., Fan, X. D., Ren, F. and Li, Z. N. 2017. Efficacy of virus elimination from apple by thermotherapy coupled with in vivo shoot-tip grafting and in vitro meristem culture. *Journal of Phytopathology* 165: 701–706.

Hu, G. J., Zhang, Z. P., Dong, Y. F., Fan, X. D., Ren, F. and Zhu, H. J. 2015. Efficiency of virus elimination from potted apple plants by thermotherapy coupled with shoot-tip grafting. *Australasian Plant Pathology* 44:167–173.

Jacyna, T. 2001. Studies on natural and chemically induced branching in temperate fruit and ornamental trees. *Acta Horticulturae* 13: 45–66.

Jacyna, T. and Puchala, A. 2004. Application of friendly branch promoting substances to advance sweet cherry tree canopy development in the orchard. *Journal of Fruit and Ornamental Plant Research* 12: 177–182.

Jauron, R. 2000. *Germination of Tree Seed.* Horticulture and Home Pest News Integrated Pest Management. www.ipm.iastate. Edu/ipm/hortnews/2000/8-11-2000/germtreeseed.html

Kacar, Y. A., Byrne, P. F. and Teixeira da Silva, J. A. 2006. Molecular markers in plant tissue culture. In: Teixeira da Silva, J.A. (ed.), *Floriculture, Ornamental and Plant Biotechnology: Advances and Topical Issues.* Vol. 2, 1st ed. Isleworth: Global Science Books Ltd., pp. 444–449.

Kaushal, N., Modgil, M., Thakur, M. and Sharma, D. R. 2005. *In vitro* clonal multiplication of an apple rootstock by culture of shoot apices and axillary buds. *Indian Journal of Experimental Biology* 43: 561–565.

Krishnan, P. R., Kalia, R. K., Tewari, J. C. and Roy, M. M. 2014. *Plant Nursery Management and Plant Nursery Management: Principles and Practices.* Jodhpur: Central Arid Zone Research Institute, 40 p.

Kumar, G. N. M. 2011. *Propagation of Plants by Grafting and Budding.* Washington State University. www.growables.org/information/documents/PlantPropagationGraftingBuddin.pdf (accessed 03.04.2019).

Li, B. Q., Feng, C. H., Hu, L. Y., Wang, M. R., Chen, L. and Wang, Q. C. 2014. Shoot regeneration and cryopreservation of shoot tips of apple (Malus) by encapsulation–dehydration. *In Vitro Cellular and Developmental Biology Plant* 50: 357–368.

Li, B. Q., Feng, C. H., Wang, M. R., Hu, L. Y., Volk, G. M. and Wang, Q. C. 2015. Recovery patterns, histological observations and genetic integrity in Malus shoot tips cryopreserved using droplet-vitrifcation and encapsulation-dehydration procedures. *Journal of Biotechnology* 214: 182–191.

Magyar, L. and Hrotkó, K. 2005. Effect of BA (6-benzyladenine) and GA4+7 on feathering of sweet cherry cultivars in the nursery. *Acta Horticulturae* 667: 417–422.

Marcon Filho, J. L., Rufato, L., Rufato, A. R., Kretzschmar, A. A. and Zancan, C. 2009. Aspectos produtivos e vegetativos de maciei-ras cv. Imperial Gala interenxertadas com EM-9. *Revista Brasi-leira de Fruticultura* 31: 784–791

Modgil, M. and Thakur, M. 2017. In vitro culture of clonal rootstocks of apple for their commercial exploitation *Acta Horticulturae* 1155: 331–335.

Montecelli, S., Gentile, A. and Damiano, C. 2000. In vitro shoot regeneration of apple cultivar Gala. *Acta Horticulturae* 530: 219–224.

Müller, D. and Leyser, O. 2011. Auxin, cytokinin and the control of shoot branching. *Annals of Botany* 107: 1203–1212.

Noor Mohammadi, Z., Fazeli, S., Sheidai, M. and Farahani, F. 2015. Molecular and genome size analyses of somaclonal variation in apple rootstocks Malling 7 and Malling 9. *Acta Biologica Szegediensis* 59(2): 139–149.

Palmer, J. W., Seymour, S. M. and Diack, R. 2011. Feathering of 'Doyenné du Comice' pear in the nursery using repeat sprays of benzyladenine and gibberellins. *Scientia Horticulturae* 130(2): 393–397.

Paprstein, F., Sedlák, J., Polak, J., Svobodová, L., Hassan, M. and Bryxiov, M. 2008. Results of in vitro thermotherapy of apple cultivars. *Plant Cell, Tissue Organ Culture* 94: 347–352.

Paprstein, F., Sedlák, J., Svobodová, L., Polak, J. and Gadiou, S. 2013. Results of in vitro chemotherapy of apple cv Fragrance. *Horticultural Science* 40: 186–190.

Pasqual, M., Norberto, P. M., Dutra, L. F., Cavalcante-Alves, J. M. and Chalfun, N. N. J. 2001. Rooting of fig (Ficus carica L.) cuttings: Cutting time and IBA. *Acta Horticulturae* 605: 137–140.

Pathak, H. and Dhawan, V. 2010. Molecular analysis of micropropagated apple rootstock MM111 using ISSR markers for ascertaining clonal fidelity *Acta Horticulturae* 865: 73–80.

Pathak, H. and Dhawan, V. 2012a. Evaluation of genetic fidelity of in vitro propagated apple (Malus × domestica Borkh.) rootstock MM 106 using ISSR markers. *Acta Horticulturae* 961: 303–310.

Pathak, H. and Dhawan, V. 2012b. ISSR assay for ascertaining genetic fidelity of micropropagated plants of apple rootstock Merton 793. *In Vitro Cellular and Developmental Biology – Plant* 48: 137–143.

Piagnani, M. C., Guglielmetti, S. and Parini, C. 2007. Identification and effect of two bacterial contaminants on apple organogenesis. *Acta Horticulturae* 38: 335–339.

Podwyszyńska, M., Kruczynska, D., Machlańska, A., Dyki, B. and Sowik, I. 2016. Nuclear DNA content and ploidy level of apple cultivars including polish ones in relation to some morphological traits. *Acta biologica Cracoviensia* 58(1): 75–87.

Ramsey, J. and Schemske, D. W. 1998. Pathways, mechanisms, and rates of polyploidy formation in flowering plants. *Annals Review of Ecology, Evolution and Systematics* 29: 467–501.

Robinson, T. L. and Miranda, M. 2014. Effect of promalin, benzyladenine and cyclanilide on lateral branching of apple trees in the nursery. *Acta Horticulturae* 1042(1042): 293–302.

Sazo, M. and Robinson, T. 2011. *The Use of Plant Growth Regulators for Vranching of Nursery Trees in NY State*. New York Fruit Quarterly.

Shahid, Q. D., Bisati, I. A., Mohammed, T. A., Abdul, W. W., Tawseef, R. B., Mohd, I. D., Ghulam, I. H., Amit, K. K. and Sajad, A. B. 2017. Effect of growth regulators and pinching on feather traits of apple nursery plants *cv*. Royal Delicious. *Vegetos – An International Journal of Plant Research* 30(2): 509–513.

Sharma, R. R. (2002). *Propagation of Horticultural Crops: Principles and Practices*. New Delhi: Kalyani Publishers.

Soni, M., Thakur, M. and Modgill, M. 2011. *In vitro* multiplication of Merton I. 793: An apple rootstock suitable for replantation. *Indian Journal of Biotechnology* 10: 362–368.

Srivastava, R. P. (1966). Research on horticultural crops at Chaubattia. *Indian Horticulture* 10: 9–11.

Theron, K. I., Steyn, W. J. and Jacobs, G. 2000. Induction of proleptic shoot formation on pome fruit nursery-trees. *Acta Horticulturae* 514: 235–243.

Upadhayay, P. K., Kharal, S. and Shrestha, B. 2020. Effect of indole-butyric acid (IBA) and wounding on rooting ability and vegetative characteristics of apple rootstock cuttings under Nepal conditions. *Journal of Agricultural Science and Practice* 5(4): 184–192.

Virscek-Marn, M., Javornik, B., Štampar, F. and Bohanec, B. 1998. Assessment of genetic variation among regenerants from in vitro apple leaves using molecular markers. *Acta Horticulturae* 484: 299–304.

Vivek, M. and Modgil, M. 2018. Elimination of viruses through thermotherapy and meristem culture in apple cultivar 'Oregon Spur-II'. *Virus Disease* 29: 75–82.

Vivian, R., Silva, A. A., Gimenes Jr, M., Fagan, E. B., Ruiz, S. T. and Labonia, V. 2008. Weed seed dormancy as a survival mechanism: brief review. *Planta Daninha*, 26: 695–706.

Wang, M. R., Chen, L., Liu, J., Teixeira da Silva, J. A., Volk G. M. and Wang Q. C. 2018a. Cryobiotechnology of apple (*Malus* spp.): Development, progress and future prospects. *Plant Cell Reports* 37: 689–709. https://doi.org/10.1007/s00299-018-2249-x

Wang, M. R., Cui, Z. H., Li, J. W., Hao, X. Y., Zhao, L. and Wang, Q. C. 2018b. In vitro thermotherapy-based methods for plant virus eradication: A review. *Plant Methods*14: 87.

Wang, M. R., Li, B. Q., Feng, C. H. and Wang, Q. C. 2016. Culture of shoot tips from adventitious shoots can eradicate apple stem pitting virus but fails in apple grooving virus. *Plant Cell, Tissue and Organ Culture* 125: 283–291.

Wertheim, S. J. and Webster, A. D. 2003. *Propagation and Nursery Tree Quality*. Wallingford: CABI Publishing.

Zhao, L., Wang, M. R., Cui, Z. H., Chen, L. and Wang, Q. C. 2018. Combining thermotherapy with cryotherapy for efficient eradication of apple stem grooving virus (ASGV) from infected apple in vitro shoots. *Plant Diseases* 102: 1574–1580.

5 High Density Plantation

Shazia Hassan, Mohammed Tauseef Ali,
Ab Raouf Malik, Rafiya Mushtaq,
Shamim A. Simnani and Khalid Mushtaq

CONTENTS

5.1   INTRODUCTION

The conventional system of planting involving seedling rootstock has a long juvenile period and is low yielding with low quality produce (Majid et al., 2018). This low productivity is due to low plant density, low yielding varieties, improper management of orchards and less technological involvement. The ongoing decline in cultivable land and escalating land and energy expenses together with the growing demand of horticulture produce has led to the concept of high density plantation. High

DOI: 10.1201/9781003239925-5

density plantation can be understood as a system of planting in which a higher number of plants can be accommodated within a unit area compared to the conventional system. High density plantation is one of the technologies for achieving a higher productivity of fruit crops. High density plantation was introduced to get maximum returns per unit of inputs and resources by efficiently utilizing vertical and horizontal space. As the high density plantation is a relatively new concept, many physical and agronomical factors such as climatic conditions, soil fertility, rootstock and scion vigor, methods for training and pruning and orchard management should be taken into consideration. The yield and quality of the produce are the two most important components of productivity. High density plantation aims to achieve the twin requirements of productivity and plant health by maintaining a balance between the vegetative and reproductive load.

5.2   MERITS AND DEMERITS OF HIGH DENSITY PLANTING SYSTEMS

- High density plantation induces precocity which results in an earlier return of the investment.
- The plants under high density plantation are easily operated and give higher returns per unit area.
- The modern horticultural techniques for input saving such as drip irrigation, fertigation, mechanical harvesting, etc. have better responsiveness in high density plantation.
- High density plantation improves fruit quality due to maximum light distribution and interception within the tree canopy.
- Harvesting and pruning from the ground or from a short stool makes efficient use of labor another advantage of using a high density plantation system (Anonymous, 2007).
- High density plantation makes possible fruit crop production mechanization. It enables effective use of water, solar radiation, fertilizers, fungicides, pesticides and weedicides.
- Even though high density orchards have many merits, the demerits should also be kept in mind. The main demerit is the orchard establishment cost as compared to traditional system of planting.

5.3   ESTABLISHMENT OF HIGH DENSITY PLANTATION

5.3.1   SITE SELECTION

Site selection is an important factor for the establishment of a successful orchard. It is important to critically assess the site before proceeding toward planting. Factors like soil conditions, drainage and weather information are the first significant steps to be taken into consideration. Soil with a 4–5 feet minimum rooting depth and with drainage characteristics of surface water is desirable. Apple trees cannot tolerate prolonged periods of waterlogged soils during the growing season. A consistent, close and clean source of water is required for irrigation and spraying; overhead frost protection and evaporative cooling is also required. The direction of the slope is another concerning factor for planting. South facing slopes increase the chance of blooming earlier while North facing slopes delays blooming; East facing slopes decreases disease potential as the early morning sun dries the foliage. The market should be close to a population with accessibility of good road conditions (Parker et al., 2007).

5.3.2   SITE PREPARATION

The orchard site prior to planting should be prepared 6 months minimum to 1 year. Further, even before the trees are planted, an ideal soil environment should be created because once the trees are in the ground, it is nearly impossible to fix certain site problems. While preparing the site, the following points should be kept in mind (Parker et al., 2007):

- pH and fertility test for soil.
- Eradicate noxious and perennial weeds at and around the orchard site.
- Arrange a proper drainage system.
- Install proper irrigation systems.
- Fumigate the soil, if required, particularly on replant sites.

5.3.3 LAYOUT

The orchard is not only affected by the environment but also by the slope of the land and tree spacing. The ideal slope of the field should be between 4% and 8% to allow cold air to drain and prevent the crop from frost damage, while slopes greater than 8% make machinery operation difficult (Berkette et al., 2007). A north-south row orientation is preferred to ensure that trees receive the maximum amount of sunlight possible throughout the growing season. North-south rows produce higher yields and higher quality fruit (Anonymous, 2009). For the layout, mark the corners of the block with stakes. Lay a straight line between the corners. Mark the place for installation of the posts. The post to post distance should be kept 9 m and row to row should be 3 m.

5.3.4 INSTALLATION OF A TRELLIS SYSTEM

Trellises are very expensive therefore must be built right the first time so they can last a generation. Trellising is a means to control the growth of the trees, enhance the fruit-bearing area of the trees, increase light interception throughout the tree canopy and improve fruit quality. The trellis should be installed before planting the trees. The main components are the end posts (14ft long, 5–6 inches in diameter, driven 4ft into ground and the length depends on the desired tree height), line posts (14ft long, 4–5 inches in diameter, driven 3–4ft deep), tieback post (8ft long, 5–6 inches in diameter, driven 4ft into ground) and wires (high tensile, class 3, 12.5 gauge, 0.1 inch galvanized steel wire). After marking the points on the ground, push the tieback post to a depth of 4ft into the ground stretching it at an angle of 10° in the direction of the wire. After fixing the tieback post, the end post should be fixed into the ground. End posts should be pounded or vibrated into at least 3–4 feet of undisturbed soil. Ideally, the post-wire-ground position should form an equilateral triangle with 60° angles which equalize the forces in the post and wire. After fixing the end posts, the line posts should be sunk into the ground at a depth of 4ft. The line posts should be kept 30ft apart from each other for the strength of the trellis. After fixing the posts, string the wires in horizontal position at 2.5ft, 5.5ft, 7.5ft and 9.5ft above the ground.

Different types of materials used for building trellises are wood, metal and plastic.

5.3.4.1 Wood

Working with wood is easy. It is best used for the trellis that is easy to move or replace. Many types of wood are used for building trellises like bamboo (light weight, strong enough to hold the weight of fruiting plants and can last up to 10 years), pine (soft wood but does not hold up well on exposure to soil, water and insects and if untreated will rot within few years), oak (hard wood that is stronger than pine, lasts longer than pine but difficult to work with), cypress (soft wood contains natural oils that preserve it from decay and insect damage) and cedar (soft wood holds up well against rot and insect damage and more expensive than other types of wood).

5.3.4.2 Metal

For durable trellises, metal might be the right choice of material. Different types of metals are available for trellises. Some commonly used metals are wrought iron (heaviest material, sturdy, can hold plenty of weight, trellis needs a prop against a wall or fence, resistant to rust, tough, durable and expensive), stainless steel (heavy, holds lot of weight, difficult to move, expensive), copper (heavy, does not form rust however corrodes over time) and aluminum (much lighter, easier to work and move, does not rust).

5.3.4.3 Plastic

Plastic is light weight and does not rot, rust or corrode.

5.3.5 Drip System Installation

Drip irrigation is a type of microirrigation system that distributes water slowly to the plant's root zone. Millions of farmers lives worldwide have changed through drip irrigation by increasing yields while conserving water and fertilizer.

The drip system of irrigation has various advantages which are as follows (Anonymous, 2009).

- Water is used at a maximum level.
- During spraying operations or on windy days water can be applied.
- The foliage is not wetted, which reduces disease problems and preventing materials of crop protection from being removed from the leaf canopy or maturing fruit.
- Water does not come into contact with the produce; therefore the risk of fruit contamination associated with poor water quality is reduced.
- It is suitable for fertigation as well.

The drip irrigation system's components (Figure 5.1) include the water source, the pump unit, the control head, the main and sub-main lines, laterals, valves and emitters and other accessories. The pump unit extracts water from the source and delivers it at the proper pressure to the pipe system. The control head is equipped with valves that regulate the discharge and pressure throughout the system. It utilizes screen filters and graded sand filters to remove suspended particles from the water. Certain control head units include a fertilizer or nutrient tank that gradually adds a measured fertilizer dose to the irrigation water during irrigation; this is one of the major benefits of drip irrigation over other methods. Water is supplied to the fields from the control head via mainlines, sub-mains, and laterals. These hoses are made of PVC or polyethylene and should be buried below ground because they degrade rapidly when exposed to direct solar radiation. Lateral pipes typically have a diameter of 13–32 mm. Emitters or drippers are used to regulate the rate at which water is discharged from the lateral to the plants. They are typically more than one meter apart and contain one or more emitters for a single plant, such as a tree. The emitters are designed in such a way that they provide a specified constant discharge that is not significantly affected by pressure changes and does not easily block (Brouwer et al., 1985).

5.3.6 Time of Planting

Planting under Kashmir region's conditions is generally done in the month of March and may vary according to agro-climatic conditions.

5.3.7 Planting Material

A number of decisions should be made prior to planting an apple orchard such as variety selection, which rootstocks to use, tree spacing and where to place pollinizers. To obtain the ideal size for the tree to maintain density in the orchard the selection of a proper rootstock is very important. Trees propagated on dwarfing rootstocks are required for high density orchards. It is not enough to have small trees; the trees must bear fruit early in the orchard's life.

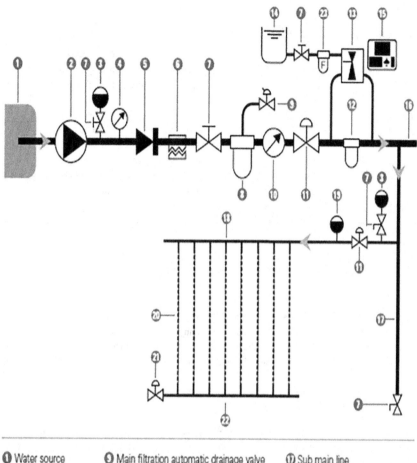

❶ Water source	❾ Main filtration automatic drainage valve	⑰ Sub main line
❷ Pumping station	⑩ Water meter	⑱ Distribution line
❸ Air valve	⑪ Hydraulic valve	⑲ Kinetic valve (vacuum breaker)
❹ Pressure gauge	⑫ Secondary filtration unit	⑳ Dripperline
❺ Check valve	⑬ Dosing unit	㉑ Flushing valve
❻ Shock absorber	⑭ Fertilizer tank	㉒ Flushing manifold
❼ Manual valve	⑮ Irrigation controller	㉓ Fertilizer filter
❽ Main filtration unit	⑯ Main line	

FIGURE 5.1   Drip irrigation system (Anonymous, 2015)

5.3.8 Rootstocks

In apple trees, rootstocks affect the rate of growth, final size, early bearing, size of fruit, yield capacity and nutritional response. Different rootstocks have different characteristics (Table 5.1) such as compatibility responses with various scions, responses to insects and diseases and behaviors in different soil conditions. Rootstocks have also shown different behaviors toward temperatures. It is therefore essential to select the right rootstock according to soil, site and variety. Many rootstocks are being used but M.9 is widely being used as dwarfing rootstock, however this rootstock has a

TABLE 5.1
Apple Rootstocks with Characteristics

Rootstock	Characteristics
M.9	Dwarf; tolerant to high soil pH; susceptible to the woolly apple aphid; resistant to collar rot
M.27, P.22	Ultra-dwarf
MM.104 and MM.106	Resistant to wooly apple aphids
MM.111	Semi-vigorous; drought tolerant
G.41	Dwarf; highly resistant to fire blight and phytopthora
G.16	Dwarf; resistant to fire blight
G.935	Winter hardy; highly resistant to fire blight and phytopthora
Northern Spy	Resistant to the wooly apple aphid
B.9, B.491, MAC.9, O.3, P.2, Novole, G.30, G.65	Resistant to crown and root rot

TABLE 5.2
Bearing Habits of Apple Varieties

Spur Type Varieties	Standard (Tip Bearer)	Mixed (Spur and Tip Bearers)
Red Chief, Oregon Spur, Golden Spur, Starkrimson, Silver Spur, Well Spur	Honey Crisp, Empire, Granny Smith, Irish Peach, Red Free, McIntosh, Maharaji, Rome Beauty, Fuji	Ginger Gold, Golden Noble, Lord Lambourne, Northern Spy, Vista Bella, Yellow Newton, Tydeman's Early Worcester

brittle and shallow root system and needs staking. It is susceptible to the wooly apple aphid but is resistant to collar rot.

5.3.9 Varieties

The selection of cultivars is the key to the profitability of the new orchard. While selecting the cultivar following considerations are important:

- Climatic suitability
- Disease resistance
- Pollination
- Harvest schedule and management
- Market prices and facilities.

Different varieties are being used although their growth habits and characteristics can also influence the high density plantation. For high density plantation, most adopted varieties are spur varieties as they have compact growth with more spurs and less shoot formation compared to standard varieties. Different varieties with their bearing habits are given in Table 5.2.

5.3.10 Pollinizers

Apple production relies heavily on pollination to be profitable. No full crop will be set unless a compatible pollinizer is used on apple varieties that are not self-fertile. The eventual size and quality of the fruit are influenced by successful pollination and the formation of a large number of healthy seeds. When pollination occurs improperly it can lead to lower yields and distorted fruit. Therefore, choosing a compatible pollinizer variety with the main variety having an overlapping flowering period is also critical. Even cultivars that have some degree of self-compatibility produce a superior crop when pollinated by a different cultivar. Consequently it becomes necessary to include suitable pollinizers when a new apple block is established. Nearly 25%–33% pollinizers should be planted in an apple orchard (Anonymous, 2017). The recommended pollinizers are 'Tydeman's Early Worcester', 'Golden Delicious', 'Spur Winter Banana', 'Lord Lambourne', 'Yellow Newton', 'McIntosh', 'Red Gold', 'King of Pippin' and 'Granny Smith'.

5.3.11 METHOD OF PLANTING

Prior to planting, pits with a depth of 60 cm are dug 2 weeks before. The first pit is dug at a distance of 18 inches from the post, and then the distance between the pits in a row is kept as 1 m. The pits should not be dug too small or deep. Too small pits might require the removal of roots and too deep pits result in the lowering of bud union too close to the ground. The pits are then filled with good loamy soil and organic matter. In the center of the pit, planting is done by scooping the soil and placing the young tree. Ideally the tree should be planted at a depth so that the bud union is 8–10 cm above the ground after settling of the soil. Loose soil is filled up in the remaining area and pressed to remove air gaps. After planting, the plants should be drenched immediately to ensure good soil-root contact. Mix the fertilizer and drench over the root area of the tree. The irrigation should be ensured for at least 10 days.

5.3.12 IRRIGATION

For any grower the economic success of a high density orchard depends upon the high quality produce with economic size. For attaining an economic size of the produce many factors contribute, and irrigation is one of them. For any newly planted high density plantation the irrigation has a critical role in improving and maximizing the growth of young newly planted trees. High density plantation economic success depends upon the yields which are significant in the 3rd, 4th and 5th years which is expected only if the growth of the trees is excellent during first 3 years. This tree growth can be optimized through adequate water supply to the plants. Moreover the irrigation is important for the uptake of different nutrients. So in order to provide adequate and timely water supply during the critical stages of tree growth drip irrigation is adopted in high density plantations. The concept of drip irrigation is preventing moisture stress rather than relieving the moisture stress, and the response to this approach is positive for the crop. At each base of the tree, drip irrigation supplies 2–8 L/hr. Water is delivered continuously (typically daily as needed) through this system. Moreover, only the rooting areas are watered, not between the rows. The irrigation schedule depends upon how much water the trees need per week and how much water the soil can hold based on water holding capacity. A crop's water requirement can be expressed as evapotranspiration (ET) which is the amount of water transpired by the plant and evaporated from the soil surface; it is expressed as millimeters or inches of water used per day. For optimum crop production, the water use must be replenished by irrigation. With a drip irrigation system, irrigation should be done daily, on alternating days or every third day as needed (Anonymous, 2009).

5.3.13 MANURES AND FERTILIZERS

Early on, the apple trees should grow quickly to reach the desired height, which is usually 10ft or 3.5 m. This aids in the accomplishment of two objectives: establishing the leader's dominance (which controls the tree's growth) as well as the growth of fruitful lateral branches or short scaffolds which will generate large early yields (Anonymous, 2009). As in the initial years of high density plantation establishment, the tree growth is very important therefore the proper quantity of manures and fertilizers becomes necessary for good quality fruits and high yields over an extended period. The nutritional requirements of the vegetative and reproductive growth stages are distinct and must be met accordingly (Ahad et al., 2018). However, before the application of manures and fertilizers it is important to test the soil for the pH and fertility so that the nutrients are applied accordingly. The pre-plant year, planting time and foliar feeding, if necessary, are the three times that will provide the necessary fertility to freshly planted apple trees. The pre-plant year is the perfect time to comprehensively incorporate organic matter, phosphorus, potassium and lime, as these are all needed to maximize orchard productivity. During the planting time the main goal is to boost as much growth as possible of both the root system and the leader for the quick establishment of the tree and bearing area; therefore the newly planted trees need adequate quantities after planting. Foliar applications are required at the time of additional need so that a nutrient deficiency can be avoided. During the foliar application the controlled concentration should be used so that it does not cause damage to the plant (Anonymous, 2009).

5.4   CANOPY MANAGEMENT

In high density orchards the growth of trees is controlled as a higher number of plants is to be accommodated per unit area. For controlling growth the trees are trained using different training systems which are different than those in standard orchards. The primary objective of training is to develop a strong framework that will support fruit production. The main motive for adopting a training system in high density orchards is to enhance the yield and quality of fruits by increasing the fruiting area and improving the sunlight penetration and diffusion. There are many different systems of training and generally the Tall Spindle is one such system that works best with the tree's natural growth and allows uniform tasks in the orchard.

5.4.1   TALL SPINDLE

This new planting system achieves the goals of very high early sustained yields and excellent fruit quality while moderating the initial investment. The Tall Spindle system has the following advantages (Anonymous, 2009):

- High early yields make it the most profitable
- Requirement of minimal labor for pruning and training once the tree is established
- Uses branch bending as a natural tree growth control
- Reduced spraying costs.

The important components of this system are:

- Highly feathered nursery trees (10–15 feathers)
- Selection of fully dwarfing rootstocks
- Minimal pruning at planting
- Branch bending to induce fruiting
- No permanent scaffold branches
- Limb renewal pruning to remove and renew branches as they get too large.

5.4.1.1   Pruning and Training Principles
- Only keep branches that are half the diameter of the leader.
- Only weak branches should be kept.
- Branches that are upright and strong in growth should be removed.
- Horizontal and pendant branches should be kept.

- Thinning cuts should be primarily made; tipping cuts should only be used in exceptional circumstances, such as forcing laterals in a branchless area.
- Recycle back to Axis or make Dutch or bevel cuts on branches older than 3–4 years.
- In the summer (22 June–15 July), tie down or weight as many branches as possible, particularly fresh succulent branches.

5.4.2 TRAINING THE NEW APPLE TREE

At planting minimal pruning is done. The highly feathered nursery trees with 10 to 15 small branches along the trunk require almost no pruning. If the tips of the feathers are damaged or dehydrated during the packaging or planting, then only the heading should be pruned.

5.4.2.1 Goals and Tasks for the Tree for Its First Year

- The leader should not be cut and should be allowed to grow as tall as possible, reaching the top wire by the 2nd year.
- Terminal growth on feathers of at least 6–8 inches should be encouraged, allowing the tree to fill its allotted space.
- Broken or damaged branches should be removed.
- Cut the feathers that are more than 50% of the diameter of leader in order to preserve the dominance of leader.
- Tie down feathers to below horizontal, especially if they are vigorous.

5.4.2.2 Training of the Tree in the 2nd Year

After the first year of planting is completed, the tree has a tall straight leader with a large number of fruitful side branches. The scaffold branches that are more than 50% the diameter of the leader should be cut. Dead/diseased/damaged branches, if any, should be removed. The scaffold branches should be tied down. Also the root suckers/shoots emerging on the main trunk should be removed in the spring.

5.4.2.3 Training of the Tree in the 3rd to 5th Years

- The shoots (more than 50% the diameter of leader) that compete with the leader should be removed. Retain the weaker shoots and remove the shoots if any top of leader. Low scaffolds should be removed as they hinder with cultural operations.
- Until the 4th or 5th year the tree should have a well-established and bearing leader. If the leader has been pulled over by the fruiting branches, then renew it to a single side branch. The scaffold branches that are large, especially in upper half of the tree, should be removed. Every year only 1–3 scaffold branches should be removed. Broken, damaged, strong uprights branches and root suckers should be cut. Also the lower branches that interfere with the cultural operations should be removed. A tall spindle tree that is 5 years old should fill its space with fruiting wood, and the sunlight should penetrate throughout the tree.

5.5 CROP LOAD MANAGEMENT

Cropping management in high density orchards during the first 4 years is crucial for sustaining a healthy balance between the vegetative development and cropping when the trees bear fruit. Young apple trees on precocious dwarfing rootstocks frequently overset in the 2nd or 3rd year, culminating in biennial bearing as early as the 4th year. This results in increased vigor in the 4th year precisely when the trees have reached the capacity of their assigned space and reduced vigor is required. Different varieties have a biennial bearing tendency which must be included into the crop loads allowed on immature trees. A crop loads of 6 fruits/cm^2 trunk cross-sectional area (TCA) for annual cropping cultivars like Gala (20–25 apples/tree in the 2nd year; 30–50 apples/tree in the 3rd year; 80–100 apples/tree in the 4th year). A crop loads of 4 fruits/cm^2 TCA for slow growing and biennial yielding cultivars such as Honeycrisp (12–18 apples per tree in the 2nd year; 20–35 apples per tree in the 3rd year; 40–70 apples per tree in the 4th year). Within each year, the lower end of the range should be chosen for trees with low vigor and the upper end for trees with strong vigor (Robinson, 2008b).

5.6   ECONOMICS

For assessing the net economic gains of adopting high density apple technology in Kashmir, a partial budgeting approach was used. High density apple systems required higher costs as compared to traditional apple plantation systems. Maintenance and human labor costs accounted $1181 and $13655 US dollars per hectare, respectively. Further, an increase in income was $16293 US dollars per hectare with an increase in apple yield to 53.6 MT per hectare; therefore the credit side demonstrated large gains totaling $1638 US dollar per hectare. Returns increased by $3094 US dollars per hectare as a result of the net change in returns. Hence, the adoption of high density apple systems might be said to have enhanced the livelihood of its growers by creating greater employment and income (Wani et al., 2021).

5.7   ORCHARD CARE

Once the high density orchard is established it requires essential care for higher production. The orchard floor should be kept clean. Weeds must be controlled throughout the season as they compete with the plant which affects its growth. The irrigation should be given regularly as the plants have limited root system. Mulches can also be used to conserve moisture in order to prevent water stress. The adequate soil fertility levels should be maintained. The insect pests should be managed properly so that they don't interfere with the tree growth. Disease management is very important for controlling diseases like scab, *Alternaria*, powdery mildew, etc. These diseases can be managed by spraying proper and registered fungicides at the proper growth stage of tree.

5.8   CONCLUSION

Apple growers around the world are looking for better apple orchard systems with higher yields, better fruit quality and lower production costs per unit of production. By involving high density plantation systems these characteristics are full filled. The Tall Spindle system's higher yield and quality can result in significant cost savings per unit of output. Planting high-priced types improves profitability more than cutting expenses. Furthermore, current initiatives to mechanize some or all of the pruning, harvesting and tree training processes may help to cut production costs and increase the profits of apple growers in the world (Robinson, 2008a). The biggest obstacle for growers adopting high density apple orchards is the initial cost of establishment and water availability. However, with the Government assistance in facilitating the finance and subsidy for initial cost of establishment for high density orchards in union territory of Jammu and Kashmir, farmers have shown a lot of interest; therefore a lot of area has been converted to high density apple orchard systems. Likewise, for other apple producing regions in the world wherein this production system is lacking among farmers, the government support can provide impetus for early adoption of this apple production technology.

REFERENCES

Ahad, S., Mir, M.M., Ashraf, S., Mumtaz, S. and Hamid, M. 2018. Nutrient management in high density apple orchards: A review. *Current Journal of Applied Science and Technology* 29(1): 1–16.

Anonymous. 2007. *Developing the Tree Fruit Industry in British Columbia*. West Georgia Street, Vancouver, BC: Ference Weicker & Company, pp. 76.

Anonymous. 2009. *Planting New Apple Orchards in Ontario*. Ministry of Agricultural Food and Rural Affairs. Retrieved from www.omafra.gov.on.ca/neworchard/english/apples/index.html

Anonymous. 2015. *Drip Irrigation Handbook; Understanding the Basics*. Netafim. V 001.02–2015.

Anonymous. 2017. *Recommendations for Successful Apple Pollination*. Department of Primary Industries and Regional Development, Government of Western Australia. Retrieved from www.agric.wa.gov.au/pome-fruit/recommendations-successful-apple-pollination.

Berkette, D., Bradshaw, D., Richards, S. and Cromwell, M. 2007. *Extension and Research for the Commercial Fruit Tree Grower in Vermount and beyond*. Retrieved from https://www.uvm.edu/~orchard/fruit/tree-fruit/tf_horticulture/ForBeginners_Apples January2007.pdf

Brouwer, C., Prins, K., Kay, M. and Heibloem, M. 1985. *Irrigation Water Management: Irrigation Methods*. Training Manual No. 5. FAO Land and Water Development Division.

Majid, I., Khalil, A. and Nazir, N. 2018. Economic Analysis of high density orchard. *International Journal of Advance Research in Science and Engineering* 7(SI-4): 821–829.

Parker, M., Unrath, C.R. and Safley, C. 2007. *High Density Apple Orchard Management*. NC State Extension Publication. Retrieved from https://content.ces.ncsu.edu/high-density-apple-orchard-management.

Robinson, T.L. 2008a. The evolution towards more competitive apple orchard systems in the USA. *Acta Horticulturae* 772: 491–500.

Robinson, T.L. 2008b. Crop load management of new high-density apple orchards. *New York Fruit Quarterly* 16(2): 3–7.

Wani, M.H., Bhat, A. and Baba S.H. 2021. Economic evaluation of high density apple (Ex-Ante) in Kashmir. *International Journal of Fruit Science* 21(1): 706–711.

6 Flowering, Pollination and Fruit Set

Amardeep Kour, Rifat Bhat, Chetak Bishnoi and Navjot Gupta

CONTENTS

6.1 INTRODUCTION

The most critical phenological phases of apples are the flowering, pollination and fruit set (and the yield components are determined in these periods). The floral conversion in the flowering

DOI: 10.1201/9781003239925-6

development is the most important phenomenon. After a certain period of vegetative growth, the plants come into flowering. For most woody species, including apples, flowering is not activated by a single environmental factor but a varied range of factors like temperature, light, nutrition, water supply and growth regulators may all govern the response. Not all potential flowering sites (bud meristems) respond to the prevailing conditions in a similar manner, as only part of the meristems truly become propagative. In apples, flower bud formation is usually used to define the entire sequence of events leading to the formation of flower buds. It can be divided into three successive developing processes (floral induction, differentiation and anthesis). Each of these steps is amenable to independent regulation. The floral induction which occurs during the period of active growth and denotes the process by which the meristem becomes committed to form flower buds. Floral differentiation refers to the morphologic change of the apex leading to the formation of an inflorescence and starts around the time of cessation of shoot growth and continues until anthesis. Anthesis is the terminating stage of the flowering process during which the stamens produce pollen and the pistil is receptive to pollination and fertilization. Effective fertilization involves a pollen transfer to the stigma, pollen germination, and active pollen tube growth, which results in fruit set. The first step of a successful apple pollination is the transfer of pollen to the stigmatic surface (typically vectored by bees) followed by an adhesion of pollen grains to the papilla cells of the stigmatic surface. The deposited pollen hydrates and germinates, and then pollen tubes penetrate the stigma and grow down the style. Pollen recognition occurs both on the stigmatic surface and within the style. Pollen source and temperature have an incredible influence on the rate of pollen tube growth. The amount of the stigmatic surface covered by the germinated pollen of apples depends on the pollen donor and the environmental temperature at the time of pollination. Most apple cultivars are reported to be either self-incompatible or semi-compatible and require cross-pollination to set fruit in marketable quantities. The effective pollination leads to fruit set. Fruit set is the change from ovary to a quickly growing young fruit which is started after successful pollination and fertilization.

6.2   FLOWERING IN APPLES

Flowering is a complex process of morphological and physiological stages under the regulation of a number of external and internal factors. The different stages of flowering towards fruit development includes flower initiation, flower differentiation, fertilization, fruit set and fruit development.

Most apple cultivars possess epigynous and hermaphroditic flowers (Kraus and Ralston, 1916) and a syncarpous gynoecium with typically five locules. Such flowers develop into pome fruit. The five carpels resemble drupes with a fleshy exocarp and mesocarp and a tough membranous endocarp (Zielinski, 1955). Two hypotheses exist concerning the nature of these tissues, the receptacular or axial, and the appendicular hypotheses. Every description of the pome involves terminology drawn from these hypotheses. An apple fruit is most commonly described as possessing a pith (fleshy part of core) and cortex (flesh outside of the core line) as per the receptacular hypothesis. According to appendicular hypothesis (Esau, 1965), the carpellary tissue (core) and the floral tube or hypanthium (flesh on the outside of the core line) are present in the fruit. The evidence for the receptacular hypothesis follows (MacDaniels, 1940): 1) Leaflike structures rarely emerge on the sides of the fruit like on the sides of a stem. 2) The primordia of the floral organs are forced up together by the expansion of an intercalary meristem. 3) Pith is uninterrupted from the spur all the way through the pedicel to the center of the fruit. 4) Rosaceous floral organs are as frequently attached to a disk as to a stem. 5) Vascular traces supply more than one organ. A sepal trace supplies one dorsal carpellary bundle, one stamen bundle and the median bundle of the sepal. A petal trace supplies bundles to three stamens, one to the petal and one lateral bundle to each sepal. The remaining vascular tissue forms the ten ventral carpellary bundles, which supply one bundle to each ovule. The dorsal and ventral carpellary bundles end in the five styles of the pistil. The evidence for the appendicular hypothesis follows (MacDaniels, 1940): 1) The basic premises are that (a) a flower is a determinate stem with appendages; (b) the differentiation of a floral organ is initiated when its vascular trace leaves the vascular cylinder; and (c) the vascular structure is the key to homologies. 2) Relative anatomy of the *Rosaceae* reveals fusion of sepal, petal and stamen traces in the floral tube (hypanthium), which merges with the compound ovary. 3) If the pome is partially constituted of receptacular tissue, the carpel traces must double back prior to entering the carpels and styles, but in fact they pass directly through the carpels to the styles. 4) Variation in cells, color or levels of oxidation of floral tube and carpel tissue take place in cross-sections of rosaceous fruits counting some species and cultivars of apples. A thorough comparative analysis of carpellary vascularization is presented by Sterling (1965) who supports the appendicular hypothesis. Additional evidence for the appendicular hypothesis comes from cytochimeras (Blaser and Einset, 1950). The apical meristem of apples has four preliminary layers which may vary in ploidy: ZX-~X-~X, ZX-~X-~X and 2x-2x2x-4x. Blaser and Einset (1950) evaluated the arrangement of 4x tissue with the distribution predicted on the basis of the receptacular and appendicular hypotheses. As per the receptacular hypothesis, most of the cytochimeral fruit inside the ring of sepal and petal bundles must be tetraploid. While in 2x-2x-4x chimeras, 4x tissue is restricted to the pith of the pedicel beneath the vestigial stem apex and to the traces to the floral organs. In 2x-2x-2x-4x chimeras, 4x tissue is present in the pith of the pedicel and in the traces to the sepals and petals. This arrangement supports the appendicular hypothesis, as per the authors. Unusual types of apple flowers have been interpreted as supporting the appendicular hypothesis. Only some apple cultivars are considered to have apetalous flowers (Brase, 1937). The flowers are comprised of five sepals; five sepal-like, green petals; no functional stamens; two whorls of pistils; a proximal whorl of five and a distal whorl of 9–10 pistils. The vascular supply to the sepals, petals and proximal carpels is just like hermaphroditic flowers. Though, the five petal bundles at the level of the distal whorl of carpels divide to appear as 15 bundles which unite as single dorsal and two ventral carpellary bundles of each of the distal carpels. Apple flowers may show nongenetic abnormalities such as extra sepals or petals or pistils replaced by a leafy shoot (Eames, 1961; Tukey, 1937). They were considered to arise from a variation in a floral meristem to a vegetative meristem. 'Vegetative' flowers were supposed to support the hypothesis that a flower is homologous with a leafy shoot (Tukey, 1937). The pedicel is included in the vascular system of a flower or fruit. A study of the pedicel of an adhering 'York Imperial' fruit revealed that a vascular cambium in the central zone of the pedicel was active from 12 to 48 days subsequent to bloom (Barden and Thompson, 1963). The completion of xylem took place 7 weeks after bloom after which mature cortical sclereids (brachysclereids) linking the vascular tissues and hypodermal collenchyma differentiated from cortical parenchyma cells. The increased diameter of the pedicel was because of the 'growth of the cortical parenchyma.' At the distal end of the pedicel the vascular tissue bifurcated into ten primary bundles which resulted in the formation of the sepal and petal bundles. The flowers of apples are formed on the terminal inflorescence. The fresh growth starts from one or more axillary meristems near the terminal inflorescence (Crabbe and Escobedo Alvarez, 1991). In the growing seasons such axillary meristems mark a series of vegetative leaves, bud scales and leaf primordia which consist of the bourse shoot which either terminates into an inflorescence or may remain vegetative (Fulford, 1966). The cycle of flower development is completed in about 9 to 10 months. There are two main types of floral buds: spur buds and axillary buds. The spur buds are produced on spurs and small shoots at the tips which start growth soon after fruit set in the former season. Axillary buds are produced in the basal leaf axils on extension shoots after the spur buds get fully developed in the season. The goal must be to promote the speedy formation of qualitative blossoms on the spurs and short terminal shoots of young trees. In some varieties, the axillary flowers contribute significantly to cropping on young trees, e.g. in 'Gala' and its sports.

6.2.1 Floral Bud Developing Sites

Fruits are born on the spurs of most of the known apple varieties. On these spurs the major leaf area of the tree from the green tip stage to complete flower is formed. After 1 month of flowering the total spur leaf area is formed. And at this time on spurs the primary stimulation of the flower bud development occurs. The leaves on the spur type cultivators comprise greater than half of the entire leaf area of the tree. The leaf count and per spur leaf area are greater in the spurred strains of 'Delicious' than in the standard varieties. In the last 90 days of the growing season each non-reproductive spur requires 100 to 150 cm^2 leaf area to induce the formation of a healthy floral bud. But certain apple cultivars under special conditions and age initiate floral sprouts over the apex of the elongated shoots and sideways in the leaf stipules. The apical or lateral floral buds are mixed and consist of primordia of non-reproductive and floral organs. Floral bud development and the formation of a flower bud is intricate phenomenon. An essential factor in its occurrence at the time of meiosis is the floral bud development genes and the formation of gametes. The initiation of floral bud formations, cellular transformations and morphological changes are involved in floral bud development. The induction of flower buds stimulates the adequately formed and 'susceptible' buds to switch from the non-reproductive to reproductive stage, or it is a process in which a formerly subsided signal is being altered to produce a new organ, namely the flower bud.

6.2.2 FLOWER BUD EMERGENCE AND DEVELOPMENT

Just from induction to the anthesis the growth of floral buds are exposed to the effects of a number of conditions which govern flower bud development.

6.2.2.1 Variety
The commencement of the histologic and morphologic changes and floral bud development is influenced by the type of variety or cultivar.

6.2.2.2 Rootstock
Rootstocks as a component of grafted fruit trees control the moment of induction of floral buds due to a mechanism linking the two elements of a tree. It results in an alteration in the time of commencement of the morphological differentiation.

6.2.2.3 Bearing Branches
Variation in the commencement of floral buds on different bearing branches has been noticed. Differentiation of buds generally initiates early on the recurrent spurs, then on the under-developed and vigorous ones, and lastly, on the shoots.

6.2.2.4 Shoot Growth
Differentiation of floral buds is related to the development of the shoots. Growth cessation is supposed to be important for flower commencement. The spurs growth discontinues 2–4 weeks following the flowering period while the shoots carry on their growth. According to Luckwill and Silva (1979), after the 2 weeks of termination of growth of the extended shoots the flower emergence in spur and bud formation in axils of 'Golden Delicious' apple take place whereas, it occurs after one month in the terminal buds of the shoots.

6.2.2.5 Cytochemical Variations
The cytochemical changes happening in the buds and their neighboring organs and tissues, influence in some way their transition from vegetative to generative state. The increased content of nucleic acids and the decreased content of IAA favor the flower bud formation. The presence of fruit on spurs results in a decrease of the nucleic acid content and increase of the nucleohistone and this has a negative effect on the floral bud differentiation (Buban and Simon, 1978). The cellular biochemical variations in buds and their adjacent structures and tissues govern their change from the vegetative to the reproductive phase. An increase in nucleic acids and the reduced amount of IAA support the floral bud development. The existence of fruit on spurs leads to a reduction of the

nucleic acid amount and enhancement of the nucleohistone which has a subsiding impact on the floral bud transformation (Buban and Simon, 1978). Fruits also decrease the amount of starch in the fifth to sixth week from complete bloom. The Sachs' hypothesis (1977) for diversion of photo-assimilates with respect to the interaction between the non-reproductive and the generative growth needs consideration. In this respect, the greater content of starch in the spurs can't be supposed as a straight reason for the floral bud commencement but just as a clear and quick sign of the path of the assimilatory processes due to small quantities of growth hormones like auxin and gibberellins in the adjoining tissues.

6.2.2.6 Hormones

Five different kinds of phytohormones (auxins, gibberellins, cytokinins, ethylene and growth inhibitors) are documented. All of them carry out a number of functions that act at once to control the activities of fruit plants. Gibberellins, in relation to flower bud formation, are regarded as the basic reason for poor floral bud development in the alternate bearers. Reduction in the flower bud formation with the goal of incapacitating the problem of biennial bearing is also noticed with the use of synthetic gibberellins up to the fourth week from bloom (Tromp, 1982). It has been summarized by different authors that auxins have an indirect but positive impact on the commencement of fruit buds at the start of the season. Auxins in seeds, at age of less than 4 weeks, draw further assimilates to the spurs. This is essential since the early, quick growth of the primordia of leaves and immature leaves sooner in the season is needed for the floral bud commencement. The gibberellins that are shifted from the seeds starting in the third to fourth week after the completed flowering period neutralize the positive effect of auxins and depress floral bud development. With enhancing shoot growth, they indirectly reduce floral bud growth. In the stage prior to the flower emergence, seeds of the alternate bearer apple varieties disperse a significantly greater number of gibberellins than seeds of the regular bearer apple varieties (Grochowska and Karaszewska, 1976).

6.2.2.7 Environmental Conditions

The flower bud commencement and growth are affected by the natural surroundings. Excess temperatures can indirectly delay the development of a bud in certain apple varieties by changing the period of the plastochrone underneath the effect of the gibberellins formed in the tips of the long growing shoots. Thermal differences with day and night ranges also result in a delaying effect on floral bud development. Cool temperatures prior to the beginning of the morphogenetic alteration may result in the depression of its onset. Heat builds from complete bloom to inception of the floral opening is not considered for differences in the beginnings of seasons. The positive weather conditions for the photosynthestic process with an extended duration subsequent to fruit picking, even in delayed maturing varieties, in some nations can result in the development of highly 'vigorous flowers' which bear a positive impact on the yield of varieties. The component of the floral buds that have initiated their growth are changed into vegetative ones. From these buds the floral parts are not developed, and bourse shoots are formed.

6.2.2.8 Excellence of the Propagative Organs

The competence of the flowers of apples for fruit setting, as well as for the type and size formed, depends upon the morphology of the clusters of flowers, which are dependent on the stage of development acquired by the floral buds at the bloom. The apple trees that possess less vigor have the production of floral buds on non-reproductive spurs begins near the beginning of summer. Accordingly, in the following spring the previously formed flowering spurs emerge as 'old' and have tiny primary leaves which promote the formation of small fruits. The primary leaves arise at the inactive buds of the productive apple trees. The flower bud development occurs late in cases of extremely vigorous trees or in the ones bearing a lot of fruit. In spring, the spurs are 'under-developed,' and possess big primary leaves and do not develop sufficient fruit. Vegetative spurs produce 'vigorous' floral buds in well managed trees over time and in the following spring the regular clusters of flowers with 5–6 blooms and a normal size of leaves grow from them. Such cluster of flowers set a lot of fruit, a huge portion of which abscises during the June drop. With respect to quantity and quality, the yield remains standard.

6.2.2.9 The Stage of the Bearing Wood

The stage of the bearing wood in apples also directs the excellence features of the flower clusters (Costes, 2003). The clusters of flowers on the previous season's wood arranged axillary differ in excellence from the clusters which are present on a 2–3-year-aged wood terminally. The flower clusters on the previous season's wood arranged axillary are reduced in size; the primary leaves are lighter in color and are small; and the flower clusters have fewer numbers of flowers which are generally smaller. They have a shorter duration of flowering time which reduces the probability of effective pollination. The development of fruits from such types of flowers is not vital enough and sets smaller fruits (Robbie and Atkinson, 1994). The bloom bunches of a 2-year-aged plant vary as per standard from the clusters present in the aged trees. The flower which are present at the apex are more dominating on the lateral flowers. This dominance varies by variety and can be improved by environmental situations and by the features of distinct years. The blossoms at apices develop more fruit in contrast to flowers which are borne laterally. With respect to size, the apical fruit dominates others. The date of harvesting and the yield affect the formation of floral appendages in the fall and in the beginning of bloom. The high crop yield in the preceding season can deform stigmas and papillas, and delayed harvesting depresses the formation of blossom clusters which results in the reduced fruit set. For fruit set and fruit development (Ferree and Palmer, 1982), primary leaves in clusters are important, and the cumulative yields of different apple varieties depend on such leaves. The 'strong' and 'weak' spurs are based on the count and primary leaves area, but these features are not satisfactory for expecting the fruit characters. The average dehydrated matter of one primary leaf, the cumulative mass of the leaves per spur, the average dry mass of one primary leaf and the specific dry mass of the primary leaves, which depends upon the intensity of isolation, are more valued. Early growth of the fruit is linked to assimilates drawn from the primary leaves. After 2 weeks of complete blossom, assimilates drawn from the primary leaves are dispersed among the budding fruit lets, leaves and plump shoots. In the next 2 weeks the amount of photosynthates allocated to the developing fruits rises significantly. After 3 weeks of blossom, the strong spurs with 9–10 primary leaves and no more than single fruit can meet the requirements of own assimilates while the less vigorous spurs, the over-shaded ones and the ones shaving higher fruit counts must derive assimilates from the rest of the branches. For the ultimate fruit development and for their ability to exhibit growth, the allocation of assimilates in these 3 weeks is quite vital. The shoots themselves exhausted the assimilates imported from the bourse shoots; afterwards the portion of assimilates distributed to the young fruits consecutively augments, and in the eighth week after flowering the two attracting cores possess approximately equal proportions in drawing the photoassimilates. The overshading, in the year previous to floral bud development as well as in the present year, reduces the photosynthetic power of the spurs by reducing the exact weight of the primary leaves, the leaf area of the bourse shoots and the specific mass of these leaves.

6.2.2.10 Pruning

Resting pruning and its effect on the development as well as the yield of the apple trees has been a goal of several experiments (Mika, 1986). Experiments on pruning in active season are highly restricted. Pruning in the active season practiced on developing vigorous apple trees can enhance floral bud development. It is advantageous for such pruning to be practiced in the beginning of summer. Under various circumstances, summer pruning does not positively affect the emergence of floral buds. Its impact is based on the features of the varieties. Pruning in summer may decrease the final count of floral buds on the trees, but it might raise the flower count for each inflorescence, thus the final count of flowers on the tree does not change. For shaping the top of the tree, pruning in the summer is preferable over pruning in the winter, but it may not always lead to enhancement of the yield. In cases of dormant pruning, reduction in the length of previous year shoots can be practiced in the varieties known for initiating floral buds on lengthy shoots with no retarding consequence on the floral bud development. Varied tree responses toward summer pruning, relevant from a number of experiments, are due to the variation in the date of pruning, way of pruning and the extent of the pruning.

6.2.2.11   Use of Fertilizers

The application of fertilizers in fruit trees must be carried out with respect to soil nutrient status, the types of varieties and rootstocks, environmental scenario, load of crop and desirable quality of fruit of fruit crops (Jonkers, 1979). It significantly influences both the floral bud formation as well as the fruit set and must be carried out with respect to soil nutrient status, types of varieties and rootstocks, environmental scenario, crop load and desirable fruit quality. At the time of planting, an application of a larger amount of phosphorus enhances the count of the emerged floral buds. It affects floral bud development indirectly by altering the amount of cytokinins developed in the roots. After cessation of terminal growth of extension shoots, the application of nitrogen encourages floral bud formation. The former application enhances growth which is not desired for the flower bud formation. Nitrogen application during the summer as a supplement to the application already given in bloom encourages the strength of ovules and stigmas in several apple varieties. Nitrogen fertilization (in alternate bearers) during spring in the 'off' year ought to be restricted, while the fertilization in autumn and in the next spring of the 'on' year should be adequate.

6.2.2.12   Irrigation

Irrigation positively affects the floral bud development in apples. With the adequate provision of irrigation and nutrients through fertigation, developing apple trees produce a greater number of floral buds. Competition between vegetative growth and reproductive growth can be reduced by uneven influences: a particular rootstock, water supply or other causes. However, the condition might be altered when development is encouraged with an immediate supply of water and nutrients, which influences metabolism in a highly balanced way and probably prevents the deficiencies in particular assimilates that are essential for flower emergence in crucial phases. By means of root function, fertigation, annual shoots expansion and the formation of the nodes in the axillary, buds might be enhanced to levels where a larger portion of the buds result in floral buds.

6.2.3   Morphological Differentiation

The apple floral bud comprises of a short axis, typically having 21 formations, pegged in the coiled arrangement (Abbot, 1977). The floral bud is comprised of bud sheaths, transitional leaves, functional leaves and bracts. On the stipules of the three bracts and the three superior leaves, axis terminates into rudiments of the 'king flower' and the rudiments of the lateral flowers. Below the group of the flowers in the axils in one or two of the leaves, the rudiment of a non-reproductive bud is present. A new flower bud is formed during the vegetation from a brouse shoot. When the formation of a non-reproductive bud is completed, the morphological transformation takes place (Buban, 1996). The changes in the rudiments of flowers in apples begin as soon as the primordia of the appendages have been initiated. The critical count of nodes is the initial sign for the change of the buds from non-reproductive into generative state. The significant count of nodes is dependent on the variety, and the numbers range from 16 to 20 (Hirst and Ferree, 1995; Luckwill and Silva, 1979). The total count of appendages formed lead to the emergence of the sepals of the flower at apices in the cluster which is also specific to the cultivar's environmental conditions and growth habits. For the development of bracts and the flower emergence in their bases, it is important for the plastochron to be reduced to 7 days. The primary visible sign of the transformation from the non-reproductive to the reproductive phase is the expansion and swelling of bud tips and they appear as dome shaped. A higher accuracy in defining the instant of a move from a vegetative state to a reproductive state is the beginning of the initiation of bract. In apples, Foster et al. (2003) suggested that the expansion of the tip of the bud followed by the dome shape of the bud can be recognized as the indicator of the transformation from vegetative to flowering phase.

6.3   POLLINATION

Apple flowers are formed in clusters of five to six on short, woody 'spurs.' The best quality fruit are producing on primary or 'king' buds and generally the buds open first. Each flower has a style, which is divided into five stigmas, and is bounded by 20 or more stamens with pollen-producing anthers. The ovary is formed with the base of the style which contains the developing seeds (ovules). A ring of petals surrounds the stamens. Nectar is produced between the bases of the stamens and the style. Pollination occurs when a bee actively collects pollen from one flower and then inadvertently transfers it to the sticky surface of the stigmas of another flower on a tree of a compatible variety. Pollen tubes grow down the style to the ovary where fertilization occurs. The flowering period for apples lasts about 9 days and is longer in cool weather and shorter in warmer weather.

6.3.1   POLLEN MORPHOLOGY

The expansion of flowers, along with anther shedding and pollen grain liberation, is called anthesis. Pollen grains are inactive, resilient components consisting of lipid deposits for germination and initial growth, but they get abruptly desiccated after anther shedding and need to imbibe in water to germinate when they reach stigmas. They are ellipsoid and have three grooves under dry conditions with three germinal furrows expanding the entire grain length. They bulge out if moist and acquire a spherical shape. The external coating of the pollen grain includes long parallel streaks and occasionally possesses minute openings on the surface. The pollen is weighty and not easily passed away by wind.

6.3.2   POLLEN GERMINATION, FERTILIZATION AND PHYSIOLOGY

For the proper ovule fertilization, pollen germination is the initial event which ultimately results in growth and development of fruits. When a pollen grain comes in contact with the floral stigmatic surface, the process of fertilization takes place. The stigma possesses a moist surface made of extracellular discharge from its papilla cells that crumple after anthesis (Sedgley, 1990). The wet pollen grain germinates on the stigmatic surface, and the initiated pollen tube starts to expand through the interstitial material of the transmitting tract (Jackson, 2003). Pollen germination on apple stigmas is developed through a chain of complex events which include proteins and other molecules. RNA, protein and polyamine concentrations inside a pollen grain stay comparatively unaffected prior to germination. They start to reduce after germination (Bagni et al., 1981). Fully developed pollen grains consist of two generative nuclei and the tube cell nucleus. When the compatible pollen grains reach the stigmas, germination is carried on with pollen tube expansion; each has a tube nucleus and two generative nuclei down each style into the ovaries (Dennis, 2003).

6.3.3   POLLINIZERS

Cross-pollination is essential for every apple cultivar with a pollinizer to guarantee production of quality fruit yields. Cultivars vary in self-productiveness. Apart from the level of self-fruitfulness, cross-pollination is a must in every plant. For adequate cross-pollination, the following conditions are necessary:

- The bloom periods of the pollinizer and main variety should coincide.
- The pollinizer should possess functional diploid pollen.
- The location of the pollinizer plant must be in the vicinity of the main plant.
- Pollinators should be functional during flowering in the orchard.
- The flowers of weeds like mustard, dandelion, etc. must not be available in abundance since they divert the pollinators away from fruit tree flowers.

Plants that supply ample compatible pollen to the main varieties in the plot are required for pollination. The closer the location of the pollinizer to the main tree, the more effective the transfer of the pollens to every flower, with the help of the bees.

Some of the cultivars have an alternate bearing habit. In the 'off' year of the pollinizer, the main variety, though a regular bearer, turns out to be biennial due to the lack of cross-pollination. Naphthalene acetic acid or ethephon treatment during the summer aids in enhancing the production of blossom clusters in the succeeding year.

Every fruit crop requires pollination in order to develop fruit. Several species are self-fertile, and there is no need for any pollinizer (e.g., peaches, some cherries, nectarines, etc.). For satisfactory cross-pollination in the case of apples, sweet cherries and pears, mixed plantings of numerous cultivars are highly required. The proportion of blossom which ought to be set depends upon the fruit crops. In the case of cherries, in which every single blossom sets an appreciable sized fruit at picking and no thinning is practiced, productivity depends on pollination; hence 20 percent to 60 percent of sweet cherry and 20 percent to 75 percent of sour cherry flowers are required to develop into an economical yield. The proportion of flowers which needs to be pollinated is quite low for rest of the fruit species where commercially the size is essential or where preventing alternate bearing is a must, e.g., apples: 2 percent to 8 percent; peach: 15 percent to 20 percent; plum: 3 percent to 20 percent. Hence, hand thinning later in the season in a number of fruit trees like peaches or chemical thinning in apples must be done as under usual conditions such crops will set huge number of fruits, which, in cases left as such, will turn out to be poor in size at harvest.

6.3.4 POLLINIZER PLACEMENT

6.3.4.1 Pollination with Compatible Varieties

Pollinizer placement is essential. If possible, each plant in a fruit farm ought to be grown in the vicinity to the pollinizer tree. The preferrable organization is in solid rows in case another commercial variety is planted as a pollinizer. One way is to arrange two alternate rows of pollinizers in between four rows of the main variety. When pollination is needed to be maximized in case of exceptions like 'Delicious' that are inclined to be less self-fertile, a pollinizer row must be planted after every two rows.

6.3.4.2 Effective Pollination Period (EPP)

The effective pollination period (EPP) is the gap from the time for pollen tube expansion to the time of ovule longevity. The longer the duration of this interval, the higher the chance of sufficient fusion of gametes and development of seed. The transfer of pollens ought to take place within 2 to 4 days after anthesis; otherwise, the embryo sac disintegrates prior to fertilization. This period varies with respect to cultivar. The pollen tube development and ultimate embryo fertilization is highly varying with temperature and its correlation to the effective pollination period.

The EPP was first set up in the middle of 1960s as means of setting the time period from the time a flower is pollinated to the time the embryo becomes non-receptive. It requires a certain time period for the pollen tube to arrive at the embryo sac after pollination. Just when anthesis takes place, the embryo has inadequate time to become receptive. In cases where the pollen tube does not arrive to the embryo prior to its degeneration, no flower set takes place. The length of the EPP

TABLE: 6.1
Effective Pollination Period Index

Mean daily temp (°C)	5.0	6.1	7.2	7.8	8.9	10.0	11.1	12.2	12.8	13.9	15.0
Pollen tube growth index (%)	8	9	10	11	12	14	17	20	25	35	50

varies with respect to the position of the flower within the cluster and cultural practices. Generally, EPP lasts 3 days to 12 days. A temperature response index was developed by Williams and Wilson to estimate the time needed for a pollen tube to get in touch with the embryo. The index is based on the regular average temperature over some days. When the index attains the value of 100 percent or exceeds it, the pollen tube ought to have arrived at the embryo and fused with the egg (Table 6.1)

As an example, the average mean temperature over the past 5 days had been 50°C, 54°C, 50°C, 52°C and 59°C. The pollen tube growth would be expected to be 14 + 20 + 14 + 17 + 50 = 115 percent, meaning the pollen tube growth would have taken slightly less than 5 days.

6.4   FRUIT SET

Fruit setting has long been considered a major problem in many fruit plants, especially in temperate fruit crops. In apples, certain varieties are completely fruitful, some are partly self-fruitful and a few are self-unfruitful. Most apple cultivars are effectively self-incompatible, or very largely so, and that fruit set usually depends on cross-pollination between genetically different cultivars. The site of incompatibility is in the style or ovary because of the physiological reactions occurring between the pollen tube and the stylar and ovarian tissue (Modlibowska, 1945). It is always safe to grow a block of single variety, which is self-fruitful, without any provision for cross-pollination, as in 'Baldwin' apple. There are many varieties, which are otherwise considered self-fruitful because they produce abundant crop in one season, that often greatly benefit if some provision of cross-pollination is made. There are many examples wherein increased fruit set, size and yield have been reported in self-fruitful types if some provisions of cross-pollination (pollinizer and pollinators) are made. Apple flowers are true hermaphrodites. Although most of the varieties have a good amount of pollen, in some varieties a fraction of the pollen is non-functional. However, it appears that all varieties have a good content of pollen, which is facilitated by environmental and genetic factors. Certain cultivars are self-productive, and many others are not. Many research workers have reported that more than 60 percent of apple varieties are self-unfruitful. Self-infertility in apples is chiefly related to a variation from the standard diploid (2n = 34) chromosomes number, although there are a lot of exceptions as the level of self-fruitfulness in apples differs to a great extent with the age and environmental conditions. Thus, the 'Jonathan' apple variety that is self-fertile in many regions of the United States is self-productive in Victoria (Australia) when raised on soils of medium productivity, but it is self-unfruitful if raised on highly fertile soils. In general, the old varieties like 'Yellow Newton,' 'Baldwin,' 'Ben Davis,' 'Grimes' and 'Yellow Transparent' are listed as self-fruitful; 'King,' 'Arkansas,' 'Missouri,' 'Pippins,' 'Rome,' 'Rhode Island Winesap,' 'Wealthy,' etc. are listed as self-unfruitful. Nevertheless, the self-fruitful cultivars have been reported to be self-unfruitful at certain localities, and the same is true with self-unfruitful varieties which bear considerable crops at certain other localities. In the wine sap group, inter-fruitfulness has been observed but is not of much significance in the production of apples. Yet, the colored strains, which have developed as chance seedlings, are generally inter-sterile with one another and their parents. Parthenocarpy takes place quite often, but true parthenocarpic cultivars are uncommon. Immature vigorous trees generally shed more of their fruits than the well-established and slightly older trees. Moreover, aged and weak trees commonly flower profusely but produce little to no fruit.

Rootstocks also play a vital role in fruiting precocity, fruit bud formation, fruit set and the yield of a tree. Dwarf rootstocks induces fruiting precocity whereas vigorous rootstocks delay in fruiting precocity (Preston, 1967). For reducing tree size, maximum precocity and yield-efficiency apple rootstocks are mainly used. Intensive planting of small trees results from dwarf rootstocks that intercept light and have less internal shading, which increase the production dry matter and fruit yield. The plants grafted on dwarf rootstocks contribute higher yield efficiencies than those plants grafted on vigorous clonal and seedling rootstocks due to a higher ratio of fruit weight to trunk and branch weight (Strong and Azarenko, 1991).

6.4.1 Percentage of Flowers that Need to Set

A grown-up standard apple tree with profuse flowering might have approximately 100,000 flowers. Just a small share of these flowers, possibly 8 to 10 percent, will ultimately grow into fruit. The greater the competition for nutrition, up to 90 percent of the flowers may abscise even if they were sufficiently pollinated. Though just one quarter of the blossoms on a profusely flowering tree develop fruit, the tree would have a significant crop load which would lead to poorly developed, unbendable fruit. A set of one flower in 20 will habitually be adequate to generate ample productivity. High yield, possibly 15 percent to 20 percent, is required on trees that have light blooms. Floral buds are produced in the summer prior to the spring in which they open into blossoms. Thus, the density of bloom usually depends on the earlier health conditions and cropping of the tree. The quantity of flowers can at times be promoted by following these steps the previous summer: enhancing leaf exposure to the sun, branch bending, trunk ringing and delivering ample amount of water, accurate fertilizer applications or both. An insignificant overset may be required as thinning after flowering may regulate cropping intensity.

6.4.2 Cross-Pollination by Honeybees

Honeybees generally transfer apple pollen from one cultivar to another variety to other. Contrary to the accepted idea, bees do not live exclusively to transfer pollen for the orchardist or the apple tree; rather, it is by accident that they bring pollen on their body surface when they are looking for nectar or pollen. Some of the honeybees just carry pollen that might be reserved as feed; others gather just nectar that has oozed out from the nectarines right on top of the ovaries. Pollen collection bees exhibit more pollinating than those that just collect nectar. Honeybees generally do not build huge surpluses of honey to stock up inside the beehive at the time of the flowering of apples. Their exploitation in fruit farms is advantageous for the cultivator by promoting fruit formation not for the production of honey. Honeybees, by chance, wipe through the gluey floral stigmas and leave some unknown pollen. Pollen gets stuck to the bee's hairy body while it is entering the orchard, then it brings the pollen to the beehive and then out to the orchard another time. Bees also lift up the pollen by striking against other bees in the hive. Bees do not travel over wider regions of the orchard; they stay to just a few trees in each journey. Their flight is negatively affected below 65°F and it ceases below 55°F. Honeybees are responsible for 90 percent of apple pollination; other bees may also provide some pollination.

Even when compatible pollinizers are accurately planted in an orchard, it still may be necessary to bring honeybees into the orchard. Bees produce their most striking results in seasons that have only a few hours of favorable weather during bloom. It is in these unfavorable seasons that apples are generally inadequate and bring the highest prices. Apple flowers are very attractive to bees because they produce more nectar than most other kinds of fruit at that season. That is why honeybees are potent pollen carriers.

6.4.3 Requirement of Effective Pest Management

'Rome Beauty' and 'Golden Delicious' both exhibit late blooming habits and will successfully pollinate one another. Their pest management needs might be different, however. 'Golden Delicious' requires mild insecticide which might be desirable to prevent skin russet.

6.4.4 TOP GRAFTING OF POLLEN SOURCE VARIETIES

If no cross-pollination is assured in an orchard at the time of planting, then another better pollen source variety can be top-grafted over the trees of the main variety. A pollen source can be introduced into an orchard more quickly by top grafting than by planting new young trees. Be cautious of eliminating the pollen source grafts during regular pruning practices.

6.4.5 FERTILIZER MANAGEMENT

Cultural practices that result in strong spurs with a proper provision of reserved foods will help to ensure good fruit set. In particular, an adequate availability of nitrogen at flowering time is important for setting fruits, though excess nitrogen leads to excessive vegetative growth and reduced fruit bud formation later in the season. When the trees are low in nitrogen, urea spray applied at the pink bud stage of flowering may help to increase the nitrogen content of the flowers and prepare them for fruit setting. Flower quality depends on a proper balance of all of the essential elements as determined by leaf analysis. An adequate supply of boron is required for the normal development of ovules and pollen grains, pollen germination and the normal growth of pollen tubes. Proper levels of zinc and copper, which are deficient in orchards, are especially important for good fruit set. Proper pruning in winter is also beneficial. Trees that grew weakly the preceding summer or were seriously damaged by insects or diseases might result in the development of unproductive crops even with abundant blooms. Even starved trees often blossom abundantly; their flowers can be very poor at setting fruit. This is why orchards that are left as such and not looked after may flower profusely but develop very little crop.

6.4.6 DISEASE MANAGEMENT

Apple scab and fire blight disease incidence infect the flowers of the developing fruits and cause maximum shedding of fruits. Thus, such diseases should be prevented by following appropriate spray schedule measures to get adequate fruit setting.

6.5 CONCLUSION

The flower bud formation of fruit plants growing in temperate regions like apples, sweet cherries, etc., is a complicated phenomenon. Versatile relations are supposed between the genetic control, the hormonal balance and the presence of sufficient amount assimilates in the plant as a whole and more precisely in the forming flower buds. The development of flower buds is linked to the appearances of the fruit tree species and cultivars, ecological conditions and agricultural practices. Excellence of the reproductive organs rest on the factors and conditions for flower bud formation, which in turn effects the quantity and quality of fruit production. The initiation and development of flower buds in apples can be successfully regulated by means of scientifically well-founded agrotechnical practices such as pruning, fertilization, irrigation and treatment with growth regulators. The EPP is the gap from the time for pollen tube expansion to the time of ovule longevity. The longer the duration of this interval, the higher the chances of a sufficient fusion of gametes and the development of seed. Fruit setting has long been considered a major problem in many fruit plants, especially in temperate fruit crops. Most apple cultivars are effectively self-incompatible, or very largely so, and that fruit set usually depends on cross-pollination between genetically different cultivars. Many cultural practices like insect pest management, fertilizers scheduling, source of pollinizers, rootstocks used, etc. affect the intensity of fruit setting.

REFERENCES

Abbott, D. L. 1977. *Fruit bud formation in Cox's Orange Pippin*. Report of Long Ashton Research Station, pp. 167–176.

Bagni, N., P. Adamo, D. Serafini-Fracassini, and V. R. Villanueve. 1981. RNA, proteins and polyamines during tube growth in germinating apple pollen. *Plant Physiology* 68: 727–730.

Barden, J. A., and A. H. Thompson. 1963. *Developmental anatomy of vascular tissues in York Imperial apple with special emphasis on the pedicel.* Maryland Agriculture Experimental Station Bulletin, p. A131.

Blaser, H. W., and J. Einset. 1950. Flower structure in periclinal chimeras of apple. *American Journal of Botany* 37: 297–304.

Brase, K. D. 1937. *The vascular anatomy of the flower of Malus domestica Borkh. f. apetala Van Eseltine.* M.S. Thesis. Cornell University.

Buban, T. 1996. *Floral biology of temperate zone fruit trees and small fruits* (Eds. J. Nieki and M. Soltesz). Akademia Kiado, Budapest, pp. 3–54.

Buban, T., and I. Simon. 1978. Cytochemical investigations in apices of apple buds with special reference to flower initiation. *Acta Horticulture* 80: 193–198.

Costes, E. 2003. Winter bud content according to position in 3-year-old branching systems of 'Granny Smith' apple. *Annals of Botany* 2(4): 581–588. doi:10.1093/aob/mcg178.

Crabbe, J., and A. Escobedo. 1991. Activites meristematiques et cadretemporel florale des bourgerons chez le pommier (*Malus domestica* cv Golden Delicious) "L" abre. *Bilogie et development Naturalia Monspeliensia* 7: 369–380.

Dennis, F. J. 2003. Flowering, pollination and fruit set and development. In: Ferree, D. C. (Ed.), *Apples botany production and uses.* CAB International, Cambridge, pp. 153–166.

Eames, A. J. 1961. *Morphology of the angiosperms.* McGraw-Hill, New York.

Esau, K. 1965. *Plant anatomy.* 2nd ed. Wiley, New York.

Ferree, D. C., and J.W. Palmer. 1982. Effect of spur defoliation and ringing during bloom on fruiting, fruit mineral level, and net photosynthesis of 'Golden Delicious' apple. *Journal of American Society of Horticultural Science* 107: 1182–1186.

Foster, T., R. Johnston, and A. Seleznyova. 2003. A morphological and quantitative characterization of early floral development in apple (*Malus x Domestica* Borkh). *Annals of Botany* 92: 199–206.

Fulford, R. M. 1966. The morphogenesis of apple buds. The development of the bud. *Annals of Botany* 30: 25–38.

Grochowska, M.J., and A. Karaszewska. 1976. The production of growth promoting hormones and their active diffusion from immature developing seeds of four apple cultivars. *Fruit Science Reproduction* 3: 5–16.

Hirst, P., and Ferree, D. 1995. Rootstock effect on the flowering of Delicious Apple I Bud development. *Journal of American Horticultural Science* 120: 1010–1017.

Jackson, J. E. 2003. *Biology of apples and pears.* Cambridge University Press, Cambridge.

Jonkers, H. 1979. Biennial bearing in apple and pear: A literature survey. *Scientia Horticulturae*, 11, 303–317.

Kraus, E. J., and G. S. Ralston. 1916. *The pollination of the pomaceous fruits. 111. Gross vascular anatomy of the apple.* Oregon Agricultural College Experimental Station Bulletin, p. 138.

Luckwill, L.C., and Silva, J. M. 1979. The effect of daminozide and gibberellic acid on flower initiation, growth and fruit of apple cv. Golden Delicious. *Journal of Horticultural Science* 54: 217–223.

MacDaniels, L. H. 1940. *The morphology of the apple and other pome fruits.* New York: Cornell University, Agricultural Experimental Station Memoir, 230.

Mika, A. 1986. Physiological responses of fruit trees to pruning. *Horticultural Reviews* 8: 337–378.

Modlibowska, I. 1945. Pollen tube growth and embryo-sac development in apples and pears. *Journal of Pomology and Horticultural Science* 21: 57–89.

Preston, A. P. 1967. Apple rootstock studies: Fifteen years' results with some M.IX crosses. *Journal of Horticultural Science* 42: 41–50.

Robbie F. A. and Atkinson C. J. 1994. Wood and tree age as factors influencing the ability of apple flowers to set fruit. *Journal of Horticultural Science* 69: 609–623.

Sachs, R. M. 1977. Nutrient diversion: A hypothesis to explain the chemical control of flowering. *Horticulture Science* 12: 220–222.

Sedgley, M. 1990. Flowering of deciduous perennial fruit crops. *Horticulture Review* 12: 223–264.

Sterling, C. 1965. Comparative morphology of the carpel in the Rosaceae. V. Pomoideae: Amelanchior, Aronia, Malacomeles, Malus, Peraphyllum, Pyrus, Sorbus. *American Journal of Botany* 52: 418–426.

Strong, D., and A. M. Azarenko. 1991. Dry matter partitioning in 'Starkspur supreme delicious' on nine root-stocks. *Fruit Variety Journal* 45: 238–241.

Tromp, J. 1982. Flower-bud formation in apple as affected by various gibberellins. *Journal of Horticultural Science* 57: 277–282.

Tukey, H. B. 1937. The occurrence of apple blossoms with prolonged central axes and its bearing upon flower morphology. *Proceedings of American Society of Horticultural Science* 35: 117–127.

Zielinski, Q. B. 1955. *Modern systematic pomology.* Wm. C. Brown, Dubuque, IA.

7 Regulations of Form-Training and Pruning

*Mohammadebrahim Nasrabadi, Mojtaba Kordrostami,
Ali Akbar Ghasemi-Soloklui, Mehdi Gararazhian and
Ali Gharaghani*

CONTENTS

7.1 INTRODUCTION

In spite of the frequent association between the concepts of pruning and training, there are significant distinctions between the two practices when it comes to timing and execution. When it comes to pruning fruit trees, it's often separated into two distinct phases: training, which occurs early in the life of a plant, and pruning, which occurs later in the life of a plant with the goal of creating a strong framework for the tree and improving its ability to develop. We'll go through the ins and outs of training and pruning in great depth in the paragraphs that follow.

The term "fruit tree training" refers to the process of creating a well-structured framework in order to produce huge amounts of high-quality fruits in the desired environment (Franzen, 2015). In addition to developing a strong structure, tree training aims to open up the canopy so light can reach all sections of the tree (Forshey et al. 1992). After an orchard establishment is complete, the primary purpose of tree training in its early years is to produce fruit as soon as possible, and appropriate tree training may have a considerable impact on the requirement for pruning trees during crown growth in the following few years as well. In part, this is because tree training takes place early in the trees' lives. As a result, one of the most important principles of fruit tree training is the selection and manipulation of scaffolds and positioning them at appropriate heights and orientations in accordance with the orchard application system used (Franzen, 2015). Bending the branches to increase

DOI: 10.1201/9781003239925-7

light penetration into the tree canopy and encouraging fruit production with the use of specific gardening activities like weights, branch spreaders, and a special rope are some of the most common methods of scaffold manipulation (Ferree and Warrington, 2003). The bending of the branch and the development and apical dominance of branches in horizontal position have been found to have no effect on shoot growth and blooming (Wareing and Nasr, 1961; Franzen, 2015). Branch bending has been shown to improve blooming and fruit production in many studies, including the ones listed in this chapter. Bending increases the density of flower buds on the branch and afterwards increases the yield in some rootstocks (Hamzakheyl, 1976; Ferree, 1994). Furthermore, when branches have an angle between 30° and 60° from the vertical, the rate of flowering increases (Franzen, 2015).

Orchard practices for fruit production also include pruning, which may be defined as "the art and science of cutting away a piece of the plant in order to achieve horticultural goals" (Franzen, 2015). Pruning, in its broadest sense, may be defined as the act of removing branches in order to establish a balance between the tree's vegetative and reproductive development and to promote fruit production and quality. This is accomplished by removing branches from the tree's special branches (Zhang et al. 2018). It is necessary to remove crisscrossed, diseased, dried-out, and broken branches from the main stem of the tree in order to enable sufficient light to reach the tree's crown and downward branches and parallel branches with a close distance from the trunk (Brasil et al. 2018; Zhanget al. 2018). Branch thinning cuts and heading cuts are the two basic methods of pruning cuts. Thinning cuts remove the branch from the main trunk; its purpose is to improve the penetration and distribution of light into the tree crown. Removing the apical portion of the branch is called a heading cut, and its purpose is to stimulate the growth of the lateral branches below the cut part (Ferree and Warrington, 2003; Franzen, 2015).

In addition to the previously mentioned methods, there are two forms of pruning: winter pruning and summer pruning. Winter pruning is used while trees are dormant and therefore, less susceptible to cold harm after cutting. Summer pruning, on the other hand, is used to improve light penetration and distribution into the tree canopy to increase the percentage of red cultivars that have a blush color (Autio and Greene, 1990; Franzen, 2015). Aside from reducing vegetative growth and increasing fruit quality, summer pruning of apple trees also improves fruit maturity and increases the number of apple tree flower buds (Ashraf et al. 2017).

Additionally, pruning can be further separated into root pruning and shoot pruning. Root pruning may significantly and quickly reduce the development of new shoots after pruning and lead to the formation of new roots by reducing the water potential of plants and rearranging photosynthetic resources (Geisler and Ferree, 1984; Franzen, 2015). It is important to consider the appropriate pruning methods and wound care as quickly as possible in order to minimize the danger of infections such as fungus or insects and to ensure optimal pruning (Badrulhisham and Othman, 2016). It is the primary goal of this chapter to discuss training and pruning concerns, as well as to provide an explanation for all of the elements described in the introduction.

7.2   PRINCIPLES OF APPLE TREE TRIMMING

In order to minimize future trimming time, it is important to begin training trees as soon as they are planted and maintain this training over the tree's whole lifespan.

The final shape of untrained apple trees will be large and dense; there will be many branches on the main trunk that will be poorly connected, and the light will not penetrate the tree interior well. To prevent this from happening, there is no choice but to train the tree early in the life by using the appropriate methods. However, in some cases it may be necessary to train the trees after they have come into the production phase (Ferree and Warrington, 2003).

In untrained trees, as the age of the tree increases the penetration of light into the interior of the canopy decrease, which leads to a reduction in fruit size, a lack of proper coloring of fruits, and a decrease in fruit sugar concentration. As these conditions progress the spurs' ability decreases and may die (Ferree and Warrington, 2003). Trees with more leaves that are exposed to light are able to produce a higher quality fruit. Removing a lot of branches during pruning can motivate the regrowth by the production of three or four branches below the portion of the branch that is removed. To prevent this from happening, it is possible to minimize it by making appropriate cuts and by cutting older wood, if possible (Roper, 1997). Whether a branch will produce fruit or a vegetative growth depends on the position of the branch (Figure 7.1). According to Figure 7.1, each branch can have three states which include: 1) a straight and upward growth, in which case they will mainly have vegetative growth and produce little fruit;2) a completely horizontal state that is perpendicular to the trunk of the tree, in which case fruit production is at its highest but vegetative growth is reduced; and 3) an angle between 30° to 60° with the main trunk (Oberhofer, 1990), in which case fruit production and growth is at its best, and the limb that has a greater angle of connection with the main trunk will be a more powerful connection than the upright limb (Figure 7.2) (Roper, 1997). According to the previous report, the best time to train the branches is when they are 3–6 inches long and their wood is soft (Roper, 1997).

7.2.1 Limb Positioning Techniques

Cutting the branches is not the only methods to train apple trees. Three other ways to do train apple trees are: 1) spread, 2) tie, and 3) weight (Roper, 1997).

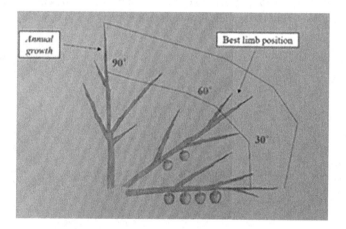

FIGURE 7.1 Effect of limbs position on fruiting and vegetative growth.

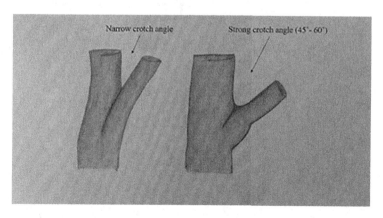

FIGURE 7.2 The connection of branches with the main trunk; branch with narrow crotch angle (left) and strong crotch angle (right).

7.2.2 SPREADING TECHNIQUE

The broad crotch angles are created by applying pressure to the branches with specific instruments such as toothpicks, small sticks, and clothespins (Figure 7.3). The optimal time to use spreaders is in mid to late June when the limbs are about 3–6 inches long. It is preferable to insert the toothpicks into the soft bark of branches and trunks and to use clothespins to hold branches in place, as shown in Figure 7.3(C), by first putting the gripping end on the trunk and then pressing the open side on the little limb to push the limbs into a horizontal position.

7.2.3 TYING TECHNIQUE

Multiple methods and materials such as twine, rope, fiberglass band, and rubber tapes can be utilized for tying the branches. For the materials like rubber tapes that decay after time, it is not essential to remove them; for the other materials, it is necessary to remove them. The rubber tapes should be used before June when the branches have grown less than 8 inches and are not too woody, otherwise it can cause damage to the branches and they will rot away before the branches become woody enough.

7.2.4 WEIGHTING TECHNIQUE

Branches may be pushed horizontally using a heavy item (Figure 7.4). The branch's position and angle are determined by where the weight is placed on the branch. An upward growth branch will

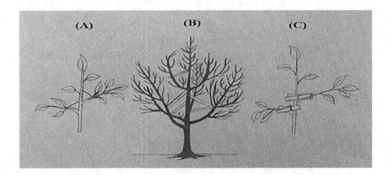

FIGURE 7.3 Spreading technique using tooth picks (A), short sticks (B), and clothespins (C).

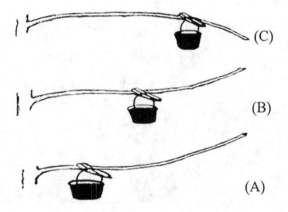

FIGURE 7.4 Weighting technique: (A) the weight is too close to the main trunk, (B) the best position of weight, (C) the weight is too far from the main trunk.

be restarted if the weight is too close to the main stem and will result in failing to accomplish the intended outcome (Figure 7.4 [A]). More arches are created, and the branch's end is pointed downward when the weight is placed too close to the main stem (Figure 7.4 [C]). Weights are a continual source of strain, and their position should be checked often to assure the correct outcome.

7.3 TRAINING THE YOUNG TREES

For the training of young apple trees, four basic tree canopy forms have been developed, including spherical, V-shaped, conical, and flat planar (Robinson, 2003). There are three main tree canopy shapes including: considered planar, angled, and vertical canopy (Musacchi, 2017). There is a good chance that new training systems will be developed throughout the world as a result of the wide variety of plant materials and environmental conditions (Lauri, 2018). The original architectural and operational characteristics of the cultivar-interstock and rootstock are critical to the garden's success. Training and pruning techniques for young trees must take into account the tree's growth tendency.

Tree planting density and dwarf rootstocks have a major impact on fruit yield in the first 10 years of an orchard's existence, resulting in greater yields. High-density planting techniques produce fruit in the second year but generate considerable fruit in the third year and are predicted to attain mature output in the fifth or sixth year. For orchards with a dense planting method, fruit production might begin in the sixth or seventh year, but full yields don't begin until the fifteenth year (Robinson, 2003).

Because several species have relatively small or large cultivars with an upward or wide growth tendency, as well as acrotonic or basitonic branches and variable internodes lengths, the techniques for cultivating and pruning young trees vary. Because contemporary apple orchards contain between 1000 and 6000 trees per hectare, and in some cases up to 10,000 trees per hectare in certain systems and cultivars, employing a specialized training technique is needed for each cultivar, planting scheme, and management method (Robinson, 2003). Only four basic forms of apple tree canopies are discussed in this chapter since it is not feasible to detail all the prevalent apple training techniques.

7.3.1 SPHERICAL-SHAPED CANOPY SYSTEM

Traditional apple orchards of the 1800s and 1900s used this kind of canopy design. The huge globe-shaped trees in the spherical system had primary trunk lengths of 1.8–2 meters and heights of 6–8 meters. Low tree density and big spreading trees were used to occupy the area allotted to each apple tree when they were planted. For great yields, a spherical form required 10–15 years of cultivation with numerous huge leaders or scaffold branches.

The spherical-shaped canopy system has the benefit of being the natural shape of apple trees, which means less pruning is needed for them (Figure 7.5). However, the canopies have a huge

FIGURE 7.5 Spherical-shaped canopy shapes.

shadow in central core that provides low fruit quality and is ineffective in terms of yields. The time it took for branches to mature under typical spherical canopies was too long. Planar, V-shaped, and conic systems supplanted spherical canopy apple orchards in the second part of the twentieth century. There are currently no apple orchards with spherical canopies in use. Compared to V-shaped or conical-shaped trees, the spherical canopy receives less light.

For spherical-shaped canopy systems
1) It should be headed back toa distance of 150–120 cm from the ground.
2) In the second year, two to five branches should be chosen from the branches produced.
3) The distance between these branches should be 20 to 30 cm.
4) The selected branches should be pruned if their length is more than 30 cm.
5) In this training system, none of the branches will continuously take the lead.
6) As soon as a branch dominates and is bigger than others, it should be headed back.

7.3.2   V-Shaped Canopy Systems

For centuries in commercial European gardens, apple trees with V-shaped canopies have been recommended for orchards with high tree density (Figure 7.6). With the goal of preventing sunburn on fruit, these canopy systems do best in situations of severe light intensities. In orchards where all the fruit must be picked from the ground, V-shaped canopies are also preferred. Pruning in a V-shape is less expensive than other methods, such as those that use numerous leaders per tree. In V-shaped systems, the best rootstocks are the dwarfing M.9 and M.27 rootstocks (Sosna, 2018). A leader branch should have a vertical angle of 60° or more for optimal fruit size (Sosna, 2018). There is a larger output in the orchards thanks to better light interception and a better quality fruit because of better light penetration to the core of the canopy. Trees with V-shaped canopies get more light than those with spherical or conical shapes. In addition, mechanical harvesting is more feasible with this canopy design than it is with others. Compared to other tree-planting technologies, the V-shaped systems allow for greater tree density.

For V-shaped canopy systems
1) The planted tree is headed at 100 cm above the budding height.
2) The emerging leaders are trained to 60° angles toward the alleyways.
3) The trees are annually pruned soon after the petals fall, starting from the fourth year following the orchard establishment.

FIGURE 7.6   V-shaped canopy systems.

7.3.3 Conic-Shaped Canopy Systems

In most commercial orchards, conic-shaped systems are the most common tree form. The majority of apple producers throughout the world have used conic-shaped systems for the past 50 years. The conic or pyramidal canopy is used to achieve a better tree form. There is no need for substantial branch and leader manipulation to grow trees in this canopy system. As a result, a variety of planting schemes have been devised, which included narrow, totally dwarf, conic-shaped trees which may be planted at high densities of 1500 to 4000 trees per hectare in either triple, double, or single row beds.

For this system, the optimum rootstock isM.9, although B.9 and G.16 stocks of the same dwarfing level are also used. With this approach, a conic form helps to distribute light more evenly by restricting the width of the tree's upper canopy. Individual tree stakes or a single-to-three-wire trellis support the trees at all times. Canopy widths are often less than 2 meters, while tree heights are typically 2 to 3 meters. However, the rows of trees in the tractor lanes are conical, and a large quantity of light energy falls between them and wastes light that might be utilized to grow fruit. Planting conic-shaped trees with wider than ideal row spacing allows farmers to employ current new equipment while planting dwarf trees.

Another downside of conic-shaped systems is that in old trees, the top branches tend to exceed the lower branches which results in too much shadow for those below them. A central leader is required in this system, but this system's open framework enables sunlight to reach the tree's interior. The horizontal scaffold branches and higher branches are positioned and regulated in size such that the shadow they throw on the lower branches is minimal. Because conic trees have less structural wood per unit of bearing surface than open-center trees, more of the leaves' products are available for fruiting. 'Golden Delicious' or 'Idared,' for example, are cultivars that may be taught to form a conic shape (Figure 7.7). There is a distinct central leader, wide-angled crotches, and a modest extension development in 'Golden Delicious' or 'Idared'. Branches from neighboring trees should not be allowed to overlap in order to avoid excessive shadowing. If a tree stands alone, its height should not surpass the distance between its branches. Once the tree reaches this height, it will take a lot of work to keep the top from shading the lower branches and depleting the tree's energy. It is less stimulating to remove one enormous branch as opposed to multiple little ones.

For conic-shape canopy systems

1) Usually, the trees are not heading cut in this system.
2) In the first growing season, the central axis closes vertically to the stake.
3) Branching occurs along the central axis.
4) When planting, it is better to use standard seedlings with a central axis with at least four lateral branches.

FIGURE 7.7 Conic-shaped canopy systems.

5) In the beginning of the summer, bring the branches (using clamps and weights) to a 60° angle to the tree trunk.
6) In the early spring, the strong lateral branch that competes with the central axis must be removed.
7) In the upper part of the tree, vegetative growth is reduced every year by replacing the central axis with a weak lower branch or bending the central axis.

7.3.4 FLAT PLANAR CANOPY SYSTEMS

The most common of the flat planar canopy systems were the espaliers grown along walls and fences. Planar canopies were developed to overcome the problems of light penetration into thick canopies. With the horizontal planar canopies, such as the T or Ebro trellis, there is a drastic reduction in light levels from the top to the bottom of the canopy. Their aim was to improve orchard labor efficiency through the use of platforms that allowed more efficient pruning, picking, and thinning of the trees in this canopy system compared with the traditional vase-shaped fruit trees of that time. With the use of platforms to position workers near their work (pruning and picking), labor efficiency can be improved by 15–20%. The success of the palmette system stimulated widespread grower adoption throughout the world.

Most of the newer forms of these systems utilize dwarfing rootstocks, which were not in the original system. The first advantage of flat planar systems is the ability to easily pick operations and mechanize orchard management. However, this has not yet been achieved (mechanized harvesting). The second advantage of planar canopies is that the thin two-dimensional canopy has good light distribution to all parts of the canopy. The planar vertical trellis systems (such as the palmette and Penn State trellis) have produced some of the best fruit quality of any system. However, the horizontal planar systems (Lincoln and Ebro) often have the poorest fruit color.

It should be noted that the labor costs to manipulate the canopies for the flat planar systems are much higher than systems that are based on a natural tree shape. Vertical planar canopies have good light exposure on both sides of the canopy. Also the vertical planar canopies must be tall or have narrow tractor alleys to optimize light interception.

7.3.5.1 Pruning Flat Planar Canopy Systems

Pruning of this tree form is simple compared with the central leader tree because light does not have to filter through the layers of canopy. Pruning consists of keeping a narrow palmette top in the north-south direction, thinning out excess limbs in the top, and removing the upright growth from the bottom whorl of scaffolds (Figure 7.8). Maintenance of the basic geometric form is easy to teach to pruners since a high level of skill is not required.

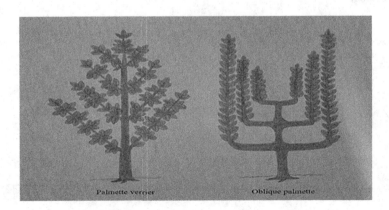

FIGURE 7.8   Flat planar canopy systems.

For flat planar canopy systems

1) This training system is based on a low initial pruning principle and on bending the most vigorous branches.
2) Usually a heading back method is not used.
3) In the initial phase of training, it is possible to bend the main axis of the tree to obtain a new leader and a very strong branch to limit the growth of the central axis.
4) Pruning involves the removal of very vigorous waterspouts when they are still in the herbaceous phase while trying to keep the branches of the current season less vigorous to ensure the formation of buds in the following year.
5) It is necessary to try to maintain a good renewal of the branches with periodic pruning to eliminate the ones that have already produced and are exhausted.
6) By shortening to 3–4 buds on shoots that are 2–3 years old, fruit set can be increased by reducing the competition between the various reproductive organs.

7.4   EFFECTS OF DIFFERENT PRUNING INTENSITIES ON APPLE TREES

Different pruning intensities affect plant growth, fruit yield and fruit quality such that the highest pruning intensity in apple trees results in higher vegetative growth than spur development, which in turn negatively affects fruit yield. Kaith et al. (2011) reported an increase in pruning severity in apple trees resulted in a decrease in fruit set and yield in 'Starking Delicious' apple trees. This effect was consistent with more vegetative growth and less spur development. In agreement with the other study (Sharma, 2014), it was observed that the heavily pruned apple trees tended to grow vigorously while maximum fruit set and yield was recorded at a light pruning intensity (25%pruning and thinning). The overall decline in fruit set with increasing pruning intensity may be related to the lowest number of flowers retained on the pruned shoots (Sharma and Banyal, 2020) and the increased foliar nutrient status (such as nitrogen and potassium content) due to the newly growing vegetative shoots (Kaith et al., 2011).

The fruit quality parameters such as fruit length/diameter ratio, fruit weight, fruit size, fruit diameter, and total soluble solids (TSS) increased significantly by pruning intensities; however, different pruning intensities had no significant effect on fruit firmness, total acidity, fruit russeting, and reducing the sugars of apple fruits (Sharma, 2014; Bound and Summers, 2000). The increase in TSS and reducing sugars in fruits of heavily pruned trees may be due to the fact that adequately illuminated leaf area on the tree allows maximum carbon assimilation due to carbohydrate supply for increased fruit quality (Bound and Summers, 2000) and the absence of other potentially competing actively growing sinks (Sharma and Banyal, 2020).

Bound and Summers (2000) also considered that the improvement in fruit size with heavy pruning could be due to 1) reducing the number of fruits at or shortly after flowering, 2) increasing resource uptake by individual fruits, and 3) decreasing negative shading effects.

7.4.1   Summer Pruning in Apple Trees

The history of traditionally applied summer pruning in apple orchards can be traced back well beyond the seventeenth century and has attracted the scientific attention of pomologists since 1863 (Marini and Barden, 1987). Scientists have claimed that summer pruning in apple trees can be done before the beginning of the growing season, up to 4 weeks before harvest (Morgan et al. 1984), or when there are leaves or flowers on the tree (Crassweller, 1999).

In general, summer pruning means removing the current season's upright and vigorous growth (water sprouts, shoots, and leaves) during the growing season (April–August) with various thinning cuts removing entire water sprouts and unwanted branches; heading back the upper, middle, and lower canopy; or pinching based on vigorously growing shoots, the age of the tree, and the cultivar and training system. Tree growth is regulated by winter pruning during the dormant season;

however, summer pruning, which affects both reproductive and vegetative growth, may also be appropriate.

Many papers show that consistent summer pruning in apple trees has an effective effect on vegetative traits such as vigor, canopy volume, trunk diameter, and leaf area. In addition, this practice can also have significant effects on fruit load (Li et al. 2003), fruit quality (Buler and Mika, 2009), and the flower bud (Rom and Ferree, 1986).

The implementation of summer pruning effectively reduces some vegetative traits such as trunk and root growth (Kikuchi et al. 1989), total crown volume (Mészáros et al. 2015), trunk circumference (Talaie et al. 2004), stem dry weight (Ferree et al. 1983), and tree height (Myers and Ferree, 1983).

The study of summer pruning in different apple cultivars showed that this technique resulted in reduced fruit characteristics including fruit weight (Bound, 2005), fruit size (Myers and Ferree, 1983; Morgan et al. 1984), total acidity (TA; Guerra and Casquero, 2010), yield (Bound, 2005), and the soluble solids content in the fruit (Taylor and Ferree, 1984; Guerra and Casquero, 2010). However, other authors have suggested that summer pruning has no effect on fruit weight (Autio and Greene, 1990), fruit size, TSS, firmness of flesh, the starch index (Li et al. 2003), yield, or flesh firmness (Greene and Lord, 1983), or they have suggested an increased percentage of large fruit (Bound, 2005) or increased fruit size (Dietz, 1984). Differences in fruit quality trait results in response to summer pruning depend on cultivars, type of summer pruning, severity of summer pruning, timing of pruning, number of leaves removed, and the amount of leaf surface removed as the main source of large photosynthetic capacity during the fruit development phase. Furthermore, the reduced leaf to fruit ratio due to summer pruning has the greatest impact on fruit quality and fruit growth.

In addition, according to general observations, summer pruning can improve light penetration and distribution in the canopy (Bhusal et al. 2017). Summer pruning also has a positive effect on fruit color (Dietz, 1984; Autio and Greene, 1990), especially on the red color of apple fruit at harvest time; this increase in the red color of apple fruits has also been observed during storage (Stover et al. 2003).

Summer pruning in apple fruits has been practiced to control postharvest physiological disorders in apples because it has a much greater effect on the quality and quantity of apple fruits during growth. Guerra and Casquero (2010) reported that summer pruning in apple fruits reduced the outer bitter pit (by 8.3% after 120 days of storage and by 10% at the end of storage) and the inner bitter pit. Reduction of bitter pit in apple fruits may be related to reduced fruit size and increased K and Mg content in fruits, i.e., increased (K+ Mg)/Ca ratio. In addition, Ashraf et al. (2018) showed that summer pruning increased fruit calcium content, leading to a reduction in physiological disorders associated with calcium such as bitter pit and cork spot. Moreover, they reported that this could be the effect of summer pruning on reducing shoot growth and available calcium for fruits, which increased the calcium content in fruits.

On the other hand, summer pruning can increase flower bud formation (Mohammadi et al. 2013), increasing the number of flower clusters (Ogata et al. 1984). An induced formation of flower buds after summer pruning due to a decrease in the vigor of trees removed apical dominance and triggered the formation of lateral buds and increased light penetration within the canopy; thus, this would be useful for flower development. However, other authors could not confirm a consistent effect of summer pruning on increasing bud flower formation as reported by Ferree and Stang (1980), Chandler (1923), and Greene and Lord (1983). This difference may be due to cultivar differences, intensity of pruning, timing of pruning, and weather conditions.

In general, there are widely varying responses to summer pruning in apple trees (Table 7.1), with many inconsistencies yet to be resolved. So some factors such as the amount and type of removal, the timing of summer pruning, the timing of evaluation, varieties, orchard management, severity of pruning, environmental conditions, vigor of the cultivar, type of rootstock or scion/rootstock combinations, and age of the tree are likely to play a role.

TABLE 7.1

Effects of Summer Pruning on Apples Trees

Pruning methods	Pruning time	Cultivars	Effects	References
Removal of water shoots and season's growth	4 weeks before harvest	Fuji	Reduced fruit weight, fruit size, fruit Length/Diameter ratios, TSS and increased fruit firmness, and russet levels	Bound (2005)
Thinning out water sprouts, unwanted branches and heading back the current season's growth	8 weeks and 12 weeks after full bloom	Red Delicious	Increased TSS, fruit color, and enhanced fruit calcium	Ashraf et al. (2018)
Eliminating all vigorous shoots above 30 cm	94 days after full bloom	Reinette du Canada	Decrease bitter pit incidence during storage, slightly lower in weight, accelerates red color, decrease both TSS and TA	Guerra and Casquero (2010)
Removing the current season's shoots after the fifth leave and removing all water sprouts	30, 60, and 90 days after full bloom	Golab and Shafi-Abadi	Control of the growth, no negative effects on productivity and fruit quality	Talaie et al. (2004)
Thinningcut or back to the first fruiting spur	30 July and 4 August	Empire	Does not affect fruit color, TSS, starch, firmness, and internal breakdown after storage, return bloom, or root growth	Li et al. (2003)
Thinning cuts (upright branches, waterspouts, and shade-causing, hanging branches were removed)	1 July through 1 Sept	Rogers McIntosh	Enhanced red coloring, increased the percentage of crop graded, decreased TSS, harvested earlier, reduce the losses to preharvest abscission	Autio and Greene (1990)
Remove excess shoots and weak pendant wood	late August (40 days before harvest)	McIntosh	No effect on the internal ethylene concentration of fruit, increased starch fruit, light penetration in canopy, fruit color, and softer fruit, and earlier onset fruit drop	Schupp (1991)

7.4.2 PRUNING SYSTEMS IN APPLE ORCHARDS AND FRUIT QUALITY

Apple tree yield and crop load depend on pruning and training system with an emphasis on light management (light interception and distribution in the canopy and optimization of yield) and balance between vegetative and reproductive components. Currently there is no internationally accepted training system for all apple cultivars worldwide as the vegetative growth and reproductive habits of individual apple cultivars depend on soil properties, vigor of rootstocks, and the weather conditions of the countries (Table 7.2) where they are grown (Gandev et al. 2016).

Gandev et al. (2016) evaluated three training systems, including the Slender spindle, Solen, and Vertical axis on vegetative and reproductive characteristics of apple cultivars. They reported that in Bulgaria, the Vertical axis training system is the best orchard system for the 'Braeburn' cultivar because the better reproductive characteristics, the highest number of apple trees per Ha, and the higher average and cumulative yields per Ha were harvested when the Vertical axis training system was used.

Fallahi et al. (2019) announced that the apple cultivar 'Aztec Fuji' with central leader training had a lower yield (39.9 kg/tree) than the tall spindle training (53.9 kg/tree) on Nic29 rootstock. They also reported that apple trees with central leader training had a denser and darker canopy due

TABLE 7.2

Effects of Different Training Systems on Apple Trees

Training system uses	Suggested Training system	Country	Cultivars	Effects	References
Slender spindle, Solen, and Vertical axis	Vertical Axis training	Bulgaria	'Braeburn'	Solen training delay the reproductive habits of the trees; Slender spindle and Vertical axis systems induce more vigorous growth compared to Solen	Gandev et al. (2016)
Central leader and Slender spindle	Slender spindle	USA (Idaho)	'Aztec Fuji'	Fruit weight was similar in both tree training; central leader had 43% higher proline content and 4% higher chlorophyll	Fallahi et al. (2019)
Geneva Y-trellis		Iran (Karaj)	'Golab-kohans', 'Fuji', 'Starkin', 'Delbarestival', 'Prime rose'	'Delbarestival' and 'Starking' are a more efficient portion, and 'Prime rose' isn't a suitable cultivar for high density planting	Talaie et al. (2011)
V slender spindle and Geneva Y-trellis	Y-trellis	Iran (Karaj)	'Golab-kohans', 'Fuji', 'Starking', 'Delbar'	'Delbarestival' and 'Fuji' are more adaptive to intensive training systems, and the training system affected tree vigor; V-system contributed to a higher density of trees	Dadashpour et al. (2019)
Solaxe and spindle training	Slightly better Solaxe	Slovenia	'Pink Rose', 'Fuji Kiku 8', 'Rubinola', 'Gala Galaxy', 'Braeburn Hillwell', 'Golden Delicious', 'Fuji Raku', 'Juliet'	No effect on alternate bearing; Similar yield with slightly better Solaxe training system; training systems do not influence the average fruit weight with the exception of the 'Golden Delicious'; fruit mass and fruit diameter	Fajt et al. (2016)
Hytec, Solen, Mikado, and spindle	Mikado	Poland	'Sampion'	Highest fruit size, leaf area, and percentage of colored apples related 'Mikado' and 'Hytec'	Buler et al. (2001)
Tall spindle, Vertical axis, and Solaxe		Brazil	'Gala'	No effect of fruit quality; most productive system in fourth leaf was Solaxe, but all of them had a similar cumulative, and the highest canopy volume was related to the vertical axis.	Kretzschmar et al. (2016)
Minimally pruned, Tall spindle, and Tall trellis	For 'Jonagold' it was minimally pruned, but for 'Fuji' it was V axis	USA (Pennsylvania)	'Jonagold', 'Fuji'	Number of fruit trees, average fruit weight, total yield tree, and cumulative yield ha for 'Jonagold' was not influenced by training system; total yield tree and cumulative yield per unit of land of 'Fuji' was minimally pruned; Tall trellis was highest labor in both cultivars.	Crassweller et al. (2018)

Training system		Location	Cultivar	Observations	Reference
Slender spindle, Geneva Y-trellis, a modified Solen training called Solen Y-trellis, V-trellis, and the new version of the V-trellis with higher densities		Canada	'Royal Gala'	Solen Y-trellis had smaller TCA; V-trellis high density trees were difficult to keep within their allotted space and for high wood removal; cumulative yield over the first three crops related to slender spindle; training system had no consistent effect on average fruit weight; Slender Spindle and V-trellis had better fruit color	Hampson et al. (2002)
Centrifugal Training and Original Solaxe	-	France	'Ariane', 'Pitchounette'	At harvest, fruit damage with aphid infestation in centrifugal trees was less than compared to the Original Solaxe ones; European red mite infestation lower in centrifugal than in original Solaxe trees; leaf scab incidence was similar in centrifugal and in original Solaxe systems, but the number of scab lesions per leaf was higher in centrifugal trees than in the original Solaxe trees in June; fruit damage at harvest was not different between systems; during the summer, scabs increased faster in centrifugal trees than in original Solaxe trees	Simon et al. (2007)
Spindle and V-trellis training	Spindle	USA (Washington State)	WA38 ('Cosmic Crisp')	Leaf is index for the V-system and the Spindle system not significantly differences; the V-system had higher light interception compared to the Spindle; Spindle trees tended to yield a slightly higher amount of apples on a kg per tree; average fruit weight and fruit number per tree than the V-system; production per hectare (34.5 MT/hectare [MT/ha]) was lower than the V-system (52.5 MT/ha)	Anthony et al. (2019)

to more shading and less fruit bud flower formation, and thus had a lower yield than tall spindle training. Talaie et al. (2011) studied the response of the different apple cultivars 'Golab-kohans' (Iranian cultivar), 'Fuji', 'Starking', 'Delbarestival', and 'Prime rose' raised in a Geneva Y-trellis system in Iran. Talaie et al. found that 'Golab-kohans' (Iranian cultivar) had the highest pH (4.07) and moisture content (85.96%). However, 'Delbarestival' showed the highest fruit length (6.13 cm), L/D (0.87), and TSS (15.77), while 'Starking' had the highest fruit weight (145.24 g), fruit diameter (6.91 cm), and ash content (0.71%). However, 'Fuji' fruit had more dry matter (20.13%), fruit firmness (13.13 kg/cm^2), and TA (0.72%).

On the other hand, in the study by Dadashpour et al. (2019) using the same cultivar, it was reported that the Geneva V training system had higher yield efficiency (0.25 kg/cm^2), dry matter, ash, and TSS than Y training system; whereas in the Y training system wider stem cross-sectional area, cumulative yield per h (28.08 T/Ha) and tree (16.72 Kg/tree), fruit diameter, fruit length, fruit weight, and fruit firmness were observed. They also found the Y training system was more suitable than the V training system for productive apple orchards in Iran (Karaj).

Fajt et al. (2016) compared the training systems for apple trees (Solaxe and Spindle training) and found that both training systems showed similar pattern of alternate fruit set in the studied cultivar. An earlier study by Guitton et al. (2012) showed that hormones could be the main factor in regulating biennial fruiting, and that flower initiation is limited by excessive yields.

Buler et al. (2001) demonstrated that the training system in hybrid tree cone system (HYTEC) and Solen decreased tree vegetative growth of trees compared to Mikado and spindle, while the highest productivity index and yield were observed in Mikado and spindle, respectively. It was also reported that the use of Mikado and HYTEC in apple orchards resulted in better light distribution in the canopy (especially in the middle part of the canopy), which in turn resulted in more colored fruits.

The Mikado training system was introduced by Widmer and Krebs (1996) from the Gutingen research station in Switzerland. In this training system, each apple tree is trained with three or four strong leader branches attached to wires. Lespinasse (1987) has also introduced the Solen (Solaxe) training system in France.

Inconsistent effects of the apple orchard training system on fruit quality and tree characteristics were also found (Table 7.2). Kretzschmar et al. (2016) compared three training systems in Brazil with cultivar 'Gala' and reported that the different training systems had no significant effect on fruit quality traits such as TSS, flesh firmness, and fruit diameter. They also observed that some vegetative traits, including internode length and tree height, were not affected by the training system. In addition, their data showed the baseline cost for each system; the Vertical axis system being the cheapest system.

Crassweller et al. (2018) studied the Vertical axis system, minimum pruning (building a base tier of permanent scaffolds with strong crotch angles), Tall spindle, and Tall trellis training systems on two cultivars at the Horticulture Research Farm in Rock Springs, Pennsylvania, USA, from 2009 to 2017. They showed that the training system had no effect on tree size within cultivars, while the highest cumulative average number of fruit trees and hectares in both cultivars was related to a minimally pruned system.

Their study also showed that there were significant differences in the response of different varieties to different training systems (Table 7.2). For example, in the cultivar 'Jonagold', the training system had no effect on total cumulative yield, while the cultivar 'Fuji' showed the highest yield per tree per hectare under minimally pruned systems compared to other training systems. The economic analysis of the data announced by Crassweller et al. (2018) also showed that the total income per unit area did not differ by the training system for either cultivar, so that the highest income and net returns were obtained with the Vertical axis system, but the highest labor costs were obtained with the Tall trellis system. Therefore, economic comparison is an important parameter for selecting the cropping system in each country.

In Canada, Hampson et al. (2002) studied the effects of five training systems including Slender spindle, Geneva Y-trellis, a modified Solen training called Solen Y-trellis, V-trellis, and the new version of V-trellis with higher density (7143 trees/ha) on the variety 'Royal Gala' grafted on rootstock M.9; they showed that Geneva Y-trellis and Solen Y-trellis systems increased the cumulative yield/ha by 11% to 14% compared to the Slender spindle training system, while the minimum fruit size and delayed color development were observed in the higher density V-trellis training system.

Surprisingly, the training system in apple orchards has effects on pests and pathogens. Previous studies suggested that these effects are related to architectural characteristics of the tree such as the physical structure and physiological response of the tree fruit (Simon et al. 2007). Unfortunately, there are only a limited number of studies reporting the relationship between the training orchard system and pest and disease injury; a preliminary study in France compared the effects of two training systems, centrifugal training and the Original Solaxe, on pest and disease infestation (Simon et al. 2007). The experimental results indicated that the highest infestation of pests such as rosy apple aphid (*Dysaphisplantaginea*) and European red mite (*Panonychus u/mi* Koch.) was observed in Original Solaxe, while the highest fruit damage by codling moth (*Cydiapomonella* L.) and disease development by apple scab (*Venturia inaequalis* Cooke G. Wint.) were associated with centrifugal training in summer. On the other hand, this is a new topic for future research on the comparison of other training systems to reduce pest and disease infestation in organic apple production

The establishment of high density trees in apple orchards has developed rapidly in the past decade, making the control of light inside the canopy and spacing between trees in the orchard a major factor in flower bud formation and fruit quality. Therefore, a training system in apple orchards can help control light to increase reproductive growth and fruit quality. In recent years, the Spindle and V-trellis training systems are the most important systems for apple orchards in Washington State. In addition, studies from Washington State have shown that Spindle and V-trellis trainings have a significant effect on light absorption and fruit yield. These results indicate that the spindle training system on rootstock Nic29 using the click-cut method (head pruning) results in uniform yields, improved light interception, and optimal thresholds for LAI for cultivar WA38 ('Cosmic Crisp') (Anthony et al. 2019).

7.4.3 EFFECTS OF ROOT PRUNING ON FRUIT QUALITY AND TREE CHARACTERISTICS

Maintaining a proper vegetative and reproductive balance is the major challenge in orchard production of fruit trees. There are many common horticultural practices to maintain a balance between shoot growth and fruiting in fruit trees (Miller and Tworkoski, 2003). Root pruning is a common technique used to reduce vegetative growth in apple trees (Geisler and Ferree, 1984) and is an extremely effective and economical method that not only contributes to assisting in dwarfing, but also stimulates new roots necessary for growth.

Root pruning results in reduced resource uptake, including water and nutrients, an imbalance of plant hormones, and other factors such as reduced availability of carbohydrates that can affect vegetative growth of fruit trees. Literature on the effects of root pruning on apple trees report that this practice is very effective in reducing shoot growth of apple trees (Poni et al. 1992). Early studies have shown that root pruning reduced terminal shoot growth (20%) and the height of apple trees (12%), and root pruning was found to significantly reduce shoot length, number of shoots, and fruit diameter compared to a control without root pruning (Khan et al. 1998; Ferree, 1992). Despite the many studies demonstrating the benefits of root pruning in controlling canopy size, there are also several studies showing the negative effects of this horticultural practice.

Some studies suggest that root pruning may reduce fruit tree yields (Ferree, 1989; Khan et al. 1998), which can be explained by the effect of root pruning on the lower leaf area index due to reduced photoassimilate supply. In addition, Vercammen et al. (2006) reported that root pruning often has negative effects on the water and nutrient uptake capacity of fruit trees due to severe damage to the root system.

Some authors suggest that fruit trees whose roots are pruned suffer from drought stress and nutrient deficiency which may affect fruit growth, fruit yield, fruit quality, and fruit size. Recently, there is a need to develop field strategies to reduce these negative effects of root pruning on the fruit size of apple trees. Among other possible managements after root pruning, it is believed that supplemental irrigation of the pruned root system and improvement of water balance can neutralize the negative effects of root pruning on fruit quantity and quality in apple orchards (Wang et al. 2014). Pavičić et al. (2009) indicate that root pruning combined with crown pruning can significantly increase yield and reduce biennial fruit set without negatively affecting fruit quality.

The literature on the effects of root pruning on productivity and yield factors such as alternate bearing, flower induction, and return is controversial. Mitre (2012) found that the best cumulative yield was obtained from a system with root pruning (98.75 t/ha) compared to a tree without root pruning (97.8 t/ha). They also reported that root pruning increased the number of bearing branches and productivity index in high density apple orchards. Similar positive effects on tree productivity after root pruning are also reported in other studies in apples (Geisler and Ferree, 1984; Sosna, 2002). On the other hand, there are also some studies (Ferree and Rhodus, 1993; Khan et al. 1998; Ferree, 1992) reporting lower yields.

Khan et al. (1998) and Ferree (1992) reported that root pruning in apples resulted in a reduction in alternate bearing and increased return bloom, flowering spurs, and generally lower fruit drop before harvest.

Consistent with this, Wang et al. (2014) found that root pruning increased return bloom after onset of the treatments and reported a significant positive effect of root pruning on return bloom. Others found little or no effect on alternate bearing (McArtney and Belton, 1992).

There are differences in responses to root pruning due to the timing of root pruning, interaction of genetic and climatic factors, and differences among cultivars independent of root pruning which could be explained from a genetic perspective. In addition, Sokalska et al. (2009) indicate that the effects of root pruning depends on soil properties, irrigation practices, rootstock type, and other factors, and a significant amount of fine roots can also be found in the lower soil layers.

Previous work by Luthi (1974) and Schupp and Ferree (1987a) recommended root pruning between March and mid-May to reduce excessive vegetative growth. In addition, some authors have shown that root pruning in April or December does not produce the desired results (Schumacher, 1975; Schumacher et al. 1978). Schupp and Ferree (1987b) also indicated that root pruning later in the growing season, in June, had no significant effect on growth and increased fruit drop before harvest in apple trees. Another study showed that root pruning reduced the intensity of biennial fruit set in apple trees and that this reduction was unrelated to the timing of root pruning.

Despite some differences in study results, the optimal conditions and timing for root pruning to reduce vegetative growth and preharvest fruit drop, as well as to improve fruit quantity and fruit quality and minimize undesirable reductions in fruit size, depend on tree vigor, cultivars, tree age, soil properties, and rootstock type. In addition, we should not forget that root pruning can be carried out only in irrigated orchards with optimal soil conditions (Ferree, 1992).

7.5 CONCLUSION

The main objective of training and pruning is to optimize the cultivation of trees with high quality fruit production while improving the regularity of fruit yield, especially in naturally alternative fruit cultivars. Training and pruning fruit trees are a basic agricultural technique and essential for balancing the vegetative and reproductive fruit tree. Selecting the best training method can influence height; create a strong scaffold of scaffold limbs capable of supporting heavy, regular annual fruit bearing and fruit quality; expose most leaf surfaces to the sun; direct tree growth so that various cultural measures such as spraying and harvesting become economical; protect the tree from sunburn; and promote early production with light interception. Training and pruning trees has a special place in the management and success of orchards. It is difficult to train and prune due to the high cost and the need to repeat every year, and, if done poorly, it will have many negative effects on production efficiency and quality of fruits produced in the next years.

Pruning and cultivating trees interacts with other horticultural practices including fertilizing, thinning fruit, pests and diseases control, and harvesting fruit.

The most important effect of training and pruning is the fruit quality because if proper training and pruning is not done, light cannot penetrate into the tree canopy and the fruits produced will not be of good quality. Finally, although training and pruning trees does not guarantee 100% success of the orchard, the success of the orchard will not be possible without proper training and pruning.

REFERENCES

Anthony, B., S. Serra, and S. Musacchi. 2019. Optimizing crop load for new apple cultivar: "WA38". *Agronomy* 9:107.

Ashraf, N., M. Ashraf, M.Y. Bhat, and L. Sharma. 2018. Paclobutrazol and summer pruning influences fruit quality of red delicious apple. *Indian Journal of Hill Farming* 10:349–356.

Ashraf, N., M. Ashraf, M.Y. Bhat, and M.K. Sharma. 2017. Paclobutrazol and summer pruning influences fruit quality of red delicious apple. *International Journal of Agriculture. Environment and Biotechnology* 10(3):349–356.

Autio, W.R. and D.W. Greene. 1990. Summer pruning affects yield and improves fruit quality of McIntosh' apples. *Journal of the American Society for Horticultural Science* 115(3):356–359.

Badrulhisham, N. and N. Othman. 2016. Knowledge in tree pruning for sustainable practices in urban setting: Improving our quality of life. *Procedia-Social and Behavioral Sciences* 234:210–217.

Bhusal, N., S.G. Han, and T.-M. Yoon. 2017. Summer pruning and reflective film enhance fruit quality in excessively tall spindle apple trees. *Horticulture, Environment, and Biotechnology* 58:560–567.

Bound, S.A. 2005. *The impact of selected orchard management practices on apple (*Malus domestica L.*) fruit quality*. PhD thesis, University of Tasmania, 152pp.

Bound, S.A. and C. Summers. 2000. The effect of pruning level and timing on fruit quality in Red 'Fuji' apple. *VII International Symposium on Orchard and Plantation Systems* 55(7):295–302.

Brasil, M. et al. 2018. Postharvest quality of fruits and vegetables: An overview. In *Preharvest modulation of postharvest fruit and vegetable quality*. Academic Press, New York, pp. 1–40.

Buler, Z. and A. Mika. 2009. The influence of canopy architecture on light interception and distribution in 'Sampion' apple trees. *Journal of Fruit and Ornamental Plant Research* 17:45–52.

Buler, Z., A. Mika, W. Treder, and D. Chlebowska. 2001. Influence of new training systems on dwarf and semidwarf apple trees on yield, its quality and canopy illumination. *Horticulturae* 557:253–260.

Chandler, W.H. 1923. *Results of some experiments in pruning fruit trees*. Willsboro, NY: Cornell University Agricultural Experiment Station.

Crassweller, R.M. 1999. *Effects of summer pruning on apples and peaches*. College of Agricultural Sciences, Penn State University. 152 pp.

Crassweller, R.M., L. Kime, and D. Smith. 2018. Orchard architecture effects on yield and economics of two apple cultivars. *Acta Horticulturae* 1281:207–212.

Dadashpour, A., A. Talaie, M. Askari-Sarcheshmeh, and A. Gharaghani. 2019. Influence of two training systems on growth, yield and fruit attributes of four apple cultivars grafted onto 'M. 9'rootstock. *Advances in Horticultural Science* 33:313–320.

Dietz, H.J. 1984. The effect of summer pruning on growth, yield and fruit quality. *Obstbau* 9:320–321.

Fajt, N., E. Komel, J. Jakopic, F. Stampar, M. Hudina, and R. Veberic. 2016. Solaxe and spindle-comparison of two apple training systems. In *XI international symposium on integrating canopy, rootstock and environmental physiology in orchard systems*, 1228105–1228108.

Fallahi, E., S. Mahdavi, C. Kaiser, and B. Fallahi. 2019. Phytopigments, proline, chlorophyll index, yield and leaf nitrogen as impacted by rootstock, training system, and girdling in "Aztec Fuji" apple. *American Journal of Plant Sciences* 10:1583.

Ferree, D.C. 1989. Growth and carbohydrate distribution of young apple trees in response to root pruning and tree density. *HortScience* 24:62–65.

Ferree, D.C. 1992. Time of root pruning influences vegetative growth, fruit size, biennial bearing, and yield of 'Jonathan' apple. *Journal of the American Society for Horticultural Science* 117:198–202.

Ferree, D.C. 1994. Early performance of two apple cultivars in three training systems. *HortScience* 29(9):1004–1007.

Ferree, D.C., S. Myers, C. Rom, and B. Taylor. 1983. Physiological aspects of summer pruning. *International Workshop on Controlling Vigor in Fruit Trees* 14(62):43–52.

Ferree, D.C. and W.T. Rhodus. 1993. Apple tree performance with mechanical hedging or root pruning in intensive orchards. *Journal of the American Society for Horticultural Science* 118:707–713.

Ferree, D.C. and E.J. Stang. 1980. Influence of summer pruning and Alar on growth, flowering, and fruit set of Jerseymac apple trees. *Research Circular Ohio Agricultural Research and Development Center* 259:4–6.

Ferree, D.C. and I.J. Warrington. 2003. *Apples: Botany, production, and uses.* Wallingford, UK: CABI Publishing, pp. 672.

Forshey, C. et al. 1992. *Training and pruning apple and pear trees.* American Society for Horticultural Science, pp. 166.

Franzen, J.B. 2015. Application and implications of rule-based pruning of apple trees. Open Access Theses. 523. Purdue University. *Purdue e-Pubs, a service of the Purdue University Libraries.*

Gandev, S., I. Nanev, P. Savov, E. Isuf, G. Kornov, and D. Serbezova. 2016. The effect of three training systems on the vegetative and reproductive habits of the apple cultivar 'Braeburn' grafted on M9 rootstock. *Bulgarian Journal of Agricultural Science* 22:600–603.

Geisler, D. and D.C. Ferree. 1984. The influence of root pruning on water relations, net photosynthesis, and growth of young Golden Delicious apple trees. *Journal of the American Society for Horticultural Science* 109:827–831.

Greene, D. and W.J Lord. 1983. Effects of dormant pruning, summer pruning, scoring and growth regulators on growth, yield and fruit quality of 'Delicious' and 'Cortland' apple trees. *Journal of the American Society for Horticultural Science* 108:590–595.

Guerra, M. and P. Casquero. 2010. Summer pruning: An ecological alternative to postharvest calcium treatment to improve storability of high quality apple cv. 'Reinette du Canada'. *Food Science and Technology International* 16:343–350.

Guitton, B., J.J. Kelner, R. Velasco, S.E. Gardiner, D. Chagne, and E. Costes. 2012. Genetic control of biennial bearing in apple. *Journal of Experimental Botany* 63:131–149.

Hampson, C.R., H.A. Quamme, and R.T. Brownlee. 2002. Canopy growth, yield, and fruit quality of 'Royal Gala' apple trees grown for eight years in five tree training systems. *HortScience* 37:627–631.

Hamzakheyl, N., D.C. Ferree, and F.O. Hartman. 1976. Effect of lateral shoot orientation on growth and flowering of young apple trees. *HortScience* 11(4):393–395.

Kaith, N., U. Sharma, D. Sharma, and D. Mehta. 2011. Effect of different pruning intensities on growth, yield and leaf nutrients status of starking delicious apple in hilly region of Himachal Pradesh. *International Journal of Farm Sciences* 1:37–42.

Khan, Z., D. McNeil, and A. Samad. 1998. Root pruning reduces the vegetative and reproductive growth of apple trees growing under an ultra high density planting system. *Scientia Horticulturae* 77:165–176.

Kikuchi, T., T. Asada, and Y. Shiozaki. 1989. Effect of summer pruning on the next season' s shoot growth of young apple trees. *Journal of the Japanese Society for Horticultural Science* 58:491–497.

Kretzschmar, A., G. Sander, R. Arruda, A. Lima, T. Macedo, A. Rufato, and L. Rufato. 2016. The effect of different training systems on vegetative and productive performance of 'Maxigala' apple tree in southern Brazil. *Acta Horticulturae* 1228:135–140.

Lauri, P.-E. 2018. Training and pruning the apple tree according to the SALSA System. 13. SENAFRUT – *Seminario nacional sobre fruticultura de clima temperado. Sao Joaquim, Brazil.* hal-01819733.

Lespinasse, J.M. 1987. Réflexions sur la conduite du pommier: une nouvelle forme le "Solen". *Arboriculture fruitière (Paris)* 399:45–48.

Li, K.T., A.N. Lakso, R. Piccioni, and T. Robinson. 2003. Summer pruning effects on fruit size, fruit quality, return bloom and fine root survival in apple trees. *The Journal of Horticultural Science and Biotechnology* 78:755–761.

Luthi, E. 1974. Die wurzelbehandlung zu triebiger baume-ein erfolg. *Thurgauer Bauer* 121:1606–1609.

Marini, R.P. and Barden, J. 1987. Summer pruning of apple and peach trees. *Horticultural Review* 9:351–375.

McArtney, S.J. and R.P. Belton. 1992. Apple shoot growth and cropping responses to root pruning. *New Zealand Journal of Crop and Horticultural Science* 20:383–390.

Mészáros, M., J. Sus, L. Laňar, and J. Náměstek. 2015. Evaluation of slender spindle form in young 'Topaz' apple orchard. *Scientia Agriculturae Bohemica* 46:167–171.

Miller, S.S. and T. Tworkoski. 2003. Regulating vegetative growth in deciduous fruit trees. *PGRSA Quarterly* 31:8–46.

Mitre, V. 2012. Effect of roots pruning upon the growth and fruiting of apple trees in high density orchards viorel MITRE, Ioana MITRE, Adriana F. SESTRAS 2). Radu E. SESTRAS. *Bulletin UASVM Horticulture*, 69, 1.

Mohammadi, A., M. Mahmoudi, and R. Rezaee. 2013. Vegetative and reproductive responses of some apple cultivars (*Malus domestica* Borkh.) to heading back pruning. *International Journal of AgriScience* 3:628–635.

Morgan, D., C. Stanley, R. Volz, and I. Warrington.1984. Summer pruning of Gala apple: The relationships between pruning time, radiation penetration, and fruit quality. *Journal of the American Society for Horticultural Science* 109:637–642.

Musacchi, S. 2017. Achieving sustainable cultivation of apples. In *Innovations in apple tree cultivation to manage crop load and ripening*. Duane Greene, University of MA, Burleigh Dodds Science Publishing, 10, pp. 219–262.

Myers, S.C. and D.C. Ferree. 1983. Influence of summer pruning and tree orientation on net photosynthesis, transpiration, shoot growth, and dry-weight distribution in young apple trees. *Journal of the American Society for Horticultural Science* 108(1):4–9.

Oberhofer, H. 1990. *Pruning the slender spindle*. Victoria, BC: Province of British Columbia, Ministry of Agriculture and Fisheries.

Ogata, R., H. Kikuchi, T. Hatayama, and H. Komatsu. 1984. Growth and productivity of vigorous 'Fuji' apple trees on M.26 as affected by summer prunning. *III International Symposium on Research and Development on Orchard and Plantation Systems*, 160157–160166.

Pavičić, N., M.S. Babojelić, T. Jemrić, Z. Šindrak, T. Ćosić, T. Karažija, and D. Ćosić. 2009. Effects of combined pruning treatments on fruit quality and biennial bearing of 'Elstar' apple (*Malus domestica* Borkh.). *Journal of Food, Agriculture & Environment* 7:510–515.

Poni, S., M. Tagliavini, D. Neri, D. Scudellari, and M. Toselli. 1992. Influence of root pruning and water stress on growth and physiological factors of potted apple, grape, peach and pear trees. *Scientia Horticulturae* 52:223–236.

Robinson, T.L. 2003. *Apple-orchard planting systems. Apples: Botany, production and uses.* CABI Publishing, Cambridge, pp. 345–407.

Rom, C. and D. Ferree. 1986. The influence of fruiting and shading of spurs and shoots on spur performance. *Journal of the American Society for Horticultural Science* 111:352–356.

Roper, T.R. 1997. *Training and pruning apple trees*. University of Wisconsin. Cooperative Extension, in cooperation with the U.S department of agriculture and Wisconsin counties.

Schumacher, R. 1975. Einfluss des Wurzelschnittes auf die Fruchtbarkeit von Apfelbäumen. Schweiz. *Zeitschr fur Obst.-und Weinbau* 111(5):115–116.

Schumacher, R., F. Fankhauser, and W. Stadler. 1978. Beeinflussung der Fruchtbarkeit und der Fruchtqualität durch den Wurzelschnitt. *Schweiz. Zeitschr. fur Obst-und Weinbau* 114(3):56–61.

Schupp, J.R. 1991. Effect of root pruning and summer pruning on growth, yield, quality, and fruit maturity of McIntosh apple trees. *Acta Horticulturae* 322:173–176.

Schupp, J.R. and D.C. Ferree. 1987a. The effects of root pruning on apples. *Compact Fruit Tree* 20:76–80.

Schupp, J.R. and D.C. Ferree. 1987b. Effect of root pruning at different growth stages on growth and fruiting of apple trees. *HortScience* 22:387–390.

Sharma, L. 2014. Effect of varied pruning intensities on the growth, yield and fruit quality of starking delicious apple under mid hill conditions of Himachal Pradesh, India. *Agricultural Science Digest–A Research Journal* 34:293–295.

Sharma, R. and A. Banyal. 2020. Effect of summer pruning intensities on growth, quality and yield of low chill peach (*Prunus persica* L. Batsch.) cv. Early Grande. *Journal of Crop and Weed* 16:210–215.

Simon, S., C. Miranda, L. Brun, H. Defrance, P. Lauri, and B. Sauphanor. 2007. Effect of centrifugal tree training on pests and pathogens in apple orchards. *IOBC WPRS Bulletin* 30:237.

Sokalska, D., D. Haman, A. Szewczuk, J. Sobota, and D. Dereń. 2009. Spatial root distribution of mature apple trees under drip irrigation system. *Agricultural Water Management* 96:917–924.

Sosna, I. 2002. Reducing vegetative growth of 'Golden Delicious' apple trees by root pruning and ways of planting. *Journal of Fruit and Ornamental Plant Research* 10:63–74.

Sosna, I. 2018. V-shaped canopies in an apple orchard from the perspective of over a dozen years of research. *Plant Research* 19:79–83.

Stover, E., M.J. Fargione, C.B. Watkins, and K.A. Iungerman. 2003. Harvest management of 'MarshallMcIntosh'apples: Effects of AVG, NAA, ethephon, and summer pruning on preharvest drop and fruit quality. *HortScience* 38:1093–1099.

Talaie, A.R., R. Fatahi, Z. Zamani, and A. Saie. 2004. Effects of summer pruning on growth indices of two important Iranian apple cultivars 'Golab'and 'Shafi-Abadi'. *Acta Horticulturae* 707:269–274.

Talaie, A.R., M. Shojaie-Saadee, A. Dadashpour, and A.-A.M. Sarcheshmeh. 2011. Fruit quality in five apple cultivars trees trained to intensive training system: Geneva y-trellis. *Genetika* 43:153–161.

Taylor, B. and D. Ferree. 1984. The influence of summer pruning and cropping on growth and fruiting of apple. *Journal of the American Society for Horticultural Science* 109:19–24.

Vercammen, J., G.V. Daele, and A. Gomand. 2006. Root pruning: A valuable alternative to reduce the growth of cv Conference [*Pyrus communis* L.]. *Acta Horticulturae* 671:533–537.

Wang, Y., S. Travers, M. Bertelsen, K. Thorup-Kristensen, K. Petersen, and F. Liu. 2014. Effect of root pruning and irrigation regimes on pear tree: Growth, yield and yield components. *Horticultural Science* 41:34–43.

Wareing, P.F. and T. Nasr.1961. Gravimorphism in trees: 1. Effects of gravity on growth and apical dominance in fruit trees. *Annals of Botany* 25(3):321–340.

Widmer, A. and C. Krebs. 1996. 'Mikado' and 'Drilling' (Triplet) – two novel training systems for sustainable high quality apple and pear production. *Acta Horticulturae* 451:519–528.

Zhang, L., A. Koc, X. Wang, and Y. Jiang. 2018. A review of pruning fruit trees. In *IOP Conference Series: Earth and Environmental Science*, IOP Publishing. 153:062029.

8 Parthenocarpy, Unfruitfulness and Alternate Bearing in Apples

M. K. Sharma, E. A. Parray, N. Nazir,
A. Khalil and Shamim A. Simnani

CONTENTS

DOI: 10.1201/9781003239925-8

8.1   INTRODUCTION

In flowering plants, there are two distinct but simultaneous fertilization events obligatory for the establishment of the seed and several ensuing accessory tissues (Hamamura et al., 2012; Wada et al., 2018). The embryo, the forerunner of the future sporophyte plant, is the product of a fertilized egg cell. The central diploid cell of a megaspore is fertilized by one of the pollen nuclei leading to the formation of a triploid endosperm, which is a nutritive tissue essential for the persistence of the embryo growth. Normally the said double fertilization is necessary for exact seed and fruit development (Dumas and Rogowsky, 2008). After the completion of the fertilization process, certain floral tissues are transformed into structures which shield and ease out dispersion of the seeds (Sotelo-Silveira et al., 2014). The ovary wall is responsible for the formation of protective structures resulting into the botanical fruits *Annona* (sugar apple) and *Prunus persica* (peach). Nonetheless, in case of accessory fruits like apples and strawberries the story is different; here the flesh of the fruit is obtained from tissues other than ovary. Some apple mutants ('Spencer Seedless', 'Rae Ime', and 'Wellington Bloomless') are famous for only producing flowers (apetalous) which are readily transformed into parthenocarpic fruits (Yao et al., 2001).

It has been found that some fruit species, like certain apple cultivars, have an innate characteristic of superseding the need of fertilization. This phenomenon is called 'parthenocarpy' which has been derived from a Greek word for 'virgin fruit'. Parthenocarpy has played a great role in the domestication of some of the oldest fruit crops. It seems to have primarily assisted in the domestication process of bananas, breadfruits and figs (Zerega et al., 2004; Kislev et al., 2006; Sardos et al., 2016).

8.2   GENETIC BASIS OF PARTHENOCARPY IN APPLES

Genetic control of parthenocarpy is not clearly understood despite the fact that seedlessness is a profoundly advantageous characteristic for several fruit cultivars. Apple fruit is a pome obtained from the floral tube, a combination between the lower part of sepal, petal and stamen tissues (Pratt, 1988). This pome contrasts significantly from the Arabidopsis fruit (silique) which is exclusively formed from ovary tissue. A couple of apple mutant varieties comprising 'Wellington Bloomless', 'Spencer Seedless' and 'Rae Ime' are famous for giving blossoms without stamens or petals, yet with two arrangements of sepals and two arrangements of carpels (Pratt, 1988; Brase, 1937). The morphology of these blossoms is determined by a single recessive gene as affirmed by hereditary investigations (Tobutt, 1994) and is like Arabidopsis PI and AP3 mutants. Moreover, the apetalous apple blossoms can grow into seedless fruit without any pollination and fertilization. Nonetheless, seeds can be obtained upon hand pollination (Pratt, 1988; Tobutt, 1994), displaying that the stigmata are functioning properly.

The floral organ identity MADS-box genes have been assembled into three unique classes by work, as per the proposed ABC model (Weigel and Meyerowitz, 1994). In synopsis for Arabidopsis, the class A gene APETALA1 (AP1) indicates sepal development; AP1 and two class B genes (APETALA3 [AP3] and PISTILLATA [PI]) together determine petal arrangement; AP3, PI and the class C gene AGAMOUS (AG) together are needed for stamen development; and AG alone controls carpel development (Weigel and Meyerowitz, 1994; Wada et al., 2018). PI and AP3 proteins are useful accomplices and structure a heterodimer. A deficiency of capacity transformation in PI or AP3 produces blossoms that contain two whorls/arrangements of sepals and twofold quantities of carpels without stamens or petals (Goto and Meyerowitz, 1994).

8.3   SIGNIFICANCE OF PARTHENOCARPY IN THE APPLE FRUIT INDUSTRY

Several parthenocarpic fruit species enjoy an extensive upper hand over seeded fruit. This is exhibited with certain types of fruits like seedless watermelon, bananas, oranges and grapes. Seedless fruits are highly preferred by the consumers in the market. Pollination is not of much significance in such fruit crops. Seedless/parthenocarpic apple plants set fruits without pollination, which diminishes reliance on honeybees, warm weather at flowering and pollinator varieties. This trait is valuable in temperate growing conditions where the movement of bees is highly restricted by the low temperature at the time of flowering resulting in reduced crop productivity. Also seedless apple cultivation will surely aid in obtaining quality fruits without the need for proper pollination. The weight of the parthenocarpic cultivar 'Spencer Seedless' is approximately 70% of the normal fruits (Chan and Cain, 1967). The great advantage of seedless apple cultivars is the reduced biennial bearing tendency as there is no inhibition of flower bud differentiation that is associated with the developing seeds (Chan and Cain, 1967). To conclude, the susceptibility of seedless apple fruits to codling moths, a significant pest of apple trees, is much lower than the normal seeded fruits (Goonewardene et al., 1984).

8.3.1 Way Forward

The quality parameters of the modern seedless apple cultivars are very poor in comparison to the normal seeded varieties. Efforts are in progress to transfer this wonderful trait of seedlessness into commercially acceptable cultivars. Nevertheless, breeding in perennial fruit trees like apples is a gradual process as far as conventional breeding is concerned and needs good time to get the job done due to the long gestation period and many generations required (Tobutt, 1994). After the in-depth study of seedlessness in different fruit crops, it has come to the forefront that the *MdPI* sequence is of special interest. Different nonconventional techniques like genetic transformation can be put in use to down regulate *MdPI* expressions by cosuppression and antisense techniques for obtaining good quality parthenocarpic apples. *Agrobacterium*-mediated transformation could also be of special significance when used together with the said approaches for developing seedless apple cultivars (Yao et al., 1999).

8.4 UNFRUITFULNESS IN APPLES

In apples, most cultivars are self-unfruitful, which means that they will set a decent harvest consistently only when pollinated by a different variety. Unfruitfulness is a serious issue in apple cultivation which is reducing returns to the growers, thus making the cultivation less profitable. Fruitfulness refers to the state of the plant when it is not only capable of flowering and fruit setting but also takes these fruits to maturity. The inability to do so is called unfruitfulness or barrenness (Dhillon and Bhat, 2011). A couple of apple cultivars are self-fruitful and can set a great yield even when planted in solid blocks; yet even these will, as a rule, profit with cross-fertilization (Sheffield, 2014). Subsequently, by and large, it is preferable to interplant with pollinizer trees. Since the pollen grains of fruit trees are sticky, they aren't conveyed by the breeze, and honeybees (especially the nectar honeybee) should be available to impact cross fertilization.

8.4.1 Causes of Unfruitfulness

Unfruitfulness in apple trees can be because of an absence of harmony between vegetative growth and fruiting. Poor natural fruit set, which is due to the troublesome climatic conditions, also leads to unfruitfulness in apples. It can likewise be because of heavy crop load prompting a hindrance of fruit bud formation and reduced yield in the following year.

Self-incompatibility additionally prompts unfruitfulness because of restriction of pollen tube growth and the abortion of undeveloped embryos. The reasons for unfruitfulness in apples can be extensively assembled into two classes as shown in the flow chart (Figure 8.1).

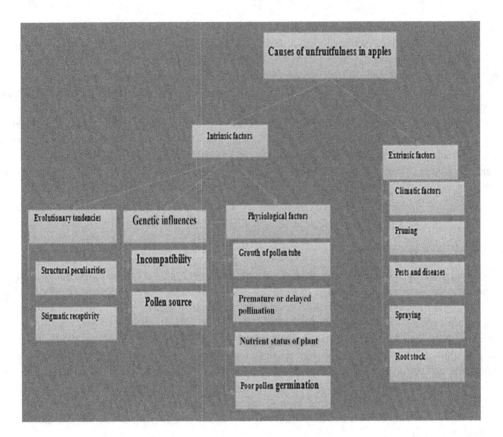

FIGURE 8.1   Causes of apple unfruitfulness.

8.4.2 INTRINSIC FACTORS

There are a number of internal factors which are associated with unfruitfulness. They have further been divided into three major categories: (a) evolutionary tendencies (b) genetic influence (c) physiological factors.

8.4.2.1 Evolutionary Tendencies

Due to evolutionary tendencies, some fruit crops cannot attain commercial fruit set by self-pollination, and apples are one of them. In such crops, cross-fertilization is a must in order to maintain the vigor of the particular tree species and obtain the consistent commercial yield. There are two evolutionary factors that lead to unfruitfulness.

8.4.2.1.1 Structural Peculiarities of Flowers

The specific morphology of the apple flower assists bee pollinators as they contact the floral nectaries and move within the flowers where they touch each of the stigmas (i.e. top working bees). In comparison to the top working bees, there are certain bees that approach the nectaries through the gap between the anthers; these are called as side working bees. But these bees have less chance to reach the stigma than top working bees. Side working behavior has been predominantly found in 'Delicious' apples (Degrandi-Hoffman et al., 1985; Schneider et al., 2002). Sometimes all the stigma surfaces within the apple flower cannot receive pollen due to the deformed gynoecium.

8.4.2.1.2 Stigmatic Receptivity

Stigma receptivity is the inherent ability of a stigma surface to aid in the pollen germination process. It is almost universally found in fruit crops that stigma is receptive at the time of anthesis. Williams (1965) studied the reproductive biology of apple flowers and found similar results. Most descriptions of the reproductive biology of apples report that stigma is receptive at anthesis.

8.4.2.2 Genetic Influence

8.4.2.2.1 Incompatibility

Most of the existing cultivated apple varieties are self-incompatible. Therefore, they need to be facilitated by proper pollinizer varieties in order to get the commercial yield (Garratt et al., 2013; Matsumoto, 2014). For profitable apple production, no less than two cross-viable cultivars with coordinated blossoming are suggested (Garratt et al., 2013; Goldway et al., 2012).

8.4.2.2.2 Pollen Source

The type of pollen source (pollinizer) has a profound influence (Table 8.1) on the pollen germination and elongation thereby affecting the overall fruitfulness of the apple trees. For example, Jahed and Hirst (2017) reported that 'Golden Delicious' and 'Delicious' pollens occupied lesser stigmatic surface in comparison to the *M. floribunda* and 'Ralph Shay' during both seasons of the investigation.

8.4.2.3 Physiological Factors

8.4.2.3.1 Growth of Pollen Tube

Pollen tube growth is an imperative for making the pollination successful. If the rate of growth is slow then there is a chance of missing the effective pollination period. The effectiveness of pollination is determined by the number of pollens that can reach the bottom of the style and fertilize the egg. In some cases pollens can travel to the egg in short duration of time, like 48 hours for 'Gala' and 'Honeycrisp' and 72 hours for 'Fuji' (Jahed and Hirst, 2017).

8.4.2.3.2 Premature or Delayed Pollination

Pollen has to be transferred on time in order to effectively fertilize the ovules. Apples, like other fruit crops, have their own unique time limit during which pollination culminates into the fruit set.

TABLE 8.1

Pollen Germination on Stigmatic Surfaces of Apple Cultivar 'Honeycrisp' as Influenced by the Pollen Source

Male	Rating of visible pollen tube growth on the stigmatic surface (0–10)^z			
	24h	48h	72h	96h
'Crab Apple'	3.6 ± 0.5 b	4.5 ± 0.5 b	1.8 ± 0.4 b	0.6 ± 0.1 a
'Delicious'	6.1 ± 0.5 a	4.5 ± 0.5 b	2.3 ± 0.4 b	0.4 ± 0.1 a
'Golden Delicious'	6.2 ± 0.5 a	7.1 ± 0.5 a	7.7 ± 0.4 a	0.6 ± 0.1 a
Significance	***	***	***	

(Source: Adopted from (Jahed and Hirst, 2017)

^{z} Rating of visible pollen tubes germinated on the stigmatic surface scaled 0 to 10 where 0 = no pollen tubes visible on the stigmatic surface; 1 = 1% to 10% stigmatic surface covered; 2 = 11% to 20%; 3 = 21% to 30%; 4 = 31% to 40%; 5 = 41% to 50%; 6 = 51% to 60%; 7 = 61% to 70%; 8 = 71% to 80%; 9 = 81% to 90% and 10 = 91% to 100% stigmatic surface was covered by pollen tubes.

*** Significant at P = 0.001.

This time period is called the 'effective pollination period' (Williams, 1965). Therefore, premature or delayed pollination leads to unfruitfulness in apples.

8.4.2.3.3   Nutrient Status of Plant

The nutrient status of apple trees has a large role to play in producing quality flowers. Plants that are not in good vigor and health produce defective pistils. The main cause for this is the exhaustion of the plant due to higher crop loads and poor nutrition management. Nutrition is that aspect of fruit production which is usually taken for granted by the orchardists. But it has a long lasting impact on getting the proper fruit set, maintaining the physicochemical parameters of the fruit and improving the pollen longevity/viability. It is an established fact that overbearing causes tree weakness due to the exhaustion of carbohydrates and reduces storage nutrition for the next season which badly affects the vegetative vigor and fruit bud differentiation for the ensuing year (Ding et al., 2017). The macronutrients like nitrogen, phosphorus and potassium play an essential part in improving the overall health of the plant. These nutrients need to be supplied in proper doses in order to enhance the transition of vegetative buds into the flower buds, increase yield and especially to maintain the fruitfulness of trees (Drake, 2002; Wargo, 2003).

8.4.2.3.4   Poor Pollen Germination

Pollen germination is greatly influenced by source of the pollen and the temperature present. Jackson (2003) and Petropoulou and Alston (1998) reported that pollen donor (pollinizer variety) and temperature have a profound impact on the percentage of the stigmatic area occupied by the germinated pollen. They further found that 'Idared' and 'Cox's Orange Pippen' had lesser germination percentages than the 'Spartan' at 14–16°C and 8–10°C, respectively. Several interrelated apple cultivars such as 'Early McIntosh', 'McIntosh', 'Macoun' and 'Cortland' do not cross-fertilize each other, and, as a result, there was no pollen germination (Delaplane et al., 2000).

8.4.3   Extrinsic Factors

8.4.3.1   Climatic Factors

Climatic factors like temperature, precipitation and solar radiation a have profound influence the apple tree physiology as well as phenology. These factors determine the fruitfulness (Pu et al., 2008; Fujisawa and Kobayashi, 2011) of an apple orchard by interfering with the flower quality beside the movement of bees. Any type of precipitation at the time of anthesis leads to unfruitfulness by constraining pollinator movement, washing pollens and increasing the incidence of diseases. Apples, as is the case with other temperate fruits, requires a certain time period without rains during blossoming for successful pollination to take place. Temperatures of as low as 4.4°C or lower in apples restricts blooming, fertilization and thereby reduces the fruit set (Jindal et al., 1993). Rising temperatures as a result of global warming are proving detrimental to the phenology of apples. Flowering is delayed in certain places and has been advancing in other places around the temperate world (Legave et al., 2013; Hoffmann and Rath, 2013). The tree's proper exposure to the light enhances the production of photosynthates thereby improving the quality of fruits. High carbohydrate reserves within the apple tree also improve the flower bud differentiation. When the light intensity is low, it leads to the fruit abscission (Wani et al., 2010).

8.4.3.2   Pruning

Pruning plays an important part in determining the fruitfulness of an apple orchard. Its impact on the apple tree depends on the season in which it is done, the severity and the kind. To improve fruit set and retain the fruits until maturity, proper and judicious pruning is imperative. It has been found that deep/severe pruning triggers the excessive vegetative growth resulting in reduced productivity (Robinson et al., 2014).

8.4.3.3 Pests and Diseases

Pests and diseases cause huge economic loss to apple production. Climatic change is creating feasible conditions for a resurgence of new pests and diseases. Apples are infected by several types of fungal, bacterial and viral diseases. Apple scab is a primary concern for apple growers. It causes premature leaf/fruit fall thereby leading to more than 70% production loss, and the percentage of production loss will continue to increase if cared for at the early stage of infection (Ogawa and English, 1991). Defoliation of apple trees caused due the diseases takes a heavy toll on tree vigor and growth and makes the tree unfruitful.

8.4.3.4 Spraying

Pesticides have a very detrimental impact on insect pollinator species like honeybees, therefore hindering the pollination process in fruit trees. Certain pesticides are also harmful to flowers and initiate abortion and reduced fruit set. The level of injury to the stigmatic surface ranges from a slight surface wrinkling to deterioration of stigmatic papillae. The pollen germination percentage and the pollen tube growth rate are significantly inhibited by the pesticide sprays resulting in reduced fruit set. Legge and Williams (1975) found that no fruit set occurred on pesticide applied flowers, and there was a decidedly substantial variation between the final fruit set on a protected cluster and an exposed one (Table 8.2).

8.4.3.5 Rootstock

The selection of proper rootstock is imperative to get a good yield in apples (Mészáros et al. 2019). Improper scion-rootstock combinations can result in the unfruitfulness of an apple orchard. Malling series (M9, M27, etc.) rootstocks are used to induce precocity in apples.

8.4.4 Management of Unfruitfulness in Apples

8.4.4.1 Balancing Reproductive and Vegetative Growth

There are different horticultural interventions/techniques available which can be put in use to strike a favorable reproductive and vegetative balance. The most significant ones for maintaining the optimum vigor and restricting excessive vegetative growth are the use of dwarfing rootstocks, proper training and pruning which give optimum crotch-angled branches, and the selection of short internode and compact scions and trees. Growth retardants are also a good option to use for slowing down the annual growth rate in apple trees. To improve fruit set and retain the fruits until maturity, proper and judicious pruning is imperative. It has been found that deep/severe pruning triggers the excessive vegetative growth which results in reduced productivity (Robinson et al., 2014). It has also been found that pears on *P. communis* and 'E.M. Quince C' rootstocks improved initial fruit thinning to one fruitlet/cluster and enhanced the final fruit set percentage of the 'Comic' cultivar (Westwood and Bjornstad, 1974). Sheban (2009) found that number of flushes per shoot was increased in mangos by adopting severe pruning technique in combination with GA_3 spray at 100 ppm. Singh and Daulta (1986) reported that pruning up to 12 buds, enhances fruit sets substantially and further reported that light pruning improves the yield of the 'Sharbati' cultivar.

TABLE 8.2

Influence of Binapacry and Captan Sprays on Fruit Set in Apples

Cluster treatment	Initial set (%)	Final set (%)
Exposed	0.30	0.30
Protected	23.30	12.60

8.4.4.2 Control of Pollination

Use of pollinizers is prerequisite to get a commercial set in self-incompatible fruit crops. Several apple varieties have been found to be partially self-fertile, particularly at a higher temperature. But in most of the cultivars, proper use of pollinators and pollinizers amplifies fruit set. Self-incompatibility in apples offers great scope for undertaking breeding and selection programs for screening self-fertile cultivars or clones in order to lessen difficulties and accomplish adequate pollination, particularly in cooler cultivation areas. In such regions, temperatures at flowering are not conducive to promote the activity of insect pollinators like bees and facilitate pollen tube development. Up until the provision of such upgraded varieties, interplating of appropriate pollinizer cultivars and the introduction of beehives at the time of flowering could be of great benefit to the orchardists to ensure commercial fruit set. Generally, it is recommended to use crab apples for pollination as they have extended flowering periods and produce high amounts of viable pollens. For ensuring commercial fruit set in apple orchards, it is preferable to have pollinizer varieties with prolonged flowering windows. The flowering time of pollinizers should synchronize with the main cultivar to obtain the better fruit productivity. Control of pollination service in orchards is one the key factors in improving the fruit fullness of an orchard. It is only possible when we have an in-depth knowledge related to early phenological behavior of pollinizers and main cultivars and are well acquainted with the foraging behavior of the honeybees. Almost all fruit crops, irrespective of the level of cross-pollination, immensely benefit from the pollinating insects such as honeybees in apples. It is a general observation that the closer the main cultivar to the pollinizer variety, the more fruit set and vice-versa. Viable pollens should reach to the stigma of the main cultivar within a specific time period known as the effective pollination period (Williams, 1965). So, synchronization of the blooming time between the pollinizer and the main cultivar should be given due attention before establishing the orchard.

8.4.4.3 Proper Nutrition

Nutrition is of utmost importance when it comes to the ensuring fruitfulness of an orchard for the long term. A balanced supply of macronutrients and micronutrients is always required for improving the carbohydrate reserves of apple trees, thereby realizing optimum fruit production. The best time for the application of fertilizers is a few days before the emergence of blossoms. As flowering is a nutrient exhaustive process, it requires a supply of nutrient proper time to favor the fruit set. It has been reported by Jackson (2003) that to enhance embryo longevity nitrogen must be applied after terminal bud formation. Nonetheless, surplus application of fertilizers and manures might cause the production of abnormal flowers.

8.4.4.3 Use of Suitable Rootstocks

The variation in the yield of any fruit cultivar is quite significant when grown on different types of rootstocks. The performance of the composite plant can be elevated by desired combinations of scion (Mészáros et al., 2019). Such a huge distinction in the production can be attributed to the difference in tolerance levels to adverse soil conditions, resistance to insect pests and in nutrient uptake potential.

8.5 ALTERNATE BEARING IN APPLES

Fruit crops experience cyclical changes in cropping: heavy fruit load ('on' crop) in one year and low fruit load ('off' crop) in the succeeding year (Figure 8.2). This cycle of 'on' and 'off' years is called alternate bearing (Goldschmidt, 2013; Sharma et al., 2015). Apples have been heavily studied for alternate bearing because of their great commercial significance. Carbohydrate exhaustion of apple trees in the years of heavy crop load is the most probable reason for alternate bearing (Jupa et al., 2021).

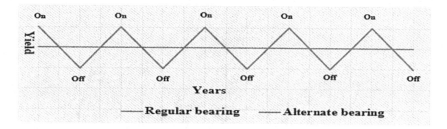

FIGURE 8.2   Cropping pattern of regular and alternate bearing apple cultivars.

Since the start of horticulture, fruit cultivators and investigators have concentrated on the aspects that seem, by all accounts, to be related to fruit bud differentiation in fruit trees. Most of the investigations on this subject have been more focused on apples in comparison to all the other fruit crops combined.

Study of fruit bud differentiation in apples is of the utmost importance for understanding the biennial bearing of apples clearly. In fact, alternate bearing is one of key factors that generated interest in scientists to comprehend the concept of fruit bud transformation in a detailed manner. This problem of biennial bearing has been documented in fruit crops since the very beginning of fruit domestication, and the condition is prevalent in every orchard some form of severity. Several varieties of apples have a tendency to develop this trait under specific conditions, and it has been seen that once the rhythm becomes consistent in an apple tree, it almost impossible to reverse the trend. Nevertheless, the tendency of alternate fruiting in different apples varies significantly with type of cultivar.

8.5.1   THEORIES OF ALTERNATE BEARING

Two theories have been put forth to understand the essence of the alternate bearing process: the starch depletion theory and the hormonal inhibition theory.

8.5.1.1   Starch Depletion Theory

The supporters of this theory believe that heavy fruit set during the 'on' year causes the depletion/exhaustion of carbohydrate reserves thereby negatively affecting the flower bud differentiation for the succeeding year (Lenahan et al., 2006; Krasniqi et al., 2013). Pallas et al. (2017) found that apple trees show strong positive correlations between yield and starch content.

8.5.1.2   Hormonal Inhibition Theory

Hormonal influence is one of the important aspects that have an important role in determining the cropping pattern of apple cultivars (Figure 8.3). Developing seeds are major source of gibberellic acid production in fruit crops. During the season of heavy crop load, there is significant production of this hormone which leads to flower bud inhibition for the proceeding cropping season (Hanke et al., 2007). In apples, a significant difference in the flower bud formation between the heavily fruiting branch and the poorly fruiting one can be noticed easily (Pallas et al., 2017).

8.5.2   MANAGEMENT OF ALTERNATE BEARING IN APPLES

Different potential strategies to minimize economic losses due to alternate bearing in apple orcharding may be broadly discussed under following headings.

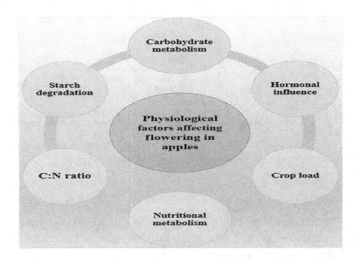

FIGURE 8.3   Factors influencing the flowering of apples.

TABLE 8.3

Classification of Apple Varieties Based on Bearing Tendency

Regular bearers	Moderate bearers	Distinctly alternate bearers
'Early Worcester', 'Golden Delicious', 'Golster', 'Jonagold', 'Rome Beauty'	'Delicious' group, 'Golden Spur', 'Granny Smith', 'Jonagold', 'McIntosh', 'Northern Spy', 'Red Gold', 'Rhode Island Green'	'Baldwin', 'Benoni', 'Boskoop', 'Cox's Orange Pippin'

8.5.2.1   Site and Cultivar Selection

According to bearing tendency, apple varieties can be categorized into three groups as follows (Jonkers, 1979; Table 8.3).

The selection of even bearing cultivars and the planning of orchards at sites with optimal climatic conditions and good soil physical properties reduces the chances of alternate bearing and other cropping anomalies. It is preferable to plant deciduous trees like apple on locations with minimum chances of frost or freeze injury.

8.5.2.2   Cultural Practices and Pollination

Alternate bearing is a natural phenomenon that the plant regulates internally. To prevent this from happening, good crop management measures are required. In general, under ideal growing conditions, bloom development is favored in perennial fruit trees. Annual trimming, consistent irrigation and maintaining ideal nutrient levels in soil and plant tissue, as well as judicious pesticide usage, can all help to ensure optimal plant growth. Honeybees are efficient apple pollinators. Pollination and fruit set are improved by placing beehives in apple orchards throughout the flowering season. Although the mentioned cultural practices do not directly minimize the danger of alternate bearing, they do lessen the chance of cropping irregularity to some degree.

8.5.2.3   Crop Load Management

This is the most important management strategy for reducing the likelihood of alternate bearing. Crop load adjustment is critical for yield regulation in the current season as well as in the following season. Thinning and pruning techniques are used to manage crop load.

8.5.2.3.1 Chemical Thinning

Heavy crop load in apples adversely affects the next year yield as a large crop load causes carbohydrate exhaustion by creating strong competition for the available photosynthates among the strong sinks/fruits (Rodrigues et al., 2019). In order to get rid of biennial bearing tendencies, certain chemical thinners such as lime sulfur (Allen et al., 2021Z), carbamates (Carbaryl and Oxymyl) (Knight, 1986), NAA (Irving et al., 1989), 6-BA (Greene et al., 1990) and Ethephon (Irving et al., 1989) are used in apples at varying concentrations for getting consistent fruit productivity.

8.5.2.3.2 Pruning

The level of severity of pruning has a great impact on the apple tree physiology (Farooqui, 2012). It has been found that light pruning improves physicochemical characteristics of fruits and vice-versa. Removal of overshaded branches by pruning increases the interception of light by the plant thereby minimizing alternate bearing tendency.

8.5.2.3.3 Hand Thinning

Hand thinning is a significant crop management technique used to lessen the crop load to the optimum level and improve the quantitative and qualitative characteristics of the fruits (Fattahi et al., 2020). Excessive crop load is manually removed in this method. Hand thinning is time-consuming, and it must be done early in the growing season to have a positive impact on alternative bearing.

8.6 WAY FORWARD

The quality parameters of the modern seedless apple cultivars is very poor in comparison to the normal seeded varieties. Efforts are in progress to transfer this wonderful trait of seedlessness into commercially acceptable cultivars. Nevertheless, breeding in perennial fruit trees like apples is a gradual process as far as conventional breeding is concerned and needs good time to get the job done due to the long gestation period and many generations required (Tobutt, 1994). After the in-depth study of seedlessness in different fruit crops, it has come to the forefront that the *MdPI* sequence is of special interest. Different nonconventional techniques like genetic transformation can be put in use to down regulate *MdPI* expression by cosuppression and antisense techniques for obtaining good quality parthenocarpic apples. *Agrobacterium*-mediated transformation could also be of special significance when used together with the said approaches for developing seedless apple cultivars (Yao et al., 1999).

8.7 CONCLUSION

Parthenocarpy, unfruitfulness and alternate bearing are interconnected occurrences in apples. Commercialization of parthenocarpic apple cultivars will minimize the occurrence of an alternate bearing cycle thereby indirectly lessening the chances of unfruitfulness. It has been reported that alternate bearing cultivars activate the process of unfruitfulness in apple cultivars. So, an in-depth understanding of said processes will not only help to regulate the cultivars but will allow to us make the timely interventions if and when required. Different nonconventional interventions like genetic transformation can be put in use to down regulate *MdPI* expression by cosuppression and antisense techniques for obtaining the good quality parthenocarpic apples. Moreover, the use of regular bearing and high yielding cultivars is a long-term way to address the problems of alternate bearing and unfruitfulness in apples. Molecular and genetic approaches can be effective and efficient techniques to deal with the aforesaid concerns in a durable way. However, concreted investigations in this direction are still needed in order to get a clear picture of the genetic and molecular basis underlying the discussed topic.

REFERENCES

Allen, W.C., T. Kon, and S.M. Sherif. 2021. Evaluation of blossom thinning spray timing strategies in apple. *Horticulturae* 7: 308–320.

Brase, K.D. 1937. *The vascular anatomy of the flower of Malus domestica Borkh. f. apetala Van Eseltine*. M.S. Thesis, Cornell University.

Chan, B.G., and J.C. Cain. 1967. The effect of seed formation of subsequent flowering in apple. *Proc. Am. Soc. Hortic. Sci.* 91: 63–68.

Degrandi-Hoffman, G., R. Hoopingarner, & K.K. Baker. 1985. The influence of honey bee 'sideworking' behavior on cross-pollination and fruit set in apples. *HortScience* 20: 397–399.

Delaplane, K.S., D.R. Mayer, and D.F. Mayer, 2000. *Crop pollination by bees*. CABI Publishing, New York.

Dhillon, W.S. and Z.A. Bhat. 2011. *Fruit tree physiology*. Narendra Publishing House, New Dehli, pp. 10–22, 93–94.

Ding, N., Q. Chen, Z. Zhu, L. Peng, S. Ge, and Y. Jiang. 2017. Effects of crop load on distribution and utilization of 13C and 15N and fruit quality for dwarf apple trees. *Sci. Rep.* 7: 14172.

Drake, S.R. 2002. Time of nitrogen application and its influence on 'golden delicious' apple yield and fruit quality. *J. Plant Nutr.* 25(1): 143–157.

Dumas, C., and P. Rogowsky, 2008. Fertilization and early seed formation. *Compt. Ren. Biol.* 331: 715–725.

Farooqui, K. 2012. *Temperate fruit production scientific fruit farming*. Sunnyvale, CA: Lambert Academic Publishing, pp. 48–385.

Fattahi, E., A. Jafari, and E. Fallahi. 2020. Hand thinning influence on fruit quality attributes of pomegranate (*Punica granatum* L. cv. 'Malase Yazdi'). *Int. J. Fruit Sci.* 20(2): 377–386.

Fujisawa, M., and K. Kobayashi. 2011. Climate change adaptation practices of apple growers in Nagano, Japan. *Mitig. Adapt. Strateg. Glob. Change* 16(8): 865–877.

Garratt, M.P.D., C.L. Truslove, D.J. Coston, R.L. Evans, E.D. Moss, C. Dodson, and S.G. Potts. 2013. Pollination deficits in UK apple orchards. *J. Pollinat. Ecol.* 11: 1–6.

Goldschmidt, E.E. 2013. The evolution of fruit tree productivity: A review. *Eco. Bot.* 67: 51–62.

Goldway, M., R. Stern, A. Zisovich, A. Raz, G. Sapir, D. Schnieder, and R. Nyska. 2012. The self-incompatibility fertilization system in Rosaceae: Agricultural and genetic aspects. *Acta Hort.* 967: 77.

Goonewardene, H.F., W.F. Kwolek, and R.A. Hayden. 1984. Survival of immature stages of the codling moth (Lepidoptera: Tortricidae) on seeded and seedless apple fruit. *J. Econ. Entomol.* 77: 1427–1431.

Goto, K., and E.M. Meyerowitz. 1994. Function and regulation of the Arabidopsis floral homeotic gene PISTILATA. *Genes Dev.* 8: 1548–1560.

Greene D.W., W.R. Autio, and P. Miller. 1990. Thinning activity of benzyladenine on several apple cultivars. *J. Am. Soc. Hort. Sci.* 115: 394–400.

Hamamura, Y., S. Nagahara, and T. Higashiyama. 2012. Double fertilization on the move. *Curr. Opin. Plant Biol.* 15(1): 70–7.

Hanke, M.V., H. Flachowsky, A. Peil, and C. Hättasch. 2007. No flower no fruit – genetic potentials to trigger flowering in fruit trees. *Genes Genom. Genomics.* 1: 1–20.

Hoffmann, H. and T. Rath. 2013. Future bloom and blossom frost risk for Malus domestica considering climate model and impact model uncertainties. *PLos One* 8: e75033.

Irving, D.E., J.C. Pallesen, and J.H. Drost. 1989. Preliminary results on chemical thinning of apple blossom with ammonium thiosulphate, NAA, and ethephon. *NZ J. Crop. Hort. Sci.* 17: 363–365.

Jackson, J.E. 2003. *The biology of apples and pears*. Cambridge University Press, Cambridge, UK.

Jahed, K.R., and P.M. Hirst. 2017. Pollen tube growth and fruit set in apple. *HortScience* 52(8): 1054–1059.

Jindal, K.K., D.R. Gautam, and B.K. Karkara. 1993. Pollination and pollinizers in fruits: Advance in horticulture. In *Fruit crops part 1*. eds. K.L. Chadha and O.P. Paruk, Malhotra Publishing House, pp. 463–480.

Jonkers, H. 1979. Biennial bearing in apple and pear a literature survey. *Sci. Hort.* 11: 303–317.

Jupa, R., M. Mészáros, and L. Plavcová. 2021Z. Linking wood anatomy with growth vigour and susceptibility to alternate bearing in composite apple and pear trees. *Plant Biol.* 23(1): 172–183.

Kislev, M.E., A. Hartmann, and O. Bar-Yosef. 2006. Early domesticated fig in the Jordan Valley. *Science* 312: 1372–1374.

Knight, J.N. 1986. Fruit thinning with carbaryl. *Acta Hort.* 179: 707–708.

Krasniqi, A.L., L. Damerow, A. Kunz, and M.M. Blanke. 2013. Quantifying key parameters as elicitors for alternate fruit bearing in cv. 'Elstar' apple trees. *Plant Sci.* 212: 10–14.

Legave, J.M., M. Blanke, D. Christen, D. Giovannini, V. Mathieu, and R. Oger. 2013. A comprehensive overview of the spatial and temporal variability of apple bud dormancy release and blooming phenology in Western Europe. *Int. J. Biometeorol.* 57: 317–331.

Legge, A.P., and R.R. Williams. 1975. Adverse effect of fungicidal sprays on the pollination of apple flowers. *J. Hort. Sci.* 50: 275–277.

Lenahan, O.M., M.D. Whiting, and D.C. Elfving. 2006. Gibberellic acid inhibits floral bud induction and improves 'Bing' sweet cherry fruit quality. *HortScience* 41: 654–659.

Matsumoto, S. 2014. Apple pollination biology for stable and novel fruit production: Search system for apple cultivar combination showing incompatibility, semi compatibility, and full-compatibility based on the S-RNase allele database. *Int. J. Agron.* Article ID 138271: 9.

Mészáros, M., L. Laňar, J. Kosina, and J. Náměstek, 2019. Aspects influencing the rootstock – scion performance during long term evaluation in pear orchard. *Hort. Sci.* 46: 1–8.

Ogawa, J.M. and English, H. 1991. *Diseases of temperate zone tree fruit and nut crops.* UCANR Publications, University of California, Division of Agriculture and Natural Resources, Oakland, CA.

Pallas, B., S. Bluy, J. Ngao, S. Martinez, A. Clément-Vidal, J. Kelner, and E. Costes. 2017. Growth and carbon balance are differently regulated by tree and shoot fruiting contexts: An integrative study on apple genotypes with contrasted bearing patterns. *Tree Physiol.* 38: 1395–1408. doi:10.1093/treephys/tpx166.

Petropoulou, S.P., and F.H. Alston. 1998. Selecting for improved pollination at low temperatures in apple. *J. Hortic. Sci. Biotechnol.* 73(4): 507–512.

Pratt, C. 1988. Apple flower and fruit: Morphology and anatomy. *Hort. Rev.* 10: 273–308.

Pu, J.Y., X.Y. Yao, X.H. Yao, Y.P. Xu, and W.T. Wang. 2008. Impacts of climate warming on phonological period and growth of apple tree in Loess Plateau of Gansu province. *Chin. J. Agrometeorol.* 29(2): 181–186.

Robinson, T., S. Hoying, M. Miranda Sazo, and A. Rufato. 2014. Precision crop load management. *Fruit Q.* 22(1): 9–13.

Rodrigues, J., D. Inzé, H. Nelissen, and N.J.M. Saibo. 2019. Source – sink regulation in crops under water deficit. *Trends Plant Sci.* 24(7): 652–663.

Sardos, J., M. Rouard, Y. Hueber, A. Cenci, K.E. Hyma, I. van den Houwe, E. Hribova, B. Courtois, and N. Roux, 2016. A genome-wide association study on the seedless phenotype in banana (Musa spp.) reveals the potential of a selected panel to detect candidate genes in a vegetatively propagated crop. *PLos One* 11, e0154448.

Schneider, D., Stern, R.A., Eisikowitch, D. and Goldway, M. 2002. The relationship between floral structure and honeybee pollination efficiency in 'Jonathan' and 'Topred' apple cultivars. *J. Hort. Sci. Biotech.* 77: 48–51.

Sharma, N., S.K. Singh, N.K. Singh, M. Srivastav, B.P. Singh, A.K. Mahato, and J.P. Singh. 2015. Differential gene expression studies: A possible way to understand bearing habit in fruit crops. *Transcr. Open Access* 3(110): 10–12.

Sheban, A.E.A. 2009. Effect of summer pruning and GA3 spraying on inducing flowering and fruiting of zebda mango trees. *World J. Agri. Sci.* 5(3): 337–344.

Sheffield, C.S. 2014. Pollination, seed set and fruit quality in apple: Studies with *Osmia lignaria* (Hymenoptera: Megachilidae) in the Annapolis Valley, Nova Scotia. Canada. *J. Pollination Ecol.* 12(13): 120–128.

Singh, D. and B.S. Daulta. 1986. Effect of pruning severity on growth, flowering, fruit set, fruit drop, the extent of sprouting and flowering at various nodal positions in Peach (Prunus persica Batsch) cv. 'Sharbati'. Haryana. *J. Hort. Sci.* 15(3–4): 200–205.

Sotelo-Silveira, M., N. Marsch-Martínez, and S. de Folter. 2014. Unravelling the signal scenario of fruit set. *Planta* 239: 1147–1158.

Tobutt, K.R. 1994. Combining apetalous parthenocarpy with columnar growth habit in apple. *Euphytica* 77: 51–54.

Wada, M., H. Oshino, N. Tanaka, N. Mimida, Y.M. Tanaka, C. Honda, T. Hanada, H. Iwanami, and S. Komori. 2018. Expression and functional analysis of apple MdMADS13 on flower and fruit formation. *Plant Biotechnol.* 35: 207–213.

Wani, I.A., M.Y. Bhat, A.A. Lone, and M.Y. Mir. 2010. Unfruitfulness in fruit crops: Causes and remedies. *African J. Agric. Res.* 5(25): 3581–3589.

Wargo, M.J. 2003. Fruit size, yield and market value of 'Gold Rush' apple are affected by amount, timing and method of nitrogen fertilization. *Hort. Technol.* 13(1): 153–160.

Weigel, D. and E.M. Meyerowitz. 1994. The ABCs of floral homeotic genes. *Cell* 78: 203–209.

Westwood, M.N. and H.O. Bjornstad. 1974. Fruit set as related to Girdling, Early Cluster thinning and pruning of 'Anjou' and 'Comice' pear. *Hort. Sci.* 9(4): 342–344.

Williams, R.R. 1965. The effect of summer nitrogen applications on the quality of apple blossom. *J. Hort. Sci.* 40: 31–41.

Yao, J.L., Y.H. Dong, A. Kvarnheden, and B. Morris. 1999. Seven MADS-box genes in apple are expressed in different parts of the fruit. *J. Am. Soc. Hort. Sci.* 124(1): 8–13.

Yao, J.L., Y.H. Dong, and B.A.M. Morris. 2001. Parthenocarpic apple fruit production conferred by transposon insertion mutations in a MADS-box transcription factor. *PNAS* 98(3): 1306–1311.

Zerega, N.J., D. Ragone, and T.J. Motley. 2004. Complex origins of breadfruit (Artocarpus altilis, Moraceae): Implications for human migrations in Oceania. *Am. J. Bot.* 91: 760–766.

9 Orchard Cultural Practices
Irrigation, Fertigation and Weed Management

*Rafiya Mushtaq, Insha Majid, Shaila Din,
Ab Raouf Malik, M. K. Sharma,
Suheel Ahmad and Tabish Jahan Been*

CONTENTS

DOI: 10.1201/9781003239925-9

9.1   INTRODUCTION

Orchard management through framed cultural practices is crucial. Cultural practices should aim to maintain the soil in good tilth and better productivity through better retention and supply of moisture and nutrients beside adequate aeration in the root zone. Apple orchards can be low maintenance or high maintenance depending on the desires of the grower, orchard densities and use of variety or rootstock. High density orchards will require more watch and ward as compared to commercial orchards. Cultural practices also include the removal of weeds and maintenance of proper orchard sanitation. Similarly, conservation of moisture and maintenance of optimum nutrient requirements in the soil and plants are important components of orchard management. Irrigation is a method of applying water to the land surface. Irrigation water may be applied to crops by flooding it on the field surface, by applying it beneath the ground surface (subsurface irrigation), by spraying it under pressure (sprinkler) or by applying it in drops near the root zone (drip or trickle irrigation). Improved irrigation systems and water management are directly linked to fertigation. Drip irrigation and other microirrigation methods are great for fertigation since they are very efficient at applying water. At various frequencies fertilizers can be inserted into the irrigation system; for example, the frequency could be once a day, once every 2 days or even once a week. Weeds are major problem in fruit orchards, costly to control, decrease yield and interfere with harvesting, pesticide sprays, and irrigation of the trees by blocking the sprinklers. The presence of uncontrolled weeds in the orchard have been seen to cause huge economic losses to the farmers by means of a reduction in quality and quality of the produce (Yudh, 2001). The amount of the nutrients and water that are directed toward the weed growth can be easily utilized for the main crop growth by means of proper management of weeds. This chapter gives detailed information on important orchard management practices like irrigation, fertigation and weed management.

9.2   IRRIGATION

Irrigation plays a vital role in crop production and is considered an important aspect of orchard management. With an optimum amount of irrigation supplied, fruits of high quality with better yields can be achieved (Litschmann, 2004; Mushtaq et al., 2021). Fruit size, fruit quality and the performance of apple crop in terms of growth, yield and productivity are highly governed by irrigation and its management (Girona et al., 2010). However, access to water resources for irrigation and other horticultural techniques will be a major concern for future as the world population grows and land resources shrink. Irrigation techniques have been studied in detail for decades and substantial progress has been made in understanding soil, plant and atmospheric water relations (Naor et al., 2001). Furthermore, the optimal irrigation and irrigation requirements that are commonly stated in the literature range widely (Kireva and Mihov, 2019) due to the variations in soil-climatic conditions, the age of the plant, the vegetative development of the trees and the applied cultivation technology.

9.2.1   CONVENTIONAL IRRIGATION IN APPLES

Apple trees require regular irrigation during fruit set and development and any mismanagement of irrigation during this critical period leads to fruit drop, reduced fruit size and quality. "The flowering, pollination and fruit set are the most critical phenological phases of the apple" (Cardoso, 2011) because yield components are determined in these periods. The method of irrigation and the vigor of rootstock is among the most important factors affecting the uptake of mineral nutrients, tree growth, fruit yield and quality attributes of apples (Fallahi, 2012). Conventionally, apple orchards placed on seedling orchards are irrigated through flood irrigation, especially during summer. This, however, leads to an ambiguity of how much irrigation water is used.

9.2.2 Microirrigation in Apples

Microirrigation is known to be one of the most perceptual ways of irrigating horticultural crops, primarily due to the higher water productivity and the possibilities for complete automation of the irrigation process with the drip irrigation methods dominating for orchard crops. Generally perennials, especially fruit species, adapt quite well to microirrigation resulting in significant water savings (Azzeddine *et al.*, 2019). "Along with all the benefits of micro irrigation in physiological and technological terms, the most important of which is the obtaining of the biologically optimal yield of high-quality fruits with significant water savings compared to traditional irrigation methods" (Kireva *et al.*, 2017). Modern microirrigation systems, which have now supplanted flood irrigation systems, have become the world's most valuable horticultural innovation. Several years after it became commercially available, the first scientific research findings on this microirrigation technology were published (Goldberg *et al.*, 1971). Because there is less surface evaporation, less surface runoff and low deep percolation, microirrigation can typically achieve high water application efficiency. Furthermore, a drip irrigation system can readily be employed for fertigation which allows for precise nutrient delivery to crops.

9.2.2.1 Drip Irrigation in High Density Apple Plantations

After World War II, drip irrigation for orchard irrigation became popular, and S. Blass in Israel created the first commercial drip irrigation system in the early 1960s. Drip irrigation is a permanent system of plastic pipes with emitters that direct water application near specific plants utilizing high-frequency application and low discharge rates. A drip irrigation system typically consists of a water source (e.g., well water, river water), a pump, a pressure controlling system, valves, piping, emitters and other accessories.

Drip irrigation has greater water saving over other systems as it has the potential to increase crop yield even with reduced irrigation water application. Also, drip irrigation can help to irrigate hilly terrains or texturally nonuniform fields. It also allows farming on flat lands to save labor and operating cost of land leveling, making furrows and ridges. Drip irrigation systems are commercially viable for apple plantations under high density systems. Water relations are considerably more crucial in high density apple orchards. Irrigation is necessary for the optimal growth of freshly planted and young apple orchards as well as the attainment of the required fruit size. The economic success of high density orchards is mostly dependent on achieving sufficient yields in the third, fourth and fifth years to cover the expenditures of establishment. Excellent tree growth over the first 3 years following planting is required to achieve the predicted high yields. One of the most common issues we find with new high density orchards is insufficient tree growth over the first 3 years. When low tree growth in the early years of a new orchard delays cropping, it is predicted that the peak investment is delayed by 20% and overall revenues are reduced by 66% throughout the orchard's 20-year life. Inadequate water delivery over the first 3 years is responsible for much of poor tree development. As a result, having a precise irrigation system for high density apple orchards is critical. Apples under high density plantations requires optimum irrigation throughout the fruit growth period. However, irrigation requirement during vegetative growth is lower compared to irrigation requirement during fruit growth and development (Mushtaq *et al.*, 2020).

9.2.2.2 Irrigation Scheduling: How Much and How Often

Determining an irrigation schedule explains how to calculate how much water your trees require per week, how much water your soil can hold based on its water holding capacity, and how to calculate the output of your irrigation system. The amount and frequency of water application affects the trees' general health. Because of these conditions, it's critical to plan irrigation so that the trees don't get too much or too little water.

9.2.2.2.1 ET-Based Irrigation

Evapotranspiration (ET)-based irrigation is a type of weather-based irrigation. The ET rate is the total loss of water from the soil surface plus transpiration from plants over a particular area in 24 hours, measured in inches per day. When using ET-based irrigation, the total ET rate is deducted from the precipitation rate to determine the irrigation system's application rate. To compute the ET rate and crop coefficient for the season using ET-based irrigation, a comprehensive set of weather parameters from a nearby weather station is required.

$$ETc = ET_0 \times Kc$$

ETc = crop evapotranspiration
ET_0 = pan evapotranspiration
Kc = crop coefficient

Various computer-based software like CropWat or PETv 3.0 can be used to calculate irrigation water requirement using various weather parameters.

9.2.2.2.2 Plant-Based Irrigation

Scheduling irrigation based on plant infrared thermal response to water status is another approach to optimize water requirements. Crop water stress index (CWSI) is one of the indicators used. 'The index is based on the difference between canopy temperature and air temperature normalized for the vapor pressure deficit of the air' (Stockle and Dugas, 1992). The water stress level in the plant informs the index of when and how much to irrigate. Also irrigation based on stem water potential is significant method to estimate water potential and therefore irrigation requirement. Various tensiometers including the latest micro tensiometers developed by Cornell University are efficiently being utilized now a days.

Additionally, the kind of rootstock used also imparts significant effects. Rootstocks with shallow roots or smaller root zones are more sensitive to water and nutrient stress since roots will not grow in dry soil. As such, a larger root zone can be encouraged by running the drip system for longer amounts of time.

9.2.2.2.3 Soil Moisture Based Irrigation

Soil moisture measurements in the orchards to be irrigated are some of the best and most reliable methods for determining water requirements in high density apple orchards. "Soil water content and soil water potential are two indicators of plant-available water used by soil-based irrigation systems" (Jiang and He, 2021). Soil moisture measurements can be taken through a range of instruments/devices available like neutron probes, time-domain reflectometry/transmissivity (TDR) sensors, capacitance sensors, tensiometers and granular matrix sensors. Through the soil moisture measurements, automation of this important practice using numerous sensors is commercially viable.

9.3 FERTIGATION

This technique was initiated by Israel in late 1960s. In 1970, Arscott was the first person who described that urea becomes more effective when applied via an irrigation system instead of hand broadcasting on soil surface. After few decades, it has become a high-tech, effective and a potent method in current horticulture resulting in increased yields with enhanced quality of the crop produces. Now 80% of the irrigated land in Israel is under fertigation.

9.3.1 ADVANTAGES OF FERTIGATION

1. The quantity and form in which nutrients are supplied is governed by critical stages of crop growth in drip fertigation.

2. Due to its high fertilizer use efficacy and low losses due to leaching, it reduces the use of fertilizers.
3. It supplies the nutrients as per the need directly to the productive root zones thus optimizing the nutrient balance in soil.
4. Nutrients are applied by using water distribution system thus reducing labor and energy costs.
5. Superior quality of product and yield is obtained.
6. Employment of accurate quantity of nutrients directly to the root zone at the right time reduces leaching of fertilizers beneath the root zone and ameliorates fertilizer use efficacy.
7. It guarantees even distribution of nutrients and water.
8. It ameliorates accessibility and uptake of nutrients by crop.
9. It prevents water and soil erosion.

9.3.2   PREREQUISITE FOR SUCCESSFUL FERTIGATION

1. The system must be designed precisely and handled efficaciously.
2. Each and every dripper must deliver the same volume of water during the irrigation period.
3. The material to be used must be free from residues or deposits.
4. In order to obtain uniform mixing of the nutrients and water in the irrigation systems, there should be minimal operational pressure variations.
5. The most suitable fertilizer corresponding to plant requirements, soil conditions and costs should be chosen.

9.3.3   PRECAUTIONS TO BE TAKEN DURING FERTIGATION

1. The fertilizers should be completely dissolved in water before fertigation.
2. Fertilizer injections should not begin until all lines are filled with water and the drippers are working.
3. Knowledge of fertilizer compatibility is necessary while mixing fertilizer.
4. Fertilizer should not be injected at the same time as pesticides.
5. The time needed to supply enough water to the field has to be more than the time needed for fertigation.
6. The quality of irrigation water must be checked before mixing.
7. Avoid over irrigation as it leads to dousing some fertilizers from the root zone.

9.3.4   FERTIGATION SCHEDULING

1. Nutrient management should be site and crop specific.
2. Nutrients should be delivered timely to fit the plant needs.
3. Management of irrigation is needed to reduce dousing of soluble fertilizers beneath the productive root zone.

9.3.5   FERTIGATION EQUIPMENT

Common equipment used in fertigation are:

1. Venturi pumps
2. Fertilizer tank
3. Fertilizer injection pump.

9.3.5.1 Venturi Pump

Venturi pump is common and affordable equipment which consists of an assembling part, a throat and a deviating part. In the system a partial void is induced to allow sucking of the fertilizers into the irrigation system through the action of venturi. The void is generated by a drop in pressure which is managed by diversion of water flow and allowing it to pass via a constriction thus increasing the velocity of flow. The fertilizer solution is sucked into the venturi due to the pressure drop via a suction pipe from the tank and subsequently entering the irrigation stream. The venturi suction rate has to be 30–120 lt hr[1] and need to be monitored; the aftermath of high pressure loss may be irregular distribution of fertilizer and water.

9.3.5.2 Fertilizer Tank

The flow of water is the basic principle in a fertilizer tank, and it is generated due to the pressure reducing valve because of the pressure gradient between the entrance and exit of the fertilizer tank. Prior to returning to the primary line, the irrigation water is diverted to flow via the tank bearing the fertilizers in a soluble solid or fluid form. The pressure is same in the primary line and the tank. However, by means of a pressure decreasing valve, an asthenic drop in pressure is catalyzed between the off take and return pipes for the tank. This causes catalyzing water to flow through the tank from the primary line causing a pour of the diluted fertilizers into the irrigation system. The uniformity of fertilizer dispensation can be problematic as the proportion of the fertilizer accessing the irrigation water charges uninterruptedly, initiating a high concentration subsequently. These tanks are procurable in the range of 90–160 liters of capacity.

9.3.5.3 Fertigation Pump

These are diaphragmatic pumps that are driven by the irrigation system's water pressure and, as such, the rate of insertion of the fertilizer solution is commensurable with water flow and can be accommodated to achieve the desired position of nutrient use to score the requisitioned level of fertilizer application. Its operational cost is low, and one of its advantages is that fertilizer injection stops automatically if the flow of water stops. Therefore it is an ideal device for fertigation purpose. The pumps suction rate ranges from 40–60 liters per hour.

9.3.6 Fertigation Methods

Depending on the crop farm management system and type of soil, two types of methods for applying fertilizers can be adopted (Sne, 2006).

9.3.6.1 Quantitative Method

A calculated quantity of fertilizer per water application is injected into the irrigation system which, conceivably, can be controlled by automatic or manual means. It is used in medium to heavy soils and application is expressed in terms of Kg/ha/day.

9.3.6.2 Proportional Method

An invariable predestined ratio is maintained between the volume of fertilizer solution and irrigation water, consequently leading to an unvarying concentration of nutrients in the irrigation water. It is generally used in light soils and is expressed in terms of concentration $g/m^3 = ppm$.

9.3.7 Characteristics of Fertilizers Used in Fertigation

1. A readily form of high nutrient content available to plants
2. Altogether water soluble at field temperature conditions
3. Rapid dissolution in irrigation water
4. Fine grade, flowable

5. Decongesting of filters and emitters
6. Insoluble content is very low (< 0.02%)
7. Content of conditioning agents should be low
8. Amicable with other fertilizers
9. Slight dealing with irrigation water
10. Water pH (3.5 < Ph < 9.0) shall not lead to a drastic change
11. Minimum corrosivity for system and control head

9.3.8 Nutrient Content of Fertilizers Suited for Fertigation

The nutrient content of fertilizers suited for fertigation is shown in Table 9.1.

9.3.9 Suitability of Fertilizers for Fertigation

According to Kafkali (2005), four factors should be considered before selecting the fertilizers for fertigation, and they are:

1. Kind of plant and its growth stages
2. Condition of soil
3. Quality of water
4. Availability of fertilizer and price.

Hagin and Lowengart-Aycicegi (1996) list key characteristics pertaining to the capability of the fertilizer to the injection technique as following:

1. Form: Depending on accessibility, comfort and profitability both liquid and soluble solid fertilizers are fit for fertigation.
2. Solubility: Fertilizers that are used in fertigation should be highly soluble. Generally the solubility of fertilizers enhances with temperature.
3. Relationship between fertilizers used in fertigation: Compatibility between the fertilizers must be checked before mixing one type of fertilizer with another in the irrigation system.
4. Corrosivity. Between metallic portions and fertilizers in the fertigation system and irrigation system, chemical reactions may occur, and it can damage metallic constituents like valves, uncoated steel pipe, injection units and filters.

TABLE 9.1
Nutrient Content of Fertilizers Suited for Fertigation

Nutrient	Compound	Nutrient content in solid fertilizers (N:P_2O_5:K_2O)
Nitrogen (N)	Urea	46:0:0
	Mono Ammonium Phosphate	12:61:0
	Ammonium Sulfate	21:0:0
	Ammonium Nitrate	33:0:0
Phosphorous (P)	Phosphoric acid	0:51:0
	Triple Super Phosphate	0:46:0
	Mono Ammonium Phosphate	12:61:0
	Di Ammonium Phosphate	18:46:0
Potassium (K)	Mono Potassium Phosphate	0:52:34
	Potassium sulfate	0:0:50
	Potash	0:0:60

9.3.10 Disadvantages of Fertigation

1. Components of fertigation, i.e., drip and water soluble fertilizer, are expensive.
2. Due to the eventuality of robbery/rodent infestation, the sustentation becomes challenging.
3. Water of high caliber is vital.
4. Blocked emitters leads to severe challenges.
5. Availability of water soluble fertilizers can be limited.
6. Acclimation of fertilizers as desired is not easy.
7. If the subsidy provided by the government is withdrawn, it can lead to a reduction in microirrigation.

9.4 WEED MANAGEMENT

The definition of weeds as given by European Weed Science Society is "any plant of vegetation, excluding fungi, interfering with the objectives or requirements of people is a weed". Traditionally, the plants that grow at a place where they are not desired that is in an area that has been disturbed or altered intentionally: "Weeds are unwanted and undesirable plants which interfere with the utilization of land and water resources and thus adversely affect human welfare". Thus, in order to reduce the competition for nutrients, water, sunlight and space and to maximize the production of quality produce, proper management of the weeds is a prerequisite in high density plantations. One of the challenges in fruit production is weed management, and it can be managed through various management strategies like manual, mechanical, cultural, biological and chemical control (Hira *et al.*, 2004).

9.4.1 Mechanical Weed Control

There is a long history regarding mechanical weed control because weeding by hand pulling is considered as the first method of managing weeds (Bond and Grundy, 2001). Nowadays, weed control is done by various sophisticated methods but still weeding hand pulling and using simple equipment is conventionally the most common method adopted for small scale productions. There are several forms of mechanical weed control ranging from handheld tools to the most advanced vision-guided hoes. Weeding with handheld tools is considered highly expensive involving large labor needs. Manual weeding may also cause damage to the shallow root system of the dwarfing trees used in high density plantations. As a result, manual weeding is sometimes replaced by several management strategies which involve the use of organic and inorganic mulches and chemical methods of weed control. The alternative methods are easier, cheaper and less time consuming.

9.4.2 Mulching

Mulching for weed control is a tried-and-true method for improving apple plant performance. Mulching helps to conserve soil moisture during dry periods (Hira *et al.*, 2004) as well as provide benefits such as suppressing weed growth (Kaur and Kaundal, 2009), reducing water runoff and soil erosion, providing cover for rodents that damage tree roots and trunks and improving water and fertilizer use efficiency (Salaria, 2009). Various plant-based materials such as straw, sawdust, wood chips, shredded paper (Rowley *et al.*, 2011), hay and crop residues and synthetic materials such as polyethylene plastic (Camposeo and Vivaldi, 2011), woven polypropylene fabric (Rozpara *et al.*, 2008) and nonwoven polyacrylic fabric (Camposeo and Vivaldi, 2011) have all been used as soil mulch in apple orchards. Currently reflective mulches, with a reflecting white or silver surface on the top and a dark black color inside to suppress the weed growth have been developed (Makus, 2007). The use of these reflective mulches has been seen to improve the photosynthetic efficiency and color development in fruits which, in turn, enhance the quality of the produce.

9.4.3   Chemical Control

Herbicides are considered the most important tool for eradicating weeds in high density orchards to reduce the damage to the shallow root system and to maximize yields (Ashiq *et al.* 2007). Weed control by means of herbicides is considered indispensable, and it is the most efficient method of controlling weeds (Kahramanoglu and Uygur, 2010). Presently, herbicides constitute about two-third of the total pesticide usage of the world (by volume). Herbicides are applied to the base of the trees. Care should be taken to direct the spray to cover the weeds and minimize the contact with the tree. Spray should be done 10–15 cm away from the tree trunk. Trunks of young trees should be covered with nonporous wraps or growth tubes to minimize contact with weeds. This is particularly important for systemic postemergence herbicides (for example, glyphosate) and contact burndown herbicides (for example, paraquat, diquat and glufosinate). For improving the herbicidal weed control, the following things should be considered.

9.4.3.1   Herbicide Selection

For efficient control of weeds, herbicides are to be applied even before their emergence from the seed or soil. Such herbicides are termed as preemergence herbicides (e.g., Oxyfluorfen). After the weed have emerged out of the soil, postemergent herbicides like gluphosate, glufosinate, ammonium, etc. are used.

9.4.3.2   Optimal Timing

The application times of herbicides depends on the type of the herbicide to be applied. Since pre-emergence herbicides are to be applied before weed emergence, they are applied early in the spring or fall before the annual weeds emerge. For postemergence herbicides, efficacy decreases as weeds grow. Thus each postemergence herbicide should be applied at the right growth stage of the weed to achieve higher efficiency.

9.4.3.3   Sufficient Coverage

In addition to the selection of the proper herbicide, it is necessary to apply the proper concentration of the herbicide with a certain amount of water per hectare in order to achieve proper coverage of the weeds. A proper nozzle type and corrected spray pattern of the nozzles helps achieve higher efficacy of applied herbicide.

9.4.3.4   Adequate Activation

Preemergence herbicides require moving into the soil where they act on the weed seeds. Thus, for their activation or efficiency a little rainfall or irrigation is necessary to help the herbicide to move into the soil profile. While applying postemergent herbicides care should be taken to add a proper surfactant that is in nonionic form to increase the uptake of the herbicide by the weeds for increased effectiveness.

Although herbicides are very effective in controlling weeds, the dosage as well as the type of herbicide varies from location to location depending upon what kind of weeds are in the orchard. Selection of the herbicide depends on several factors like tree age (some herbicide labels restricts use to established orchards), the desired duration of weed control, rates of herbicide and how growers figure the cost and management of additional postemergent control. Therefore, the selection and use of herbicides in fruit orchards should be done wisely. Preemergent herbicides such as oxyflourfen, simazine, diuron and terbacil control a greater range of annual broadleaf weeds. Most preemergence herbicides do not control perennial weeds so postemergence chemicals are used to control those weeds. Rankova and Zhivondov (2012) observed higher efficiency of herbicide in controlling the weeds when a combination of oxyfluorofen (as preemergent) and glyphosate (as postemergent) were used. It was also reported that oxyfluorfen was effective for a period of about 120 days for weed control. Weed control by means of herbicides has also lead to enhanced yield and improvement of quality and more color development in fruits. Herbicide usage enhances the production as well as productivity of the orchards resulting in higher economic returns (Din *et al.*, 2020).

9.4.4 WEED SPECIES

The type of the weed species present at a particular place depends on variety of factors, among which environmental conditions, particularly soil and climate, play a very crucial role. The species composition is also determined by the techniques of cultivation such as irrigation and fertigation regimes. A variety of weed species infest orchards, including annuals, biennials and perennials (Derr and Chandran, 2000). The major weed flora consisted of *Poa annua, Setaria glauca, Cynadon dactylon, Brachiaria spp., Sorghum helepense, Veronica biloba, Capsella bursa-pastoris, Convolvulus arvensis, Portulaca oleraceae, Polygonum tubulosum* and *Cyperus rotundus.* The details of different weed species that can be seen in high density apple orchards are presented in Table 9.2.

TABLE. 9.2
Predominant Weed Species in High Density Orchards of Kashmir

Botanical Name	Family	English Name
Poa annua	*Poaceae*	Annual bluegrass
Setaria glauca	*Poaceae*	Foxtail
Echinochloa colona	*Poaceae*	Awnless barnyard grass
Cynadon dactylon	*Poaceae*	Bermuda grass/Doab grass
Sorghum helepense	*Poaceae*	Johnson grass
Brachiaria spp.	*Poaceae*	Signal grass
Erigeron canadensis/Conyza Canadensis	*Asteraceae*	Horseweed
Veronica biloba	*Plantaginaceae*	Common speedwell
Malva neglecta	*Malvaceae*	Dwarf mallow
Fumaria parviflora	*Papaveraceae*	Fineleaf fumitory
Capsella bursa-pastoris	*Brassicaceae*	Shepherd's purse
Taraxacum officinalis	*Asteraceae*	Dandelion
Sinapis arvensis	*Brassicaceae*	Wild mustard
Plantago minor	*Plantaginaceae*	Ribwort plantain
Plantago major	*Plantaginaceae*	Broadleaf plantain
Fragaria indica	*Rosaceae*	Woodland strawberry
Convolvulus arvensis	*Convolvulaceae*	Field bindweed
Amaranthus viridis	*Amaranthaceae*	Amaranth/Pig weed
Gallinsoga parviflora	*Asteraceae*	Gallant soldier/quick weed
Solanum nigrum	*Solanaceae*	Black nightshade
Portulaca oleraceae	*Portulaceae*	Common purslane
Trifolium repens	*Fabaceae*	White clover
Trifolium pretense	*Fabaceae*	Red clover
Polygonum tubulosum	*Polygonaceae*	Knotgrass
Ageratum conyzoides	*Asteraceae*	Bill goat weed/Floss flower
Papaver dubium	*Papaveraceae*	Long headed poppy
Vicia sativa	*Fabaceae*	Common vetch
Cyperus rotundus	*Cyperaceae*	Nut sedge

Source: Din, 2017

9.4.5 INTEGRATED WEED MANAGEMENT

Although the use of herbicides has led to an easy and cost-effective management of weeds in high density orchards with enhancement of tree growth and fruit yield, the negative effect of herbicide usage cannot be neglected (Lisek, 2014). The continuous use of chemicals has always been detrimental to the environment and human health. The excessive use of herbicides can lead to a decrease or the complete destruction of weed biomass, weed biodiversity and soil quality (Jiang et al., 2016; Robinson et al., 2002; Yu et al., 2015). Additionally, the continuous use of herbicides leads to development of herbicide resistance in weeds (Pieterse, 2010) and favors an uprising of soil degradation (Polverigiani et al., 2014, 2018). In replacement to chemical weed management, various methods such as reduced or zero tillage, growing living mulch, mowing, using plastic mulch, distributing organic mulch, the physical control of weeds (i.e., steaming and flaming) have been studied, but all these methods have been found to have certain negative results (Granatstein and Kupferman, 2008).

However, the negative effects of these methods can be minimized by integrating advanced shallow tilling tools and herbicide usage at a reduced level. This will also optimize orchard biodiversity and reduce the risk factors associated with herbicide usage (Mia et al., 2020). Therefore, research is ongoing to find a more sustainable weed management strategy that reduces the weed competition but also has a reduced effect on weed biodiversity without compromising fruit production and quality. Currently, priority is given to maintaining long-term sustainability of orchards and requires an integrated approach for management of weeds (Neri, 2013; Granatstein and Kupferman 2008; Ponzio et al., 2013).

9.5 CONCLUSION

Adoption of orchard cultural practices without breeding overdue energy to enhance the quality and yield of apple fruit like microirrigation, fertigation and integrated nutrient management are of prime importance in orchard management. Irrigation plays an important role in crop production and is considered an important aspect of orchard management. Fertigation systems offer numerous benefits over conventional soil fertilization in terms of vegetative growth, yield, quality and productivity of fruit crops along with substantive saving of fertilizers and high water-use efficacy. Considering the limited potential of resources (water and fertilizers), it has become essential to espouse such technologies to avoid the demand and stress in the future. In order to enhance the irrigation volumes and units of fertilizers for the growers, water application rate research requires an advance level of exploration. A sustainable orchard floor management strategy needs to be adopted for controlling weed population and enhancing production, productivity and profitability of fruit trees. In addition to weed control, care should be taken that the management strategy will have reduced negative effects on weed biodiversity, the environment and human health. Integrated weed management approaches can play a crucial role in the overall improvement of soil quality by means of reducing of soil erosion and improving humification by means of increasing organic matter content in the soil and achieving effective weed control for enhanced orchard productivity.

REFERENCES

Ashiq, M., Sattar, A., Ahmed, N. and Muhammad, N. 2007. *Role of Herbicides in Crop Production.* Unique Enterprises 17-A, Gulberg Colony, Faisalabad, Pakistan.

Azzeddine, C., Philippe, M., Ferreira, M.I., Houria, C., Chaves, M.M. and Christoph, C. 2019. Scheduling deficit subsurface drip irrigation of apple trees for optimizing water use. *Arabian Journal of Geosciences* 12: 74.

Bond, W., Grundy, A.C. 2001. Non-chemical weed management in organic farming system. *Weed Research,* 41: 383–405.

Camposeo, S. and Vivaldi, G.A. 2011. Short-term effects of de-oiled pomace mulching on a young super high density olive orchard. *Scientia Horticulturae* 129(4): 613–621.

Cardoso, L.S. 2011. *Modelagem aplicada à fenologia de macieiras 'Royal gala'e 'Fuji suprema'em função do clima, na região de Vacaria, RS. 166 f.* Tese de doutorado. Universidade Federal do Rio Grande Do Sul, Porto Alegre, Brasil.

Derr, J.F. and Chandran, R.S. 2000. Chemical control of weeds. In: *2000 Spray Bulletin for Commercial Tree Fruit Growers.* Virginia Cooperative Extension Publications, pp. 92-106.

Din, S. 2017. *Weed Management Studies in Apple (Malus × domestica) under High Density Orchard System.* M.Sc. Thesis, Sher-e-Kashmir University of Agricultural Science and Technology of Kashmir (J&K).

Din, S., Wani, R.A., Pandith, A.H., Majid, I., Nisar, S., Nisar, F. and Jan, S. 2020. Economic analysis of various weed management strategies under high density apple (*Malus × domestica*) orchard system. *International Journal of Current Microbiology and Applied Science* 11: 2179–2185.

Fallahi, E. 2012. Influence of rootstock and irrigation methods on water use, mineral nutrition, growth, fruit yield, and quality in 'Gala' apple. *HortTechnology* 22(6): 731–737.

Girona, J., Behboudian, M.H., Mata, M., Del Campo, J. and Marsal, J. 2010. Exploring six reduced irrigation options under water shortage for 'Golden Smoothee' apple: Responses of yield components over three years. *Agricultural Water Management* 98: 370–375.

Goldberg, S.D., Rinot, M. and Karu, N. 1971. Effect of trickle irrigation intervals on distribution and utilization of soil moisture in a vineyard. *Soil Science Society of America Journal* 35: 127–130.

Granatstein, D. and Kupferman, E. 2008. Sustainable horticulture in fruit production. *Acta Horticulturae* 767: 295–308.

Hagin, J. and Lowengart, A. 1996. Fertigation for minimizing environmental pollution by fertilizers. *Fertility Research* 43: 5–7.

Hira, G.S., Jalota, S.K. and Arora, V.K. 2004. *Efficient Management of Water Resources for Sustainable Cropping in Punjab.* Technical Bulletin. Department of Soils, Punjab Agricultural University, India, 20pp.

Jiang, G., Liang, X., Li, L., Li, Y., Wu, G., Meng, J., Li, C., Guo, L., Cheng, D., Yu, X., *et al.* 2016. Biodiversity management of organic orchard enhances both ecological and economic profitability. *PeerJ* 4: e2137.

Jiang, X. and He, L. 2021. Investigation of effective irrigation strategies for high-density apple orchards in Pennsylvania. *Agronomy* 11(4): 732. https://doi.org/10.3390/agronomy11040732

Kafkafi, U. 2005. Global aspects of fertigation usage. *Fertigation Proceedings, International Symposium on Fertigation,* Beijing, China. 20–24 September 2005. pp. 8–22.

Kahramanoglu, I and Uygur, F.N. 2010. The effects of reduced doses and application timing of metribuzin on redroot pigweed (*Amaranthus retroflexus* L.) and wild mustard (*Sinapis arvensis* L.). *Turkish Journal of Agriculture and Forestry* 34: 467–474.

Kaur, K. and Kaundal, G.S. 2009. Efficacy of herbicides, mulching and sod cover on control of weeds in plum orchards. *Journal of Weed Science* 41(12): 110–112.

Kireva, R. and Mihov, M. 2019. Impact of watering regimes on apple yields under various meteorological conditions and micro irrigation. *Mechanization in Agriculture* 1: 35–38.

Kireva, R., Petrova-Branicheva, V. and Markov, E. 2017. Drip irrigation of apples at a moderate continental climate. *International Research Journal of Engineering and Technology* 9: 642–645.

Lisek, J. 2014. Possibilities and limitations of weed management in fruit crops of the temperate climate zone. *Journal of Plant Protection Research* 54: 318–326.

Litschmann, T. 2004. The importance of irrigation continues to grow. *Farmer* 20: 9.

Makus, D. 2007. Use of fabric and plastic barriers to control weeds in blackberries. *Journal of Subtropical Plant Science* 59: 95–103.

Mia, M.J., Massetani, F., Murri, G. and Neri, D. 2020. Sustainable alternatives to chemicals for weed control in the orchard: A review. *Horticultural Science* 47: 1–12.

Mushtaq, R., Sharma, M.K., Ahmad, L., Krishna, B., Mushtaq, K. and Mir, J.I. 2020. Crop water requirement estimation using pan evaporimeter for high density apple plantation system in Kashmir region of India. *Journal of Agrometeorology* 22(1): 86–88.

Mushtaq, R., Sharma, M.K., Mir, J.I., Mansoor, S., Mushtaq, K., Popescu, S.M., Malik, A.R., El-Serehy, H.A., Hefft, D.I., Bhat, S.A. and Narayan, S. 2021. Physiological activity, nutritional composition, and gene expression in apple (Malus domestica Borkh) influenced by different ETc levels of irrigation at distinct development stages. *Water* 13(22): 3208.

Naor, A., Hupert, H., Greenblat, Y., Peres, M. and Klein, I. 2001. The response of nectarine fruit size and mid-day stem water potential to irrigation level in stage III and crop load. *Journal of the American Society for Horticultural Science* 126: 140–143.

Neri, D. 2013. Organic soil management to prevent soil sickness during integrated fruit production. *IOBC WPRS Bulletin* 91: 87–99.

Pieterse, P.J. 2010. Herbicide resistance in weeds: A threat to effective chemical weed control in South Africa. *South African Journal of Plant Soil* 27: 66–73.

Polverigiani, S., Franzina, M. and Neri, D. 2018. Effect of soil condition on apple root development and plant resilience in intensive orchards. *Applied Soil Ecology* 123: 787–792.

Polverigiani, S., Kelderer, M. and Neri, D. 2014. Growth of 'M9' apple root in five central Europe replanted soils. *Plant Root* 8: 55–63.

Ponzio, C., Gangatharan, R. and Neri, D. 2013. Organic and Biodynamic Agriculture: A Review in Relation to Sustainability. *International Journal of Plant Soil Science* 2: 95–110.

Rankova, Z. and Zhivondov, A. 2012. Ecological approach for weed control in young peach plantation. *Acta Horticulturae* 962: 419–42.

Robinson, R.A. and Sutherland, W.J. 2002. Post-War changes in arable farming and biodiversity in Great Britain. *Journal of Applied Ecology* 39: 157–176.

Rowley, M.A., Ransom, C.V., Reeve, J.R. and Black, B.L. 2011. Mulch and organic herbicide combinations for in-row orchard weed suppression. *International Journal of Fruit Science* 11(4): 316–331.

Rozpara, E., Grzyb, Z.S. and Bielicki, P. 2008. Influence of various soil maintenance methods in organic on the growth and yielding of sweet cherry trees in the first years after planting. *Journal of Fruit and Ornamental Plant Research* 16: 17–24.

Salaria, A. 2009. Horticulture at a glance. In *Handbook for Competitive Exams*. Volume 2. Jain Brothers, East Park Road, Karol Bagh, New Delhi pp. 234–235.

Sne, M. 2006. Micro irrigation in arid and semi-arid regions. In *Guidelines for Planning and Design*. Ed. by S.A. Kulkarni. ICID, International Commission on Irrigation and Drainage, New Delhi, India.

Stockle, C.O. and Dugas, W.A. 1992. Evaluating canopy temperature-based indices for irrigation scheduling. *Irrigation Science* 13: 31–37. https://doi.org/10.1007/BF00190242

Yu, C., Hu, X.M., Deng, W., Li, Y., Xiong, C., Ye, C.H., Han, G.M. and Li, X. 2015. Changes in soil microbial community structure and functional diversity in the rhizosphere surrounding mulberry subjected to long-term fertilization. *Applied Soil Ecology* 86: 20–40.

Yudh, R. 2001. Losses due to weeds in India. *The Indian Express*, 17th June.

10 Nutrition and Orchard Manuring Practices

Mumtaz A. Ganie, Ab Shakoor Khanday,
Aabid H. Lone, Yasir Hanif Mir,
Aanisa Manzoor Shah, Suheel Ahmad,
Ab Raouf Malik

CONTENTS

10.1 INTRODUCTION

Apple is an important temperate fruit crop worldwide. Nutrition plays a vital role in determining yield and the quality of apples. As per the criteria of Arnon and Stout (1939), 17 nutrient elements are indispensable for completing the life cycle of higher plants. Nutrients required by plants in higher quantities are called macronutrients like carbon, hydrogen, oxygen, nitrogen, phosphorous, potassium, calcium, magnesium and sulfur. In addition to these nutrients, iron, copper, manganese, zinc, boron, chlorine, molybdenum and nickel are required in lesser quantities by plants and are called micronutrients or trace elements. Apples are perennial in nature and have more nutritional requirements than annual crops (Srivastava and Singh 2008). Balanced nutrition, on one hand, plays a role in increasing yield and improving the quality of a fruit, and, on the other hand, it impacts the sensitivity of trees to biotic and abiotic stress.

Management of nutrition is the application of a judicious quantity of fertilizer, proper timing and mode of nutrient supply to enhance production (Johnson 2011). For an efficient nutrient management system of apples, the study of various processes is a prerequisite, such as the availability of nutrients in the soil, the physicochemical properties of soil, the moisture content in soil, the uptake of nutrients from the soil, the mobility of nutrients in soil and plant, the retranslocation of nutrients from leaves to spurs, the amount of nutrients required for good quality fruit, the fertilizer source and the fertilizer use efficiency. Physicochemical properties of soil like pH, electrical conductivity, cation exchange capacity, bulk density, particle density, organic carbon, etc. and the availability of nutrients in the soil are evaluated by soil testing. Foliar analysis is carried out to assess the uptake of nutrients from the soil. Critical limits of nutrients in the soil as well as apple leaves are exploited to define whether the nutrients available are deficient or adequate, and those findings provide fertilizer recommendations. The mobility of nutrients in soil gives the idea of the mechanism of the uptake of a particular nutrient like root interception or diffusion or cation exchange. Mobility of the nutrients in the plant systems is studied through nutrient partitioning exploiting tracer technique. The mobility of nutrients in the plant system, i.e., xylem and phloem, suggests whether to use soil application or foliar application in cases of a deficiency of a nutrient element. For those nutrient elements which are phloem mobile, soil application addresses the deficiency issues. Foliar application of nutrients is advised in cases of phloem immobile nutrients. Fertilizer use efficiency is considered when recommending a particular fertilizer to meet the demand of crops. Integrated nutrient management augments overall apple trees in terms of plant growth, yield and fruit quality (Verma and Chauhan 2013). The objective of nutrient management in apples is to promote flowering and fruit set, reduce fruit drop and ultimately harvest high yield of a quality crop.

10.2   NUTRIENT ACQUISITION FROM SOIL

Improved water availability, efficient soil management, plant breeding for fertilizer responsive varieties and effective nutrient scheduling (Foresight 2011) are potent strategies for optimizing crop nutrient acquisition under diverse agro-climatic conditions (Kraiser et al. 2011; Garnett et al. 2009; Henry et al. 2010). Site-specific adoption of these management options is important not only for improving fertilizer use efficiency but also for sustainable soil health management and food security. Fertilizer use efficiency has a paramount influence on crop growth and productivity. The process of nutrient and water acquisition is facilitated by the epidermal tissue extensions, the root hairs. Root hairs upsurge root surface area for efficient exploitation of soil nutrient reserves and their acquisition. Nutrient acquisition by root hairs, the function of root hairs as prominent sites of H^+-ATPase activity and their pivotal role in nutrient uptake have been demonstrated and proven at both physiological as well as molecular levels (Gilroy and Jones 2000).

Root proliferation, transporter function, mobility enhancement of barely available nutrients through their binding with root exudates, symbiosis with other organisms, mass flow and diffusion are involved in the acquisition of nutrients through roots (Jungk 1996). However, there is a wide variation (0–80% of total uptake) in the quantity of a nutrient acquisition through root hairs which depends on plant species, differences in root hair development, quantity, mobility and availability of a nutrient in soil (Jungk 2001).

Root characteristic improvement through breeding and a better understanding of root nutrient uptake under diverse soil moisture regimes and fluctuating nutrient availability conditions will help to frame the strategies for nutrient acquisition optimization under different agro-climatic situations for better crop production. The proliferation of roots is crucial for the uptake of ions, having reduced mobility or physical isolation.

Nitrate, sulfate, calcium and magnesium are drawn to the root by mass flow, and phosphate, potassium, ammonium and micronutrient metals are mainly drawn through diffusion. However, the nutrients that mainly move by mass flow can also move by diffusion under certain conditions (Kage 1997). Nitrogen in leguminous, ectomycorrhizal and ericoid in mycorrhizal plants and phosphorus and zinc in arbuscular mycorrhizal plants are acquired via microbial symbioses (Smith and Read 1997). Iron in high pH soils and phosphorus by proteoid roots are obtained through a rhizosphere mobilization processes (Barber 1995; Marschner 1995).

Mass flow and diffusion of nutrient elements depend on the water holding capacity and hydraulic conductivity of soil, which affects the movement of water and nutrient elements to the surface of roots from bulk soil. The transpiration pull that causes the mass flow of soluble nutrients is triggered by the water potential gradient from the atmosphere to the bulk soil (Jungk 1996). Root hairs improve water and nutrient acquisition through a reduction in the matric potential gradient near the root soil interface (Carminati et al. 2017). The ion influx creates a gradient of dissolved ions within the soil solution toward the root surface that results from the localized exhaustion of the ions at the root surface (Jungk 1996).

Optimum nitrogen uptake under adverse agro-climatic conditions is one of the important priorities for enhancing root acquisition efficiency (Kraiser et al. 2011; Garnett et al. 2009; Gojon et al. 2011; Kant et al. 2011; Xu et al. 2012; Bingham et al. 2012). Under normal soil moisture conditions, most of the nitrogen absorption takes place as nitrate, and the process involved is dependent on the supply of dissolved nitrate ions in the soil water; in dry soils, a limited hydraulic conductivity hinders the nitrate uptake and thus necessitates effective root morphology for nitrate acquisition. The role of the diffusion in root nitrogen uptake relies on the ionic form of the nutrient, as effective soil diffusion coefficients for nitrate (3.26×10^{-10} m^2 s^{-1}) and ammonium (2.70×10^{-12} m^2 s^{-1}) differ significantly (Miller and Cramer 2005). As such, the contribution of diffusion flux to nitrate acquisition by the root is greater than that of ammonium. The rate of nitrate and ammonium diffusion is regulated by fertilizer application and soil temperature which influence their concentrations in soil and their rates of diffusion, respectively (Miller and Cramer 2005; Cramer et al. 2009).

Unlike nitrate, which moves in soil both through mass flow as well as diffusion, phosphate is highly immobile. Mass flow accounts for as 1–5% of a plant's phosphorus requirement; the growing roots intercept only half of what is delivered through mass flow (Lambers et al. 1998). To fulfill the plant demand for phosphate, a substantial quantity of phosphorus must reach the plant through diffusion. But the rate of phosphate diffusion in soil is very low $0.3–3.3 \times 10^{-13}$ $m^2 s^{-1}$ (Clarkson 1981), especially in dry soil (Bhadoria et al. 1991). Exploiting the mass flow for improving the delivery of phosphorus will decrease the water-use efficiency due to its higher use. So enhanced root proliferation, increased root hair density and root hair length will improve the phosphorus uptake through root acquisition by a plant especially under low soil phosphorus conditions (Lambers et al. 2006; Miguel et al. 2015). Root architecture improvements can be undertaken for improving the delivery of other nutrients also, especially when their level is low in the soil and the conditions for mass flow and diffusion are not optimal.

Due to variations in inherent nutrient content, microbial diversity and chemical and physical conditions of soil, there is a vast scope of nutrient acquisition investigations (Dodd and Ruiz-Lozano 2012). Intricacies of relationship between the diverse factors impacting nutrient acquisition is difficult to quantify through a particular approach. Through a holistic approach a better understanding of the interaction between root and soil components impacting nutrient acquisition through roots is possible. Use of hydrophilic gel systems (Lin et al. 2019) and simulation modelling (Postma 2017) for better understanding of the nutrient delivery, root acquisition of nutrients and their uptake is encouraged.

10.3 SOIL NUTRIENT AVAILABILITY

Nutrient availability is a function of the nutrient concentration in the soil solution, rate of diffusion within the soil and interactions with other soil minerals. The Soil Science Society of America defines available nutrients as: (1) the quantity of soil nutrients in different compounds available to plants or compounds that could possibly be convertible to such forms during the growing season, and (2) the materials of legally defined available nutrients in fertilizers assessed by specified laboratory procedures that in most nations serve as the legal basis for guarantees (Fageria et al. 1991).

Nutrient availability is determined by widely validated tests and protocols employed to know plant response to different nutrient management strategies. The availability of nutrients in soil solutions varies with variation in soil water content, soil depth, pH, cation exchange capacity, redox potential, amount of soil organic matter and microbial activity, season and fertilizer application. Due to low soil moisture constraint in dry areas, nutrient availability in topsoil and their delivery to roots progressively decreases during the growth season in dry areas. Concentration of nutrients, particularly nitrogen, in arable soils supporting high yielding crops is high, however phosphorus concentration, which is exceedingly low in these soils as compared to other nutrients, changes over time especially when fertilizer is added. Organic ligands improve PO_4^{3-} accessibility through an anion-exchange phenomenon, which involves substituting PO_4^{3-} from binding sites. Capillary movement of water from moist subsurface to dry topsoil can improve the K^+ availability and absorption from dry topsoil. The availability of the micronutrients Mn^{2+}, Fe^{2+}, Zn^{2+} and Cu^{2+} in soil solutions is primarily determined by soil pH, redox potential and soil organic matter and can vary throughout the season under temperate conditions with a peak concentration in early summer. A reduction in pH or redox potential and chelation by low molecular weight organic compounds increase Mn^{2+}, Fe^{2+}, Zn^{2+} and Cu^{2+} concentrations.

10.4   NUTRIENT UPTAKE

Nutrient uptake is the process by which plants acquire elements required for growth and is accomplished through ion exchange in the soil solution. As compared to cereal crops, apple farming is more efficient in sustaining the soil nutrient balance due to minimal nutrient loss through harvest and pruning, significant nutrient recycling through decomposition of leaf litter and nutrient conservation in the woody parts of trees. Apple trees uptake maximum nitrogen between 37 and 81 days following full bloom, and, subsequently, the uptake begins to decline (Zanotelli et al. 2014). Annual nitrogen removal from a high-density apple orchard through fruits harvest, leaf fall and pruning varies from 25 to 60 kg ha^{-1} for various cultivars (Neilsen and Neilsen 2002), and the removal is expected to be lower in traditional orchards. Apple trees utilize N fertilizer inefficiently due to their lower rooting densities (Neilsen and Neilsen 2002). Apple fruit is a good K sink and accumulates K from 0.55 to 0.8 kg K Mg^{-1} fruit dry weight (Zavalloni et al. 2001). Potassium uptake is dependent on fruit yield and ranges from 80 to 100 kg K ha^{-1} under a fruit yield of 40 to 60 Mg ha^{-1}, respectively. Because of its strong phloem mobility, K allocation to fruit is consistent from fruit set to maturity (Zavalloni et al. 2001). Scandellari et al. (2010) found that 74 kg Ca ha^{-1} are absorbed annually by the apple cultivar 'Gala' with allocation of tree structure (11%), fruits (4%), leaves (60%) and pruned wood (25%).

10.5   NUTRIENT CONCENTRATION IN APPLE FOLIAGE

The concentration of nutrients in the foliage is the expression of the soil nutrient availability and is considered a concrete approach to check the soil fertility status (Havlin et al. 2012). Tissue testing in conjugation with soil testing has often proved a very effective tool in adjudging nutrient disorders and formulating recommendations (Memon et al. 2005). Depending upon sampling techniques, analytical methods and management practices, foliar nutrient concentration varies substantially region-wise under diverse agro-climatic conditions (Kenworthy 1961). Mainly survey approaches have been exploited to define the optimum leaf nutrient ranges or critical nutrient levels for apple (Kumar 1991; Upadhayay and Awasthi 1993). The knowledge of seasonal fluctuations in leaf nutrient levels is critical for a proper recommendation of subsequent fertilizer applications and will play a theoretical and basic role in practical production processes. With the development of leaf nutrient assessment and its extensive use for determining tree nutrient status, nowadays orchard fertilization practices are mainly governed by exploiting critical leaf nutrient approaches or specific range approaches and are now consistently regulated by comparing leaf analysis results to the optimal range of leaf nutrient concentrations (Table 10.1). However, successful nutrient management in apple orchards still necessitates a thorough understanding of nutrient demands in terms of quantity and time.

TABLE 10.1

Optimum Leaf Nutrient Concentration in Apples

Nutrient	Optimum concentration	Reference
Nitrogen	2.33%	Kenworthy (1961)
Nitrogen	2.4–2.8%	Bould (1966)
Nitrogen	1.7–2.5%	Shear and Faust (1980)
Phosphorus	0.23%	Kenworthy (1961)
Phosphorus	0.2–0.25%	Bould (1966)
Phosphorus	0.1–0.4%	Tukey and Dow (1979)
Phosphorus	0.15–0.30%	Shear and Faust (1980)
Potassium	1.53%	Kenworthy (1961)
Potassium	1.3–1.6%	Bould (1966)
Potassium	1.2–1.9%	Shear and Faust (1980)
Calcium	1.40%	Kenworthy (1961)
Calcium	1.0–1.6%	Bould (1966)
Calcium	1.5–2.0%	Shear and Faust (1980)
Magnesium	0.40%	Kenworthy (1961)
Magnesium	0.26–0.30%	Bould (1966)
Magnesium	0.3–0.35%	Tukey and Dow (1979)
Magnesium	0.25–0.35%	Shear and Faust (1980)
Sulfur	0.01–0.10%	Shear and Faust (1980)
Manganese	30–100 ppm	Bould (1966)
Manganese	25–100 ppm	Tukey and Dow (1979)
Manganese	25–150 ppm	Shear and Faust (1980)
Iron	40–500 ppm	Tukey and Dow (1979)
Iron	40–500 ppm	Shear and Faust (1980)
Zinc	15–20 ppm	Bould (1966)
Zinc	15–200 ppm	Shear and Faust (1980)
Copper	6–40 ppm	Tukey and Dow (1979)
Copper	5–12 ppm	Shear and Faust (1980)
Boron	20–60 ppm	Shear and Faust (1980)

10.6 MOBILITY OF NUTRIENTS IN SOIL

Nutrient mobility is the capacity of nutrients to move within a specific system. Knowledge of nutrient mobility aids in determining the causes of nutrient deficiency. In order to be available to plants for uptake and consumption, nutrients in the soil must exist in ionic form, with the exception of B, which is available to plants in a nonionic form. The mobility of a nutrient is affected by the nature of charge as well as its valency. Seventeen elements are essential for plant growth. Carbon, hydrogen and oxygen are non-mineral and the remaining 14 are minerals. Carbon, oxygen and hydrogen enter the plants via leaves and roots as carbon dioxide and water. The remaining 14 elements must be dissolved in soil solutions for their uptake by the plants via roots. Mineral elements are further subdivided into macronutrients (primary and secondary) and micronutrients. Macronutrients are those required in relatively large quantities (>1.0% of the dry mass), whereas micronutrients are needed in small quantities (less than 0.01% of the dry mass).

Cation exchange capacity (cmolkg^{-1}), defined as the capacity of soil to exchange the ions between the soil solution and soil exchange complex, plays an important role in nutrient mobility within the soil. The higher the CEC, the more the cationic nutrients are held by the soil. Since soils have very low anion ion exchange capacity (< 5.0 cmolkg^{-1}), the anions are held loosely at the soil exchange complex and are readily washed out of the soil. The mobility of nutrients in soil differs slightly from that in plants, but it is critical to understand why deficiency of some nutrients inhibits growth more than others. Nutrient mobility, in particular, has an impact on how we should manage its availability in the soil through fertilizer application. N fertilizer can be broadcasted or incorporated into the soil with similar impact on crop growth because it is quite mobile in the soil. Plants can take up nitrogen as either nitrate (NO_3^-) or ammonium (NH_4^+), however, NO_3^-, a negatively charged ion, is not retained by the soil tightly on exchange sites. NH_4^+ is positive and is held by cation exchange sites and hence is rendered less mobile. However, phosphorus as (HPO_4^{2-}, $H_2PO_4^-$) is relatively immobile in most soils; therefore, it is either banded or drilled. Boron as BO_3^-, sulfur as SO_4^{2-} and chlorine as Cl^{-1} are mobile in soil, whereas N (NH_4^+), K^+, Ca^{2+}, Mg^{2+}, Cu^{2+}, Fe (Fe^{2+}, Fe^{3+}), Mn^{2+}, Zn^{2+} and Mo (MoO_4^-) are relatively immobile.

Mobility of all the nutrients in the soil is dependent on pH, temperature, moisture, soil texture, soil organic matter content, type of clay minerals and metal hydroxides. In addition, some of the immobile nutrients are more mobile than others. For example, NH_4^+, K^+, Ca^{+2}, and Mg^{+2} are more mobile than the metals (Cu^{+2}, Fe^{+2}, Fe^{+3}, Mn^{+2}, Ni^{+2}, Zn^{+2}). Soil application of mobile elements should be more frequent than immobile nutrient application because the mobile nutrients are more easily taken up or leached. Except for NH_4^+, which is immediately transformed to NO_3^-, the immobile nutrients can be applied to meet crop needs and to enrich the soil nutrient reserves for the next cropping cycle.

10.7   MOBILITY OF NUTRIENTS IN APPLE FOLIAGE

The uptake of the nutrients from the soil takes place through root diffusion, root interception (Foth and Ellis 1997) or ion exchange (Havlin et al. 2012). Afterwards nutrients move to the growing plant parts through the xylem quite easily. The movement of nutrients from leaves to the sink (fruits) occurs through the phloem. The concentration of the particular nutrient in the fruit tissue depends upon the phloem loading of that particular nutrient or, in other words, the phloem mobility. Knowledge of nutrient mobility helps to diagnose plant nutrient deficiency symptoms. If older growth is affected, then a mobile nutrient is more likely responsible for the deficiency. However, if new growth shows the deficiency, the plant is more likely deficient in an immobile nutrient. Nutrient deficiency symptoms appear when the plant is at maximum growth or at times of high nutrient demand.

Nitrogen is taken by plants in two forms, nitrate (NO_3^-) and ammonium (NH_4^+). Both the forms are highly mobile in plants. Phosphorus is absorbed by plants in the forms of HPO_4^{2-} and $H_2PO_4^-$ and both forms are mobile. Potassium taken in the form of K^+ by plants is also mobile. Among secondary nutrients, only magnesium is mobile in plants. The micronutrients which are taken by plants in cationic forms like iron, copper, manganese, zinc and cobalt are all immobile in plants except nickel (Ni^{2+}), which is mobile. Boron is the only nutrient element taken by plants in nonionic form, and it is immobile in plants.

Boron is mobile in the phloem of the plant species like apple which is rich in ol-compounds like sorbitol, mannitol, dulcitol, etc. (Brown and Hu 1996). Many researchers have demonstrated that *Prunus*, *Pyrus* and *Malus* do not accumulate high levels of boron in their leaves but develop twig die-back and gum exudation as a result of boron toxicity (Hansen 1955; Woodbridge 1955; Maas 1984; El-Motaium et al. 1994). Marginal leaf burn and/or chlorosis of the oldest leaves, a typical symptom of boron toxicity, is frequently absent in *Prunus*, *Malus* or *Pyrus*. Nevertheless, the toxicity of boron has been observed in apple trees which received a higher concentration of boron sprays at leaf margins and tips (Ganie 2016) suggesting phloem boron immobility in apple trees. Therefore, boron may be only partially mobile in the phloem of apple trees. However, further research is needed to understand the behaviour of phloem boron mobility in sorbitol-rich species like apple.

10.8   PROCEDURE OF SOIL SAMPLING IN APPLE ORCHARDS

- Divide each field into uniform blocks based on slope and cropping history (Dow et al. 1991).
- Even if a field appears to be uniform, divide it into several blocks and sample them separately.

- Take soil sample cores (at least five cores per 0.8 to 2 ha) from the entire field in a W-shaped or zigzag pattern.
- Mix the cores thoroughly, remove leaves, organic debris, pebbles, roots and other unwanted material.
- Sample by foot increments to a depth of 2 to 4 feet or deeper if restrictive layers are encountered.

Note: Don't take soil samples from or near unusual spots.

10.9 SOIL SAMPLING IN ESTABLISHED ORCHARDS

- Soil sampling once in 3–5 years is adequate.
- Annual soil sampling may be done in newly established orchards for establishing a fertility management schedule.
- To assess the available nutrient status over a long time, samples should be taken during the same season, i.e., fall (Daniel and William 2016).
- To assess the soil nitrate content, sampling should be done in early summer before the peak nitrogen uptake by the trees.
- Divide the orchard into uniform blocks based on slope and crop performance, age, variety and drainage.
- Remove the residue, grass or organic debris from the sampling spot.
- Take the samples with the help of an auger or probe.
- Take the sample halfway between the trunk and the drip line and within the wetting zone of the emitter (Figure 10.1.).
- Cores are taken from the entire orchard in a W-shaped or zigzag pattern.
- Mix the cores thoroughly, remove leaves, organic debris, pebbles, roots and other unwanted material
- Sample by foot increments to a depth of 3 feet.
- To obtain an accurate estimate of the nutrient availability, between 15 and 20 cores should be taken from each block for a composite sample.
- One sample per tree is generally taken. Within each block, make sure to sample different orientations relative to the trunk.
- Collect the samples in a clean plastic bucket. Avoid galvanized or rubber buckets as they may contaminate the soil sample with zinc (Thom et al. 2003).

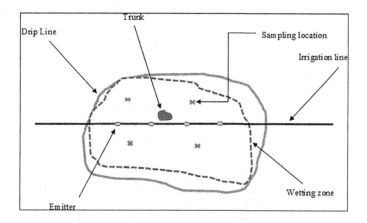

FIGURE 10.1 Sampling location under orchard trees.

Source: Figure idea adopted from Daniel and William (2016)

10.10   LEAF SAMPLING IN APPLE ORCHARDS

Leaf analysis is used to measure the amount of nutrients in perennial crops and is usually considered a better indicator of nutrient status compared to soil tests.

Procedure of Leaf Sampling in Apple Orchards

- The general time for collection of leaf sample is between 15 July and 15 August.
- First of all, divide the orchard into uniform blocks.
- In each block select 25 apple trees randomly in such a way that it represents the whole block (Holland et al. 1967).
- Take four leaves from each selected tree, one leaf each from the middle of the current season growth on all four sides of the tree at chest height (Holland et al. 1967).
- To overcome the shading effect, select leaves which are fully exposed to the sun.
- Composite leaf samples should be collected separately for different age groups and varieties.
- Avoid distorted, chlorotic, infected leaves.
- Cut the petiole of the sampled leaves.
- Wash the sampled leaves with tap water followed by distilled water.
- Sun dry the washed leaves on tissue paper.
- Pack the sample in a paper bag and oven dry at 65–70°C until constant weight is achieved.

Grind the oven-dried sample and analyze in the laboratory after wet digestion or dry ashing.

10.11   NUTRIENT DIAGNOSTIC APPROACHES

An imbalance of essential nutrients causes a disturbance in the metabolism of plants which subsequently decreases their performance in terms of growth, development and productivity. Therefore for successful fruit production, assessment of nutrient balance in orchard soil as well as foliar plant parts is significantly important. For this purpose, several approaches like soil testing, sufficiency range, critical leaf nutrient concentrations, ratios of nutrients in the soil and plant and the diagnosis and recommendation integrated system (DRIS) approach can be exploited. But there are certain limitations to each of the above mentioned approaches. Soil testing gives the details of nutrient availability in soil which does not always correlate with the nutrient content in the plant. The uptake of the nutrients from the soil depends upon a multitude of factors including the plant's own nutritional needs. Plant analysis not only helps in plant nutritional diagnosis but is also considered a practical tool for formulating fertilizer recommendations. Critical leaf nutrient concentration is soil and crop specific, which is often exploited to diagnose nutritional deficiencies and toxicities (Munson and Nelson 1990). Nonetheless, critical nutrient concentration takes into account the concentration of a single nutrient only and does not take in to account the nutrient balance, i.e., the concentration of one nutrient with respect to others (Mourao-Filho 2005). This problem was more or less solved by checking nutrient ratio which take into account the concentration of one nutrient in comparison to other. The nutrient ratio approach is considered a more reliable diagnostic tool compared to critical nutrient concentration (Dampney 1992). The nutrient ratio concept, although better than the critical nutrient approach, still takes into account the balance of only one nutrient (e.g., N) with respect to other (e.g., K) and cannot simultaneously explain the potential imbalances with other essential nutrients. The flaws in nutrient ratios can be overcome by employing the technique of the Diagnosis and Recommendation Integrated System (DRIS) (Beaufils 1973). In the DRIS approach, multiple nutrient ratios are compared in diagnosing a nutritional problem (Mourao-Filho 2005). Substantially, the DRIS diagnostic approach is considered an effective evaluation method for testing nutritional diagnosis (Nachtigall and Dechen 2007); however, it is comparatively more difficult than traditional methods like critical nutrient concentration, soil testing and sufficiency range approach

(Prado 2008). The DRIS nutrient diagnostic approach is more promising in perennial crops in which calibration tests are impractical.

The DRIS nutrient diagnostic approach has advantages over the rest of the methods because it is based on nutritional ratios instead of mean levels of a single nutrient (Wadt 2009). Deviations in the nutrient ratios are measured, and ultimately DRIS indices are developed which are employed to check nutrient imbalance (Machado-Dias et al. 2011). DRIS presents continuous scale and easy interpretation and permits nutrient arrangement (from the maximum deficient up to the most excessive). DRIS detects nutrient imbalance even when none of the nutrients is below the critical level. Finally, it offers a chance to establish the total plant nutritional balance through a nutritional imbalance index.

10.12   FERTILIZER MANAGEMENT IN APPLES

Fertilizers are pivotal in securing crop production, and the estimates suggested they currently support 40–60% of total crop production globally (Johnston and Bruulsema 2014). Fertilizer use in soil, depending upon management, could have either a positive or negative impact on soil and plant health and ecosystem functions. When managed properly, they not only increase the crop production but also maintain the ecological integrity of cropping systems. Improper management could lead to nutrient mining, yield losses, pest damages and enormous environmental hazards. Therefore, a successful nutrient management program must not only be profitable but also sustainable. The 4R nutrient stewardship concept, developed by the fertilizer industry worldwide, is based on universal scientific principles which act as a guide to fertilizer best management practices. The Rs here refer to applying the right fertilizer source, at the right rate, at the right time and in the right place. While the scientific principles governing this concept are universal, practical implementation is site specific. Therefore, there is no common nutrient management program that will work in every situation.

To develop a nutrient management plan for an apple orchard, its life cycle is separated into three age categories.

   I. Pre-planting years
  II. Non-bearing years
 III. Bearing years

10.12.1   Pre-Planting Years
At this stage, soil tests are conducted to determine the various physicochemical characteristics and the availability status of essential nutrients. Based on the results, fertilizers and other amendments are applied, mostly through broadcasting, and pH is also adjusted as per the requirement.

10.12.2   Non-Bearing Years
The initial growth, immediately after transplanting, of apple plants is mainly supported by the nutrient reserves within them and owing to the damaged root system the nutrient uptake from the soil is delayed for some time. Further, dry fertilizers aren't applied at this stage as that could damage the root system. Fertilizers are only applied after buds are broken and roots have taken hold in the soil. The rate of application is determined by soil test values.

10.12.3   Bearing Years
When developing fertilization programs for apple orchards, nutrient supply from the soil, plant nutrient demand and relationship between the nutrient supply and demand during the course of the season are taken into consideration. There are certain critical phenological stages in apples which have higher nutrient requirement. Flowering, fruit set, and June drop (when exponential fruit growth, root and

vegetative shoot growth occur simultaneously) generally represent the stages with the highest nutrient demand (Hamid et al. 1988). The events during these phenological stages have a significant bearing on the final yield, storage life, size and quality of fruit (Ganeshamurthy et al. 2019). Therefore, most of the fertilizers are applied early in the season, beginning with bud break up to about 6 weeks after fruit set. The baseline supply required thereafter is met through native soil reserves.

Nitrogen is required in large quantities for canopy development and fruit growth early in the season. Its requirement decreases toward the end of the season during the fruit quality development and wood maturity. In addition to the soil application of nitrogenous fertilizers in multiple splits during the spring, foliar applications are also sometimes recommended to satisfy the early season nitrogen demand of the tree or buildup the tree nitrogen reserves after harvest in the fall season. However, care must be taken to apply fertilizer at the right concentrations to prevent damage to tender foliage early in the season. The fall season N foliar application is also recommended; however, it must be delayed until at least 20–30% of foliage has turned yellow to avoid delay in leaf senescence.

Phosphorus could be applied in the fall or spring season while soil application of potassium is recommended at the bud break stage and 3 weeks after fruit set. Depending upon the soil test value, P and K recommendations are determined on the basis of maintenance, build-up or drawdown philosophies.

Calcium can be applied to apple trees through soil, foliage or a combination of both, depending up on the situation. If soil test value for calcium falls in the 'low' category, soil applications are recommended. However, in situations where the possibilities of calcium uptake through roots are limited, the foliar application is preferred. Care must be taken that calcium is applied directly to the fruit since the phloem mobility of calcium is very limited. Calcium, like most other nutrients, is mainly required early in the season during the cell division phase. Its uptake is progressively reduced as the fruit grows.

Boron and other micronutrients, if required, are preferably applied through foliar application. All these nutrients are applied during critical phenological stages as mentioned earlier. Boron is ideally applied at the pink bud and/or full bloom stage to promote fertilization.

Soil-applied fertilizers in apples are mostly broadcasted into the canopy region about 2 feet away from the main trunk up to the drip line and then incorporated with soil thoroughly. Only phosphorus, in view of its greater soil fixation, is applied in bands.

Foliar applied sprays, depending on the situation, are applied during active growing or the fall season. Care must be taken to apply these nutrients at the right concentrations under optimum meteorological conditions to achieve maximum benefit.

In addition to inorganic fertilizers, the application of compost and well decomposed farmyard manure is also recommended during the fall or spring seasons to promote soil health and over all plant growth. The rate of application is determined by soil organic matter content and the age of the tree.

10.13   CONCLUSION

The main aim in managing nutrition is to combat either deficiency or toxicity that may hinder normal plant functioning. Nutrient management includes nutrient availability, nutrient demand and nutrient supply sources and methods of fertilizer application. Improving nutrient use efficiency is vital both from an economic perspective as well as an environmental point of view.

Among the 17 essential nutrients, carbon, hydrogen and oxygen are supplied through air and water while the rest are supplied to plants from the soil through roots. A robust soil testing facility is essential to support a balanced use of nutrients to attain a high yield of a good quality crop in a sustainable manner. Soil plant continuum with respect to nutrient dynamics needs to be exhaustively studied in perennial crops like apples to unveil the processes involved in the movement of nutrients to plant roots and their subsequent absorption. Intensive research is needed on framing geographical information system (GIS)-based fertility maps of the apple growing areas, development and refinement of soil test methods and fertilizer recommendation systems involving sites' specific nutrient management and decision support systems based on simulation models.

REFERENCES

Arnon, D. I. and P. R. Stout. 1939. The essentiality of certain elements in minute quantity for plants with special reference to copper. *Plant Physiology* 14: 371–375.

Barber, S. A. 1995. *Soil nutrient bioavailability: A mechanistic approach*, 2nd ed. New York: Wiley.

Beaufils, E. R. 1973. Diagnosis and recommendation integrated system (DRIS). *Soil Science Bulletin* 1: 1–132.

Bhadoria, P. B. S., J. Kaselowsky, N. Claassen, and A. Jungk. 1991. Phosphate diffusion coefficients in soil as affected by bulk density and water content. *Zeitschrift für Pflanzenernährung und Bodenkunde* 154: 53–57.

Bingham, I. J., A. J. Karley, P. J. White, W. T. B. Thomas, and J. R. Russell. 2012. Analysis of improvements in nitrogen use efficiency associated with 75 years of spring barley breeding. *European Journal of Agronomy* 42: 49–58.

Bould, C. 1966. Leaf analysis of deciduous fruits. In *Fruit nutrition: Temperate to tropical* (Ed. N. F. Childers). New Brunswick, NJ: Horticultural Publication, Rutgers, The State University, pp. 651–682.

Brown, P. H., and H. Hu. 1996. Phloem mobility of boron is species dependent: Evidence for phloem mobility in sorbitol-rich species. *Annals of Botany* 77: 497–505.

Carminati A., J. B. Passioura, M. Zarebanadkouki, M. A. Ahmed, P. R. Ryan, M. Watt, E. Delhaize. 2017. Root hairs enable high transpiration rates in drying soils. *New Phytologist* 216: 771–781.

Clarkson, D. T. 1981. Nutrient interception and transport by root systems. In *Physiological processes limiting plant productivity*, Ed. C. B. Johnson. London: Butterworths, pp. 307–330.

Cramer, M. D., H. J. Hawkins, and G. A. Verboom. 2009. The importance of nutritional regulation of plant water flux. *Oecologia* 161(1): 15–24.

Dampney, P. M. R. 1992. The effect of timing and rate of potash application on the yield and herbage composition of grass grown for silage. *Grass and Forage Science* 47: 280–289.

Daniel, G. and R. William. 2016. *Soil Sampling in Orchards*. Davis: University of California. http://geisseler.ucdavis.edu/Guidelines/Soil_Sampling_Orchards

Dodd, I. C. and J. M. Ruiz-Lozano. 2012. Microbial enhancement of crop resource use efficiency. *Current Opinion in Biotechnology* 23: 236–242.

Dow, A. I., F. A. Rushmore, A. R. Halvorson, and R. B. Tukey. 1991. *Orchard soil sampling*. Available at http://content.wsulibs.wsu.edu/cahe_arch/html/eb1595/eb1595.html

El-Motaium, R., H. Hu, and P. H. Brown. 1994. The relative tolerance of six Prunus rootstocks to boron and salinity. *Journal of American Society for Horticultural Science* 119: 1169–1175.

Fageria, N. K., V. C. Baligar, and C. A. Jones. 1991. *Growth and mineral nutrition of field crops*. New York, Basel, Hong Kong: Marcel and Dekker, Inc.

Foresight. 2011. *The future of food and farming: Challenges and choices for global sustainability* (Ed. Government Office for Science). London: Department of Business, Innovation and Skills.

Foth, H. D. and B. G. Ellis. 1997. *Soil Fertility*, 2nd ed. Boca Raton, FL: Lewis CRC Press LLC, 290 p.

Ganeshamurthy, A. N., R. D. Singh, K. S. Shashidhar, and T. R. Rupa. 2019. Fertilizer best management practices for perennial horticultural crops. *Indian Journal of Fertilisers* 15(10): 1136–1150.

Ganie, M. A. 2016. *Establishment and application of diagnosis and recommendation integrated system norms for apple (cv. Red Delicious) in district Kulgam*. PhD diss., Sher-e-Kashmir University of Agricultural Sciences and Technology of Kashmir.

Garnett, T., V. Conn, and B. N. Kaiser. 2009. Root based approaches to improving nitrogen use efficiency in plants. *Plant, Cell and Environment* 32(9): 1272–1283.

Gilroy, S. and D. L. Jones. 2000. Through form to function: Development and nutrient uptake in root hair. *Trends in Plant Science* 5(2): 56–60. ISSN 1360–1385. https://doi.org/10.1016/S1360-1385(99)01551-4.

Gojon, A., G. Krouk, F. Perrine-Walker, and E. Laugier. 2011. Nitrate transceptor (s) in plants. *Journal of Experimental Botany* 62(7): 2299–2308.

Hamid, G. A., S. D. van Gundy, and C. J. Lovatt. 1988. Phenologies of the citrus nematode and citrus roots treated with oxamyl. *Proceedings of the Sixth International Citrus Congress* 2: 993–1004.

Hansen, C. J. 1955. The influence of the rootstock on injury from excess boron in Nonpareil almond and Elberta peach. *Proceedings of American Society for Horticultural Science* 65: 128–132.

Havlin, J. L., J. D. Beaton, S. L. Tisdale, and W. L. Nelson. 2012. *Soil fertility and fertilizers: An introduction to nutrient management*, 7th ed. New Delhi: PHI Learning Private Limited, pp. 298–361.

Henry, A., N. F. Chaves, P. J. Kleinman, and J. P. Lynch. 2010. Will nutrient-efficient genotypes mine the soil? Effects of genetic differences in root architecture in common bean (Phaseolus vulgaris L.) on soil phosphorus depletion in a low-input agro-ecosystem in Central America. *Field Crops Research* 115(1): 67–78.

Holland, D. A., R. C. Little, M. Allen, and W. Dermott. 1967. Soil and leaf sampling in apple orchards. *Journal of Horticultural Science* 42(4): 403–417.

Johnson, J. 2011. 4Rs right for nutrient management. *Natural Resources Conservation Service*. Available at www.nrcs.usda.gov

Johnston, A. M. and T. W. Bruulsema. 2014. 4R nutrient stewardship for improved nutrient use efficiency. SYMPHOS 2013, 2nd International Symposium on Innovation and Technology in the Phosphate Industry. *Procedia Engineering* 83: 365–370.

Jungk, A. O. 1996. Dynamics of nutrient movement at the soil-root interface. In *Plant roots: The hidden half*, 2nd ed. (Eds. Y. Waisel et al.). New York: Marcel Dekker Inc., pp. 529–556.

Jungk, A. O. 2001. Root hairs and the acquisition of plant nutrients from soil. *Zeitschrift fur Pflanzenernährung und Bodenkunde* 164: 121–129. https://doi.org/10.1002/1522-2624(200104)164:2<121::AID-JPLN121>3.0.CO;2-6.

Kage, H. 1997. Relative contribution of mass flow and diffusion to nitrate transport towards roots. *Zeitschrift fur Pflanzenernährung und Bodenkunde* 160: 171–178.

Kant, S., Y, M. Bi, and S. J. Rothstein. 2011. Understanding plant response to nitrogen limitation for the improvement of crop nitrogen use efficiency. *Journal of experimental Botany* 62(4): 1499–1509.

Kenworthy, A. L. 1961. Interpreting the balance of nutrient elements in leaves of fruit trees. In *Plant Analysis and Fertilizer Problems* (Ed. T. Renther). Washington, DC: American Institute of Biological Sciences, pp. 28–43.

Kraiser, T., D. E. Gras, A. G. Gutiérrez, B. González, and R. A. Gutiérrez. 2011. A holistic view of nitrogen acquisition in plants. *Journal of Experimental Botany* 62(4): 1455–1466.

Kumar, S. 1991. *Comparative studies on nutrient ranges in apple leaves by orchard surveys and solution culture*. PhD diss., Dr. Y. S. Parmar University of Horticulture and Forestry, Nauni, Solan (H.P).

Lambers, H., F. S. Chapin III, and T. L. Pons. 1998. *Plant physiological ecology*, pp. 540. New York: Springer.

Lambers, H., W. S. Michael, D. C. Michael, J. P. Stuart, and E. J. Veneklaas. 2006. Root structure and functioning for efficient acquisition of phosphorus: Matching morphological and physiological traits. *Annals of Botany* 98(4): 693–713. https://doi.org/10.1093/aob/mcl114.

Lin, M., S. Yichao, S. Oskar, Y. Bin, K. Timothy, S. Egner, M. Vahid, R. Kara, B. G. Lind, V. Vincenzo, and C. Ludovico. 2019. Hydrogel-based transparent soils for root phenotyping in vivo. *Proceedings of the National Academy of Sciences* 116(22): 11063–11068.

Maas, E. V. 1984. Salt tolerance of plants. In *Handbook of plant science in agriculture* (Ed. B. R. Christie). Bota Raton: CRC Press, pp. 57–75.

Machado-Dias, J. R., P. G. S. Wadt, D. V. Perez, L.M. Silva, and C. O. Lemos. 2011. DRIS formulas for evaluation of nutritional status of Cupuacu trees. *Revista Brasileira de Cienciado Solo* 35: 2083–2091.

Marschner, H. 1995. *Mineral nutrition of higher plants*, 2nd ed. London: Academic Press.

Memon, N., K. S. Memon, and Z. Hasan. 2005. Plant analysis as a diagnostic tool for evaluating nutritional requirements of bananas. *International Journal of Agriculture and Biology* 7: 824–831.

Miguel, M. A., J. A, Postma, and J. P. Lynch. 2015. Phene synergism between root hair length and basal root growth angle for phosphorus acquisition. *Plant Physiology* 167: 1430–1439.

Miller, A. J. and M. D. Cramer. 2005. Root nitrogen acquisition and assimilation. *Plant Soil* 274: 1–36.

Mourao-Filho, F. A. A. 2005. DRIS: Concepts and applications on nutritional diagnosis in fruit crops. *Scientia Agricola* 61: 550–560.

Munson, R. D. and W. L. Nelson. 1990. Principles and practices in plant analysis. In *Soil testing and plant analysis* (Ed. R. L. Westerman). Madison: Soil Science Society of America, pp. 359–387.

Nachtigall, G. R. and A. R. Dechen. 2007. Testing and validation of DRIS for apple tree. *Scientia Agricola* 64: 288–294.

Neilsen, D. and G. H. Neilsen. 2002. Efficient use of nitrogen and water in high-density apple orchards. *Hort Technology* 12: 19–25.

Postma, J. A., C. Kuppe, M. R. Owen, N. Mellor, M. Griffiths, M. J. Bennett, J. P. Lynch, and M. Watt. 2017. OpenSimRoot: Widening the scope and application of root architectural models. *The New Phytologist* 215(3): 12741286. https://doi.org/10.1111/nph.14641

Prado, R. M. 2008. *Nutricao de plantas*. Jaboticabal, Universidae de Sao Paulo, pp. 407.

Scandellari, F., M. Ventura, P. Gioacchini, L. VittoriAntisari, and M. Tagliavini. 2010. Seasonal pattern of net nitrogen rhizodeposition from peach (*Prunus persica* (L.) Batsch) trees in soils with different textures. *Agriculture, Ecosystems and Environment* 136: 162–168.

Shear, C. B. and M. Faust. 1980. Nutritional ranges in deciduous tree fruits and nuts. *Horticulture Review* 2: 142–163.

Smith, S. E. and D. J. Read. 1997. *Mycorrhizal symbiosis*, 2nd ed. San Diego: Academic Press.

Srivastava, A. K. and S. Singh. 2008. Analysis of citrus orchard efficiency in relation to soil properties. *Journal of Plant Nutrition* 30: 2077–2090.

Thom, W. O., G. J. Schwab, L. W. Murdock, and F. J. Sikora. 2003. *Taking soil test samples*. University of Kentucky Cooperative Extension. Available at www.ca.uky.edu/agc/pubs/agr/agr16/agr16.pdf

Tukey, R. B. and A. I. Dow. 1979. Nutrient content – fruit trees in Washington. Technical Bulletin of Washington State University. *Pullman* 28: 2.

Upadhayay, S. K. and R. P. Awasthi. 1993. Leaf nutrient ranges of plus apple trees in Himachal Pradesh. *Indian Journal of Horticulture* 50(2): 97–119.

Verma, M. L. and J. K. Chauhan. 2013. Effect of integrated nutrient application on apple productivity and soil fertility in temperate zone of Himachal Pradesh. *International Journal of Farm Sciences* 3(2): 19–27.

Wadt, P. G. S. 2009. Analise foliar comoferramentapararecomendacao de adubacao. In *Congresso Brasileiro de Ciencia do Solo 32: 50*. Fortaleza, Anais Fortaleza: Sociedade Brasileira de Ciencia do Solo.

Woodbridge, C. G. 1955. The boron requirements of stone fruit trees. *Canadian Journal of Agricultural Science* 35: 282–286.

Xu, G., X. Fan, and A. J. Miller. 2012. Plant nitrogen assimilation and use efficiency. *Annual Review of Plant Biology* 63: 153–182.

Zanotelli, D., M. Rechenmacher, W. Guerra, A. Cassar, and M. Tagliavini. 2014. Seasonal uptake rate dynamics and partitioning of mineral nutrients by bourse shoots of field-grown apple trees. *European Journal of Horticultural Science* 4: 203–211.

Zavalloni, C., B. Marangoni, D. Scudellari, and M. Tagliavini. 2001. Dynamics of uptake of calcium, potassium and magnesium into apple fruit in high density planting. *Acta Horticulturae* 564: 113–122.

11 Organic Apple Production and Prospects

Rajni Rajan, M Feza Ahmad, Jatinder Singh,
Kuldeep Pandey and Ab Waheed Wani

CONTENTS

11.1   INTRODUCTION

Over the last few years there has been a steady growth in global efforts to establish more self-sustaining agricultural systems, as well as the necessity of organic fruit production (Yussefi, 2004). Organic agriculture is a long-term strategy for food production since it encompasses all agricultural systems that encourage ecologically, socially, and economically acceptable food production (IFOAM, 2000; UNEP, 2011). According to recent research from the Research Institute of Organic Agriculture (FiBL) and IFOAM Organics International, organic farmland increased 14.7% from 2014 to 2015, totaling 50.9 million hectares (Willer and Lernoud, 2017). Over the last two decades, the worldwide organics industry has expanded significantly, and consumer demand for organic products is expanding internationally with around 80 billion Euros ($92 billion USD) spent annually (IFOAM, 2018). The organic apple contributes 38.75% of total organic fruit production (FiBL, 2020).

Organic management approaches prohibit the use of pesticides and fertilizers in favor of manures, compost, biostimulants, traps, and other biological control measures (Peck et al., 2006; Jönsson, 2007). In recent years there has been a growing interest in organic fruit farming due to a clear trend of reducing the negative environmental effects of fruit crops (Baranski et al., 2017). The use of synthetic inputs such as pesticides and fertilizers are prohibited in organic farming. As a result, the system is particularly demanding in terms of disease and pest management as well as plant mineral nutrition, all of which are critical variables in excellent and balanced fruit tree development and output (Kowalczyk et al., 2012). However, organic apple production is often associated with reduced crop yields when compared to conventional production techniques owing to the lack of chemical fertilizers and pesticides (Seufert et al., 2012; Orsini et al., 2016). To close the production gap, the organic agriculture industry is continually looking for novel agroecological approaches to include in crop system management. Biostimulants are regarded as one of the most creative and promising options for increasing the sustainability and profitability of this system (Povero et al., 2016).

Organic apple production has grown dramatically over the last decade as a result of high demand, new technology, and price premiums indicating that it is beneficial for producers. Organic apple production has grown quickly over the last decade owing to growing customer demand and improved technologies for controlling chronic pests and issues (Kirby and Granatstein, 2008). As a result of the growing number of organic customers, organic agricultural land and the organic market have expanded (FiBL-IFOAM, 2014). Also, organic apple cultivation consumes less energy, produces fewer greenhouse emissions, and is more profitable than conventional apple farming.

However, in most nations organic apple production is still fairly low owing primarily to inadequate pest and disease control with organic alternatives (Jönsson, 2007) and a lack of other acceptable technology (McArtneyand Walker, 2004), which limits the success of organic apple orchards. It will take time to establish an organic apple producing infrastructure. Organic standards call for a minimum of 3 years which reflects the considerable changes that must occur before an organic system may begin to function correctly. This time will need a real commitment to understanding the many approaches involved, particularly in terms of how plants are nourished and how to influence biological processes (McCoy, 2007).

11.2   CURRENT STATUS OF ORGANIC APPLE PRODUCTION

The apple is a very important tree fruit all across the world. In 2019, the value of dessert and culinary apple output in the United Kingdom (UK) was £183 million (Department of Environment, Food & Rural Affairs, 2020a). The organic product sales in the United States topped $43 billion in 2015 with produce accounting for 13% of total sales (Organic Trade Association, 2016). However, under current USDA-NOP rules, cultivating apples organically in the eastern United States can be difficult due to a lack of adequate biological, cultural, or chemical controls for the enormous insect, disease, and weed burden faced in that region (Cromwell et al., 2011; Williams et al., 2016). Furthermore, crop load management options in organic agriculture are limited since the USDA-NOP restricts the use of synthetically produced plant growth regulators, which are employed for this purpose in conventional systems (U.S. Department of Agriculture, 2016).

11.3 MARKET OUTLOOK FOR ORGANIC APPLES

The rising demand for organically grown agricultural and horticultural commodities has piqued the interest of conventional farmers who view the organic market as a viable outlet for their goods. This interest in organics is also evident in the apple sector; however growing traditional apple cultivars using organic systems is challenging because of pest and disease challenges. Organic apples are one alternative for producers seeking greater returns, and viable production strategies must be created to meet customer demand for organic products (Anon, 2008). It is also evident in the apple industry where a perceived price premium for organically grown apples entices apple producers to consider organic farming. As a result of the growing number of organic customers, organic agricultural land and the organic market have expanded (FiBL-IFOAM, 2014). Organic apple production is expanding globally and is expected to continue increasing as prominent food merchants in the United States and Europe offer more organic apples to suit customer demand (Granatstein and Kirby, 2007). Australia and New Zealand also have important markets for organic apples. There is potential to supply organic apples in addition to the existing conventional apple export sector. Destinations with high organic market growth are promising in the United Kingdom, the United States, and Europe. However, these prospects are vulnerable to competitive pressures from organic apple supplies from New Zealand, Chile, Argentina, and South Africa (Anon, 2008).

11.4 ORGANIC APPLE PRODUCTION STRATEGIES AND METHODS

Organic fruit production is the development of fruit crops in which the majority of the inputs are from contemporary farming but do not include synthetic insecticides, pesticides, or fertilizers. Crop residues, crop rotations, animal manures, off-farm organic wastes, green manures (biomass recycling), and the IMP method, as well as biological pest management, are all considered in this production. Aside from this, many other factors contribute to the maintenance of tilth and soil productivity, as well as the provision of plant nutrients (biofertilizers) and weed control. However, in this technique, the ultimate aim is ecological equilibrium (environment), which is not affected. The overall goal is to achieve crop quality and yields that are sustainable. Andrews et al. (2001) concluded that organically grown apples are firmer and sweeter. They found greater antioxidant activity in such apples, although Peck et al. (2006) found no differences in acid and sugar levels. Various apple varieties may now be grown organically and sold in the market as organic fruit. Keeping these important considerations in mind, organic apple production should adhere to the following guidelines given in Figure 11.1.

11.4.1 SITE AND SOIL SELECTION FOR ORGANIC APPLE FARMING

Like all other crops, apples also respond to good soil with plant health and productivity. Apple fruit can be cultivated in most soil types. Hence selection of the site and soil is a very critical step in the cultivation of organic apples. However, enriched with organic matter, loamy natured soil with well-drainage features is supposed to be the best for apple plantations. Spray drift and chemical contamination can be avoided at sites remote from normal manufacturing regions. Windbreaks may be necessary for windy locations not just to reduce spray drift but also to protect crops from wind influence and damage. To avoid contamination, buffer zones between conventional and organic areas must be established.

Several soil profiles should be studied before planting the orchard to identify possible root depth and whether rooting constraints may be rectified. Land that has previously been utilized for crops other than apples is the best place to start an organic orchard. Apple replant disease can cause poor tree development if apple trees are planted in an old apple orchard. This soil-borne disease that damages the tree's roots can limit vegetative growth and orchard yield. This disease tends to be more prevalent in lighter, sandy soils. It is better to do a soil test to know the status of the soil

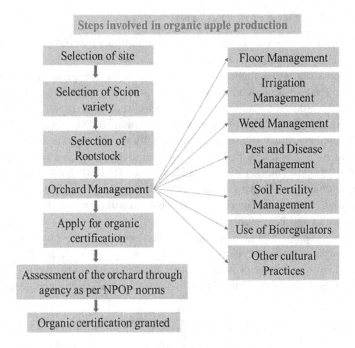

FIGURE 11.1   Steps involved in organic apple production.

concerning supply and deficiencies of nutrients. It is also necessary to obtain higher fruit production. Adjustments or alterations to the subsoil layers can be made if needed. Organic matter is responsible for feeding lots of microorganisms and providing them suitable medium to multiply and develop. Such soils are of prime importance in organic cultivation. These are correspondingly able to stimulate the action of microorganisms.

11.4.2   SITE PREPARATION/MODIFICATION

The gardener has no choice when it comes to soil selection. However, soil qualities may be altered, which aids in reducing weed intensity and disease consequences. Soils with a good balance of nutritional components and a lot of organic matter are thought to be ideal for apple farming. The following strategies can be used to modify the site/soil characteristics.

1. Before planting, several strategies such as green manuring or cover cropping can be used to boost the nutritional value of the soil. It also involves using compost, manures, and organic wastes.
2. The pH range for tree fruits is 5.5 to 6.5;6.5 is optimal. Most nutrients are easily available for absorption by the roots at this level. Increasing the pH of the soil to approximately 6.5 will also boost the activity of soil microorganisms resulting in the quicker breakdown of additional organic matter and a higher availability of nutrients to plants.
3. Plan the planting such that as many new trees as possible are planted between the existing tree rows and/or tree sites. Replace the soil that was taken from the planting hole with fresh topsoil from the unplanted land.
4. Use a more vigorous rootstock but make sure the tree spacing is appropriate for this rootstock when the trees emerge from the poor soil and restart regular growth after around 5 years.

11.4.3 Soil Preparation

Plant nutrition and health are critical in averting significant disease and pest issues. A suitable mix of macro and micronutrients is required to achieve and maintain healthy tree development. As a result, soil study of the planting location should be performed 1 to 2 years before planting. Once the trees begin to bear, soil and leaf analysis should be performed on a regular basis to ensure that soil and leaf nutrient levels remain optimal. Soil chemical or heavy metal residues must not exceed organic norms. Possible sources of unwanted pollution or excessive nutrients in irrigation water should be considered (Anon, 2008).

11.4.4 Soil Organic Matter

Soil organic matter is made up of humus, which is resilient to additional fast decomposition, and organic compounds, which decompose rather quickly. Organic matter levels in orchard soil should be greater than 3%. Growing green manures or spreading decomposed organic matter or manure prior to planting can improve soil organic matter content.

11.4.5 Plant Selection

Plants in the nursery should be healthy and not more than 1 or 2 years old. The should have lateral branches. Trees should be dormant, free of frost or another injury, and have never been exposed to drying conditions between digging and planting. Trees that have broken dormancy may not suffer too much if they are otherwise healthy until warm, dry weather arrives soon after planting.

11.4.6 Selection of Apple Cultivar

In organic gardening, selecting a cultivar to cultivate is of the utmost importance and thought. Cultivar selection should be based on local climate, soil, and market demand, as well as transport and storage capacity. Other characteristics like as size, scent, shape, color, flavor, and flesh texture are also essential. Certain other aspects are also considered, such as the tree's behavior and its requirements. Apple trees come in a variety of sizes and forms. In general, trees may grow up to 30–40 feet tall and can take a long time (6 years) to start bearing. However, smaller trees are recommended. This also necessitates the adoption of certain rootstocks. Furthermore, cultural activities such as pruning and training techniques influence tree size. Choose cultivars with horizontal branching patterns since they are easier to prune and manage.

11.4.7 Application of Organic Nutrients and Fertilizers

To cultivate apple crop organically, we can substitute chemical fertilizer with farm yard manure (FYM), compost, yard waste, etc. Seaweed extracts/fish emulsion or compost tea may be used as a foliar application, but it should be preferably applied during the spring season (when new growth and flowering) occurs. Late application of these fertilizers can interfere with the hardening-off process. It is important to note that over-fertilization can lead to excess and succulent growth and thus subject to more sucking pests and diseases.

Due to the perennial nature of the crop (apple), the nutrient supply approach should be long-lasting and holistic. Such an approach can meet nutrient requirements by taking into account both yield and quality. Fertilization has a significant effect on the fruit color, flavor, and shelf life of apples. Above all, fertilization includes the application of the macronutrients and micronutrients (trace elements) but in judicious amounts. Imbalanced fertilization can lead to negative impacts or deficiencies of a particular element. In apple plantations, it is difficult to rectify nutritional imbalances up to a certain extent.

Early in the year, the rapid growth of new leaves, flowers, and immature fruit may induce nitrogen stress on the tree, which is characterized by a light green color of the foliage. Foliar treatment of seaweed extracts has also been recommended to reduce nutrient stress. The boron content in apple trees is crucial for the establishment of good fruit set. Boron levels in soil can be increased by applying borax to the soil. Rock phosphate is an excellent supplier of phosphorus especially for young trees. Phosphorus is found in roughly 26% of bone meal. Potassium sulfate, a good potassium source, contains 55% potassium. Wood ashes include around 6% potassium, 23% calcium, and traces of other elements (Braun and Craig, 2008).

Organic fertilizers are applied three times during the growing season. The first application is in early spring, before flowering, around mid-April, especially in temperate areas. The second is 1 month later and after flowering is over. The third application is at the end of June.

11.4.8 WEED MANAGEMENT FOR ORGANIC FARMING

For higher and quality production, it is necessary to keep the apple orchard clean and weed free. It also involves techniques that help to avoid organic matter loss. Broadly spreading weeds are a major challenge in organic production since there is no application of chemical weedicide. They are responsible for the reduction in fruit size and yield. Effective weed control involves components like cover crops, soil solarization, and mulches and should be implemented. Weeds can hinder the growth of young apple trees by consuming nutrients, water, and light. For the first 3 years, a weed-free strip 1.2 m wide along the tree row or as rings around trees is preferable. Hand or mechanical cultures can be used to do this, but care must be given to prevent harming superficial roots. Weeds also influence apple tree development, especially in the growth and bearing phase. In dwarf and semi-dwarf cultivars of apples, weeds must be suppressed from the tree trunks ranging 2–4 feet in all directions. If any ground cover is cultivated, minimum competition should be posed to the plantation, especially during the autumn and rainy season. Hence, there is a need in organic fruit production for a more sustainable strategy to weed control that is compatible with organic protocols that emphasizes cautious usage rather than suppression (Sooby et al., 2007; Granatstein and Sánchez, 2009).

Effective weed control may include following:

1. Manual weeding and mechanical cultivation
2. Mowing
3. Use as a pasture or for animal grazing
4. Using biodegradable, plastic, or other synthetic mulches.

11.4.9 MULCHING

Time and type are important aspects for effective mulching. If mulch is applied as pre-planting, its layer must not be too thick to permit roots of seedlings to enter into soil surface. Mulching may also be applied to standing crops in orchards. But it is most useful when established between the rows, straight around the plant stem or evenly spread in the orchard. Mulching may be removed in the early season so that soil can be warm but subjected to crop requirements. It is established that mulching should be placed 6 inches away from the stem during dormancy to avoid vole/termite damage. If the layer (mulch) is not very thick, then seeds or seedlings of apples could be directly propagated or planted in a sandwiched manner between the mulching material.

Organic soil management approaches in Washington State that included the addition of bark mulch and composted poultry manure resulted in reduced soil bulk densities and enhanced biological soil qualities (Glover et al., 2000). A single layer of bark mulch decreased the requirement for pesticides to manage weeds in the tree row for several years. The mulch, on the other hand, was expensive to spread and would most certainly need to be redone every 3 to 5 years (Peck et al., 2010). Mulches are preferable around trees after the third year. Straw mulch has been tested with promising results. To reduce the likelihood of apple scab development and increase the color development of the fruit, it is important to choose a cover crop system that supports the apple trees with adequate N, resulting in a leaf dry matter content of around 2.1%. Mulching cover increased the soil water content, but no other effects were observed. Mulching grass increased leaf N, shoot growth, and yield while the proportion of well-colored fruit decreased (Kuhn et al., 2009).

11.4.10 Fruit Thinning

Fruit thinning may be necessary for some years to get optimum fruit size and quality and to prevent biannual bearing. Extra flower buds are plucked by hand in Europe, particularly with spur type varieties, just as they bloom. Spur pruning has been demonstrated to be beneficial in regulating crop load in Atlantic Canada. Manual fruit thinning is time-consuming, although it might be useful in small-scale cultural operations.

Thinning should ideally take place when the apples are 10 to 13 mm in diameter or 7 to 14 days following bloom. AAFC's blossom thinning studies show that lime sulfur sprayed during bloom has modest thinning action. In trials conducted in the United States, the combination of fish oil with lime sulfur improved the thinning effect. Several laboratories across North America and Europe are conducting research on experimental mechanical thinning. With some success, high pressure water jets from cannons with four nozzles have been utilized.

Recently, various thinning studies with potentially organically suitable chemicals were carried out. Lime sulfur (calcium polysulfide), a chemical allowed in organic apple production under EU law, was discovered to be effective as an apple thinning agent. Emulsions of vegetable oils might likewise be used for this purpose (ZhiGuo and YouSheng, 2001). Crop load regulation is one of the more difficult management concerns in organic apple production. Lime sulfur sprays have been used successfully as blossom thinners (Stopar, 2004; Kelderer, 2004).

11.4.11 Summer Pruning

Summer pruning is essential for enhancing fruit color, quality, and storage life. It entails removing some of the current year's growth to open up the canopy and allow for improved light penetration and crowding reduction. Summer pruning can begin after terminal bud growth is seen, which is normally around early to mid-August (Braun and Craig, 2008).

11.4.12 Orchard Floor Management

Orchard floor management is a critical component of fruit production with even greater relevance under organic production (Bloksma, 2000; Tinsley, 2000). The primary goals of orchard floor management are to maintain and/or improve soil fertility, physical and biological qualities, nutrients, and water supply. This mechanism alters the soil's microbial makeup (Yao et al., 2005). Growing grass to make mulch is the foundation of organic orchard floor management.

Typically, preferred species are established that contribute various system functions including generating biomass/organic matter (roots and tops) that feeds the soil biological activity as the foundation for sustaining soil conditions, nutrient availability, and organic matter cycling; building soil structure and improving resilience to soil compaction and erosion; providing competition to suppress problem weeds; facilitating rapid decomposition of diseased tissue, etc. This system contributes to various functions such as producing biomass/organic matter that feeds soil biological activity, nutrient availability, organic matter cycling, building soil structure and refining resilience to soil compaction and erosion, suppressing problem weeds, facilitating the rapid decomposition of diseased tissue, and so on.

11.4.13   BIO STIMULANTS

The enhanced phenolic component content, total anthocyanin, and antioxidant capability of treated apples demonstrated that biostimulants had a favorable influence on fruit functional features. Biostimulants stimulate the secondary metabolism of treated plants resulting in improved fruit quality, attractiveness, and nutritional value. Research on the effect of bio stimulants on organic apple production discovered that alfalfa protein hydrolysate, seaweed extracts, and B-group vitamins increased the final red hue of apple 'Jonathan,' hence increasing their market potential. The study also found that zinc-containing biostimulants are helpful in lowering the occurrence of physiological diseases in cold-stored apples, lending credence to the hypothesis that this element has a favorable function in the construction and resilience of cell walls and membranes at the fruit level. Given the existing lack of an efficient methods for organic fruit postharvest management, this result might have a substantial impact on the strategies now employed for organic apple conservation and marketing.

11.4.14   BIOLOGICAL CONTROL AGENTS

Most of the time it is hard to observe any beneficial effects of plant growth promoting rhizobacteria (PGPR) and arbuscular mycorrhizal fungi (AMF) under organic farming because some organic amendments already provide several microbial communities or may promote indigenous microbial communities. A study confirmed that the strategy of artificial tree inoculation with AMF + PGPR inoculation should be used mainly for improving the nutritional status of trees in terms of nitrogen and magnesium but only on soils where a deficit of potassium is not observed (Przybyłko et al., 2021). On the other hand, studies reported that low-input practices used in organic management systems and lower nutrient supplies can improve activities of soil biota in which essential components are AMF (Sas-Paszt and Gluszek, 2007; Parniske, 2008).

11.4.15   DISEASE AND PEST MANAGEMENT

Managing diseases organically is often more sustainable in the long run since they have fewer negative environmental and human health implications. Scab, powdery mildew, apple replant disease, canker, and brown rot/blossom wilt are the most common diseases in organic apple farming. The suggested organic management strategies vary according to the disease (Shuttleworth, 2021). A warmer world and more unpredictable weather patterns, on the other hand, may favor disease range extensions while raising plant stress and vulnerability (Shuttleworth, 2021). Hence, growing conventional apple types in organic systems is challenging due in part to pest and disease challenges.

11.4.15.1   Apple Scab

Apple scab is the most commercially significant apple disease in the world. The infection mechanism and disease cycle are well understood, particularly the participation of the over-wintered sexual stages on dead leaves and a secondary infection of fruit and foliage at the asexual stages (Berrie, 2019). Mycelia and conidia can survive the winter in the host tree's buds and twig and produce new inoculum the following spring (Passey et al., 2017). It is a serious apple disease in Australia and costs the industry up to $10 million per year in control measures and lost productivity. Several varieties show promise as high-quality apples that can be cultivated in both organic and conventional methods (Zeppa et al., 2002).

11.4.15.2   Apple Replant Disease (ARD)

ARD occurs in young plants when they are planted in a previously established apple orchard (Mao and Wang, 2019; Winkelmann et al., 2019). Hewavitharana et al. (2020) revealed that 4-year or longer crop rotations to a non-woody crop reduces pressure from ARD-causing organisms. Pan et al. (2017) observed that a short-term rotation of *Allium fstulosum* and Trichoderma was helpful in lowering ARD, whereas Mao and Wang (2019) discovered that a mixed cropping system with spring onions resulted in reduced populations of ARD. Throughout the propagation process, it is crucial to develop ARD-free rootstocks. ARD therapies have also been described as brassicaceous additives, charcoal, seaweed, and fermented organic materials.

11.4.15.3 Others

The other diseases are less concerning for a number of reasons. Apple trees may withstand modest amounts of powdery mildew without suffering major yield loss. However, if typical temperatures rise, powdery mildew might become a significant illness. Mildew forecasts the quantity of sporulating mildew in 3–4 days. The mildew risk index developed can be used to help in decision-making and mildew control. Cankers develop slowly which provides time to address the issue before losses occur. Fire blight hasn't been a huge concern in the past, but it's spreading swiftly and offers a real hazard as temperatures rise.

11.4.15.4 Brown Rot

Postharvest losses, particularly due to Brown rot induced by *M. fructigena*, might reach up to 22% (Berrie and Holb, 2014; Holb, 2019). The prevalence of brown rot in organic orchards is much greater due to the higher risk of insect damage (Holb and Scherm, 2007). Late-ripening varieties have more severe brown rot than early-ripening varieties (Holb and Scherm, 2007, 2008). A disease warning system for *M. fructigena* has been developed in organic apple cultivation (Holb, 2009; Holb et al., 2011).

11.4.15.5 Control Measures for Diseases

Host resistance, rootstock and scion cultivar selection, tree planting location, biological control agents, proper soil amendments, cultural management, postharvest treatments, disease modeling, and forecasting are all examples of control measures. The difficulty with organic approaches is that the outcomes are frequently more varied and unpredictable than when conventional chemicals are used. Baker et al. (2020) explained inclining implementation of biological approaches, including biostimulants, biopesticides, pheromones, and biological control is a common priority for sustainable agriculture practitioners and leaders, including those working in organic cultivation and integrated pest management. Growers should also have a solid grasp of disease ecology in order to employ organic control strategies more effectively. Early warning systems based on disease forecasting models that offer timely information on apple scab infection seasons have the potential to minimize the usage of fungicides (Hindorf et al., 2000; Jamar et al., 2008b). Furthermore, the following approaches are very significant for successful management.

Sprays

Some serious diseases of apples like fire blight and apple scabs can be managed up to a certain extent with the application of Bordeaux mixture or liquid copper soap. Sandskär (2003) also concluded apple scab is a major fungal disease caused by *Venturia inaequalis*. These sprays may not prove harmful for plantations if applied at a judicious amount and at right time during growth and development.

Use of solar energy

It should be practiced at pre-planting time, i.e., before cultivation. Before the actual transplanting of seedlings, the orchard soil should be exposed to sunlight to destroy microorganisms (soil-borne pathogens). After a certain period of time, the soil may be covered by a polyethylene sheet for better results.

11.4.15.6 Organic Treatment for Certain Insects

11.4.15.6.1   Apple Coddling Moth

This insect is responsible for massive infestation of apple plantations. The larvae of the coddling moth bore into flesh or cores and cause damage. Cultural methods including foliar application of products that contain kaolin clay and paraffinic oil are also recommended for control. For cultural control, sort, collect, and dispose of all dropped fruits and remove all infested apples by hand before the larvae develop as adults.

11.4.15.6.2   Apple Maggots

To control apple maggots by organic methods, hang 1–2 red, spherical sticky traps soaked with an attractant formulation. Some other techniques also provide good results like the use of pyrethrum products, diatomaceous earth, and sometimes rotenone.

The application of *Bacillus thuringiensis* (Bt.) also proved an effective solution as it is a safe microbial pesticide that is able to target caterpillar pests. A spray of Bt. Spp. at the apple's flowering stage is able to suppress winter moth caterpillars. In terms of insect pest control, research shows that pest-free apples may be grown organically in New York, but organic producers would likely need at least a 400% sales premium over conventional growers due to the higher costs and poorer yield associated with organic pest management (Rosenberger and Jentsch, 2006).

11.5 ORGANIC APPLE HARVESTING

It has been proven that apples that have been naturally ripened contain more vitamin C than apples that have been artificially ripened. Apples should be collected during the color-changing stage, i.e., pink or mature green. If the fruit is harvested at its peak maturity, the quality of the product will be higher, and the yield will be higher since the loss will be lower. Manual picking is preferred when the food is to be processed, whereas machinery harvesting is preferred when the fruit is to be consumed fresh. Storing the apple at 15–25°C increases the fruit's shelf life, which is 1–2 weeks, after which the fruit's outer peel wrinkles and becomes mushy.

11.6 POSTHARVEST MANAGEMENT

All postharvest operations must follow organic standards. The primary objective is to avoid contamination with prohibited substances and to ensure that organic and conventional goods are kept separate. Risk assessment HACCP (Hazard Analysis Critical Control Point)-based quality assurance methods are effective for designing protocols and assessing organic-compliant processes. In most cases, companies using existing HACCP-based quality assurance (QA) systems discover that very minor changes are required to fulfill organic criteria.

11.7 ORGANIC STANDARDS, REGULATIONS AND CERTIFICATION

Several countries set organic farming laws to offer a defined structure for the production of organic goods. For example, the European Union (EU) complies with a goal to meet customer demand for reliable organic products while also creating a fair environment for growers, distributors, and marketers. All organic food sold in the EU must comply with the EU organic regulation. The Organic Farmers & Growers CIC, the Organic Food Federation, the Soil Association Certification Ltd., the Biodynamic Association Certification, the Quality Welsh Food Certification Ltd, and the Organic Farmers and Growers (OF&G; Scotland) are all approved organic bodies in the UK (Shuttleworth, 2021). In India, the National Programme for Organic Production (NPOP) manages all certification processes. Figure 11.2 and Figure 11.3 show the general process of organic certification and field inspection in India.

FIGURE 11.2   Process of organic certification in India.

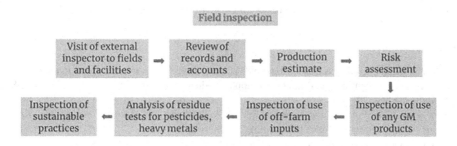

FIGURE 11.3   Field inspection for organic certification in India.

11.8   CHALLENGES IN ORGANIC APPLE PRODUCTION

The fact that a system is organic does not imply that it will be viable in the long run. A productive orchard is required for a sustainable orchard. According to a few writers, organic agricultural methods are less efficient, entail greater health risks, and yield less than traditional or integrated systems (McArtney and Walker, 2004). According to Reganold et al. (2001), fruit price premiums of 12–14% are required for the organic system to reach the conventional system's break-even point, reflecting the higher costs associated with organic cultivation. Organic apple farming methods in New Zealand sometimes yield less than conventional or integrated systems, however, market premiums for organic apples have compensated for decreased output (McArtney and Walker, 2004). Lower and more variable organic apple orchard yields are most likely due to inadequate crop load management, higher insect and weed pressures, and nutrient deficits (Peck et al., 2006; Jönsson, 2007).

11.9   BENEFITS OF THE ORGANIC APPLE INDUSTRY

In comparison to conventional systems, organic systems generate somewhat lower apple yields with higher BC ratios, less greenhouse gas emission ratios, and efficient carbon usages. A study in Turkey reported that organic apple production can reduce nonrenewable energy inputs and chemical fertilizers while also protecting the agroecological balance (Ekinci et al., 2020). Organic agriculture is thought to accomplish more sustainable practices by avoiding environmental consequences such as biodiversity loss (Tscharntke et al., 2005). There have been several instances of good correlations between organic farming and biodiversity, including natural enemy abundance (Bengtsson et al., 2005; Winqvist et al., 2011).

11.10   CONCLUSION

Organic cultivation practices are becoming more popular in long-term sustainability of fruit growing, particularly in the apple industry. Several synthetic chemicals are banned or their restricted use has been announced by almost all regulatory bodies. Moreover, more often consumers want food free from pesticides. It will be much better if such management solutions are adopted during cultivation which have nonsignificant or no effect toward the environment and ecology. Organic farming protects biodiversity and soil fertility, eliminates soil erosion, and lowers hazardous runoff pollution of the water supply. Organic food tastes better and is less likely to be tainted with pesticides, according to anecdotal evidence and common sense. However, when it comes to the claim that organic food is more nutritious than nonorganic food, most experts believe that additional study is needed. Fortunately, these studies are becoming more common, and several recent research studies have concluded that organic food may have higher quantities of certain minerals and antioxidants, as well as lower levels of nitrates and pesticides. An ideal organic apple producing infrastructure with organic certification reflects the considerable changes and reform of the global organic market of apples.

REFERENCES

Andrews, P., J. Fellman, J. Glover, and J. Reganold. 2001. Soil and plant mineral Nutrition and fruit quality under organic, conventional and integrated apple production systems in Washington State, USA. *Acta Horticulturae* 564: 291–298.

Anonymous. 2008. *Growing Organic Apples – World Class Production Systems for New Australian Apple Varieties*. Published by Union Offset Printing, Canberra.

Baker, B. P., T. A. Green, and A. J. Loker. 2020. Biological control and integrated pest management in organic and conventional systems. *Biological Control* 140: 104095.

Baranski, M., L. Rempelos, P.O. Iversen, and C. Leifert. 2017. Effects of organic food consumption on human health; the jury is still out! *Food and Nutrition Re*search 61: 1287333.

Bengtsson, J., J. Ahnström, and A.C. Weibull. 2005. The effects of organic agriculture on biodiversity and abundance: A meta-analysis. *Journal of Applied Ecol*ogy 42(2):261–269.

Berrie, A. 2019. Disease monitoring and decision making in integrated fruit disease management. In: *Integrated Management of Diseases and Insect Pests of Tree Fruit*. Burleigh Dodds Science Publishing, Cambridge, pp. 201–232.

Berrie, A.M. and I. Holb. 2014. Brown rot diseases. In *Compendium of Apple and Pear Diseases and Pests*, 2nd ed. T. B. Sutton, H. S. Aldwinckle, A. M. Angelo, and J. F. Walgenbach, eds. American Phytopathological Society, St. Paul, MN, pp. 43–45.

Bloksma, J. 2000. Soil management in organic fruit growing. In: *Conference 'Organic Fruit Opportunities and Challenges'*, Ashford, Great Britain.

Braun, G. and B. Craig. 2008. *Organic Apple Production Guide for Atlantic Canada,* 3rd ed. Agriculture and Agri-Food Canada, Publication, Canada, p. 24.

Cromwell, M.L., L.P. Berkett, and H. M. Darby. 2011. Alternative organic fungicides for apple scab management and their non-target effects. *HortScience* 46:1254–1259.

Department of Environment, Food & Rural Affairs 2019. 2020a. Agriculture in the United Kingdom, pp. 1–157. https://assets.publishing.service.gov.uk/government/uploads/system/uploads/attachment_data/file/950618/AUK-2019-07jan21.pdf

Ekinci, K., V. Demircan, A. Atasay, D. Karamursel, and D. Sarica. 2020. Energy, economic and environmental analysis of organic and conventional apple production in Turkey. *Erwerbs-Obstbau* 62(1):1–12.

FiBL. 2020. https://statistics.fibl.org/visualisation.html

FiBL-IFOAM. 2014. *The World of Organic Agriculture*, FiBL, Frick, Switzerland; IFOAM, Bonn, Germany.

Glover, J. D., J.P. Reganold, and P. K. Andrews. 2000. Systematic method for rating soil quality of conventional, organic, and integrated apple orchards in Washington state. *Agriculture, Ecosystems and Environment* 80:29–45.

Granatstein, D. and E. Kirby. 2007. The changing face of organic tree fruit production. *Acta Horticulturae* 737:155–162.

Granatstein, D. and E. Sánchez. 2009. Research knowledge and needs for orchard floor management in organic tree fruit systems. *International Journal of Fruit Sci*ence 9:257–281.

Hewavitharana, S., T. DuPont and M. Mazzola. 2020. *Apple Replant Disease; WSU Tree Fruit IPM Strategies.* Washington State University. http://treefruit.wsu.edu/crop-protection/disease-management/apple-replant-disease/.

Hindorf, H., I.F. Rövekamp, and K. Henseler. 2000. Decision aids for apple scab warning services (*Venturia inaequalis*) in Germany. *OEPP/EPPO Bulletin* 30:59–64.

Holb, I. 2019. Brown rot: Causes, detection and control of Monilinia spp. affecting tree fruit. In *Integrated Management of Diseases and Insect Pests of Tree Fruit.* Burleigh Dodds Science Publishing, pp. 103–150.

Holb, I.J. 2009. Fungal disease management in environmentally friendly apple production–a review. In *Climate Change, Intercropping, Pest Control and Beneficial Microorganisms*, pp. 219–292.

Holb, I.J., B. Balla, F. Abonyi, *et al.* 2011. Development and evaluation of a model for management of brown rot in organic apple orchards. *European Journal of Plant Pathology* 129:469–483.

Holb, I.J. and H. Scherm. 2007. Temporal dynamics of brown rot in different apple management systems and importance of dropped fruit for disease development. *Phytopathology* 97:1104–1111.

Holb, I.J. and H. Scherm. 2008. Quantitative relationships between different injury factors and development of brown rot caused by Moniliniafructigena in integrated and organic apple orchards. *Phytopathology* 98(1):79–86.

IFOAM. 2000. *IFOAM Basic Standards*, International Federation of Organic Movements, Tholey-Theley, Germany.

IFOAM. 2018. *Consolidated Annual Report of IFOAM-Organics International*, IFOAM: Bonn, Germany.

Jamar, L., B. Lefrancq, C. Fassotte, and M. Lateur. 2008b. A "during-infection" spray strategy using sulphur compounds, copper, silicon and a new formulation of potassium bicarbonate for primary scab control in organic apple production. *European Journal of Plant Pathology* 122: 481–492.

Jönsson, Å. H. 2007. *Organic Apple Production in Sweden: Cultivation and Cultivars.* 33p. Ph.D. diss., Swedish Univ. of Agricultural Sciences, Balsgård.

Kelderer, M. 2004. Organic apple growing: Legal roles, production statistics, special feature about the cultivation. *Compact Fruit Tree* 37: 82–85.

Kirby, E. and Granatstein, D. 2008. *Status of Organic Tree Fruit in Washington State, 2008.* Online. Center for Sustaining Agric and Natural Resources, Washington State Univ., Wenatchee, WA.

Kowalczyk, W., D. Wrona, and S. Przybyłko. 2012. Content of minerals in soil, apple tree leaves and fruits depending on nitrogen fertilization. *Journal of Elementology* 22: 67–77.

Kuhn, B. F. and H. L. Pedersen. 2009. Cover crops and mulching effects on yield and fruit quality in unsprayed organic apple production. *European Journal of Horticultural Sciences* 74:247–253.

Mao, Z. and Y. Wang. 2019. Apple replant disease: causes and management. In: *Integrated Management of Diseases and Insect Pests of Tree Fruit.* Burleigh Dodds Science Publishing, Cambridge, pp. 39–49.

McArtney, S. J. and J. T. S. Walker. 2004. Current situation and future challenges facing the production and marketing of organic fruit in Oceania. *Acta Horticulturae* 638:387–396.

McCoy, S. 2007. Organic apples a production guide. Department of primary industries and regional development, Western Australia, Perth. *Bulletin* 4715.

Organic Trade Association. 2016. *Organic Industry Survey.* Organic Trade Association, Washington, DC. https://ota.com/what-ota-does/market-analysis/organic-industry-survey/organic-industry-survey?oprtid=012G0000001BAsuIAG&caid=701G0000000yqzN.

Orsini, F., A. Maggio, Y. Rouphael, and S. De Pascale. 2016. Physiological quality of organically grown vegetables. *Scientia Horticulturae* 208:131–139.

Pan, F., L. Xiang, S. Wang, et al. 2017. Effects of short-term rotation and Trichoderma application on the soil environment and physiological characteristics of *Malus hupehensis* Rehd. seedlings under replant conditions. *Acta Ecologica Sinica* 37(5):315–321.

Parniske, M. 2008. Arbuscular mycorrhiza: The mother of plant root endosymbiosis. *Nature Reviews Microbiology* 6:763–775.

Passey, T. A. J., J. D. Robinson, M. W. Shaw, and X. M. Xu. 2017. The relative importance of conidia and ascospores as primary inoculum of Venturia inaequalis in a southeast England orchard. *Plant Pathology* 66(9):1445–1451.

Peck, G. M., P.K. Andrews, J.P. Reganol, and J. K. Fellman. 2006. Apple orchard productivity and fruit quality under organic, conventional, and integrated management. *HortScience* 41:99–107.

Peck, G. M., I. A. Merwin, M. G. Brown, and A. M. Agnello. 2010. Integrated and organic fruit production systems for 'Liberty' apple in the Northeast USA: A systems-based evaluation. *Hort Science* 45:1038–1048.

Povero, G., J. F. Mejia, D. Di Tommaso, A. Piaggesi, and P. Warrior. 2016. A systematic approach to discover and characterize natural plant biostimulants. *Frontiers in Plant Science* 7:435.

Przybyłko, S., W. Kowalczyk, and D. Wrona. 2021. The effect of mycorrhizal fungi and PGPR on tree nutritional status and growth in organic apple production. *Agronomy* 11:1402.

Reganold, J.P., J.D. Glover, P.K. Andrews, and H.R. Hinman. 2001. Sustainability of three apple production systems. *Nature* 6831:926–930.

Rosenberger, D. and P. Jentsch. 2006. Evaluation of organic pest controls and fruit thinning on multiple apple cultivars. *New York State IPM Program Reports* 1–16.

Sandskär, B. 2003. *Apple Scab (Venturia Inaequalis) and Pests in Organic Orchards*. Ph.D dissertation, Agraria 378, SLU, Sweden. ISBN 91-576-6416-1.

Sas-Paszt, L. and S. Głuszek. 2007. Rolakorzeniorazryzosfery we wzro 'scieiplonowaniuro' slinsadowniczych. *Post,ep. NaukRoln* 6:27–39.

Seufert, V., N. Ramankutty, and J. A. Foley. 2012. Comparing the yields of organic and conventional agriculture. *Nature* 485:229–232.

Shuttleworth, L.A. 2021. Alternative disease management strategies for organic apple production in the United Kingdom. *CABI Agriculture and Bioscience* 2:34.

Sooby, J., J. Landeck, and M. Lipson. 2007 *National Organic Research Agenda*. http://ofrf.org/publications/pubs/nora2007. pdf.

Stopar, M. 2004. Thinning of flowers/fruitlets in organic apple production. *Journal of Fruit and Ornamental Plant Research* 12:77–83.

Tinsley, A. 2000. Nutrition of trees in organic systems. In *Conference 'Organic Fruit Opportunities and Challenges'*, Ashford, Great Britain.

Tscharntke, T., A. M. Klein, A. Kruess, I. Steffan-Dewenter, and C. Thies. 2005. Landscape perspectives on agricultural intensification and biodiversity-ecosystem service management. *Ecology Letters* 8(8):857–874.

UNEP, Organic Agriculture. 2011. *A Step Towards the Green Economy in the Eastern Europe, Caucasus and Central Asia Region: Case Studies from Armenia, Moldova and Ukraine*. United Nations Environmental Programme, Geneva, Switzerland.

U.S. Department of Agriculture. 2016. *Organic Regulations*. U.S. Dept. Agr. Natl. Organic Program, Washington, DC.

Willer, H. and J. Lernoud. 2017. *The World of Organic Agriculture. Statistics and Emerging Trends*. Research Institute of Organic Agriculture (FiBL) & IFOAM-Organics International, Frick and Bonn, Switzerland.

Williams, M. A., J. G. Strang, and R. T. Bessin, et al. 2016. An assessment of organic apple production in Kentucky. *Horticulture Technology* 25:154–161.

Winkelmann, T., K. Smalla, W. Amelung, et al. 2019. Apple replant disease: Causes and mitigation strategies. *Current Issues of Molecular Biology*: 89–106.

Winqvist, C., J. Bengtsson, J. Aavik, et al. 2011. Mixed effects of organic farming and landscape complexity on farmland biodiversity and biological control potential across European. *Journal of Applied Ecology* 48(3):570–579.

Yao, S., I.A. Merwin, G. W. Bird, G. S. Abawi, and J. E. Thies. 2005. Orchard floor management practices that maintain vegetative or biomass groundcover stimulate soil microbial activity and alter soil microbial community composition. *Plant Soil* 271:377–389.

Yussefi, M. 2004. Development and state of organic agriculture worldwide. In: IFOAM, Willer H., Yussefi M. (eds.), *The World of Organic Agriculture. Statistics and Emerging Trend*. Verlagsservice Wilfried Niederland, Koeningstein, p. 167.

Zeppa, A., S. Dullahide, A. McWaters, and S. Middleton. 2002. Status of breeding for apple scab resistance in Australia. *Acta Horiculturae* 595: 33–41.

ZhiGuo, J. and D. YouSheng. 2001. New uses of vegetable oils in fruit production. *Good Fruit Grower* 52(5): 59–62.

12 Major Diseases and Their Management

*Rovidha S. Rasool, Deelak Amin, Sumaira-Hamid,
Asha Nabi, Sana B. Surma, Moin Qureshi,
Bilal A. Padder, Nazir Ahmad Ganai and
Mehraj D. Shah**
* Corresponding author

CONTENTS

DOI: 10.1201/9781003239925-12

12.1 INTRODUCTION

Agriculture, the mother of all the cultures, plays an important role in the development of human civilization. It is the backbone of the economy in many countries, including India. Agriculture has been major contribution in the development of these countries. Agricultural production and productivity are outcomes of a complex interaction of soil, seed/planting material and agrochemicals. Therefore, it is important to make good decisions for the management of all these inputs. The focus on enhancing production without taking into account the environmental consequences has resulted in environmental degradation. In agriculture, diseases and insect pests are the most serious issues which affect the crop yield as well as its quality. To prevent the crops from severe losses caused by these diseases, insect pest and disorders, accurate diagnosis and quick treatment are essential. Diseases can establish on different plant parts including the fruit, stem, leaf and root. Fruit diseases result in major economic losses in the agricultural industry around the world.

Apple (*Malus* × *domestica* Borkh.) is a temperate fruit widely cultivated in almost all hilly areas of temperate and cold arid conditions around the world. India is the second largest country in terms of area and the seventh largest in terms of apple production in the world. Major apple growing areas in the country are Uttarakhand, Jammu and Kashmir and Himachal Pradesh and more than 80 percent of the total apple production in the country is produced by Jammu and Kashmir alone. Apples have grown over last several decades in the same habitat serving as a permanent home for the multitude of diverse insect pests and pathogens (diseases). Apple is a major crop of Jammu and Kashmir and the backbone of its economy. However, its productivity, as well as its quality, is affected by many factors including diseases caused by fungi, bacteria and viruses, etc. The major diseases responsible for the significant losses caused by the fungi, bacteria and viruses are discussed below.

12.2 DISEASES CAUSED BY FUNGAL PATHOGENS

Apple scab, *Alternaria* leaf blotch, *Marssonina* leaf spot, powdery mildew, sooty blotch, fly speck, white root rot, apple collar/crown rot and cankers are the most prevalent diseases of apple and are caused by different fungal pathogens.

12.2.1 APPLE SCAB

Apple scab is one of the most dangerous diseases that is prevalent on apple and occurs in almost all apple growing regions throughout the world. The disease is a serious threat in cool and humid areas with frequent rains in spring and early summer seasons. The major predisposing factors for the disease are temperature, moisture (wetness) and relative humidity. Moderate to severe infection occurs at 8.5°C to 21.4°C with leaf wetness up to 9 hours; a relative humidity ranging from 60 to 70 percent is considered favorable for the apple scab development. The causal organism is Teleomorph *Venturia inaequalis* (Cooke) Wint; Anamorph *Spilocaea pomi* Fr.

The fungus encounters both imperfect and perfect stages in nature in its parasitic and saprophytic forms. In its perfect stage, pseudothecia are formed on over-wintered fallen leaves on the orchard floor. The pseudothecia are 90 to 150 μm in diameter, spherical in shape, and dark brown to black in color with a short-beak that has a distinct ostiole surrounded by the several single-celled bristles. Asci (55.0–75.0 × 6.0–12.0 μm) are separated by pseudoparaphysis about 50 to 100 in a perithecium which are cylindrical, fasciculate and bitunicate, each containing eight ascospores. Ascospores are yellowish to tan in color and unevenly bi-celled with the upper cell shorter, wider and slightly more pointed than the lower cell. Conidia are borne singly on short and erect conidiophores and are yellowish-olive in color, single celled, occasionally bi-celled and ovate to lanceolate in shape (Reddy, 2014).

12.2.1.1 Symptoms

Symptom appearance starts on the lower side of leaves after the bud opening or green tip stage. Later, both the leaf surfaces are covered by the pathogen. The lesions occurring on the leaf surface are olivaceous velvety in color which later turn brown to black with indefinite margins. On the older leaves, lesions may form a convex surface with a corresponding concave area on the other side of the leaf. Symptoms also occur on the fruits and lesions of the young fruits are olivaceous green which later turns to black in color. On the mature fruits, the lesions are black and corky which later cause cracking of the fruit surfaces. In case of apple scab, symptoms also occur on the pedicel and the lesions are same as those on the leaves but are somewhat broader (MacHardy, 1996).

12.2.1.2 Disease Cycle

Apple scab pathogen has one sexual cycle and several asexual cycles per year. The fungus *V. inaequalis* over-winters on the dead fallen leaves in which microscopic black flask-shaped fruiting bodies called pseudothecia are developed from late autumn to spring on the orchard floor. Sexual spores known as ascospores mature with the opening of blossom buds. Ascospores that are carried by air currents from the source of inoculum to new tissues result in primary infection. Spore germination occurs soon after landing on the wet leaves, and symptoms appear 9–17 days after germination.

Secondary infection is caused by conidia (asexual spores) that are produced at relative humidity of 70 percent on the primary lesions. Conidia are discharged throughout the season and can continue in spring, summer and until autumn. Pseudothecial initials are formed in late autumn beneath the lesions and remain there as over-wintering structure (MacHardy, 1996).

12.2.1.3 Disease Management

The disease can be managed through the integration of both cultural and chemical practices.

12.2.1.3.1 *Cultural*

1) Ensure collection and destruction of fallen leaves, mummified fruits, pruned snag and dead wood.
2) Remove crowded branches for proper aeration and sunlight to avoid high humidity conditions in the microclimate of the fruit trees.
3) Keep the orchard area clean from weeds, bushes and wild plants to avoid excessive humid conditions in the rainy season.
4) Ensure proper irrigation during hot, dry periods to avoid stress.
5) Plant scab resistant cultivars like 'Firdous', 'Shreen', 'Akbar', etc.

12.2.1.3.2   Chemical

The dosage, time of application and type of fungicides to be used against the disease play a significant role in the success of chemical management of various fungal foliar and fruit diseases of apples particularly. A comprehensive spray program comprising of 7–8 sprays of both protectants and systemic fungicides during the growing season has been recommended and adopted by the farmers in apple orchards in Jammu and Kashmir. These diseases can easily be managed by adopting the recommended fungicidal spray schedule given by the SKUAST-K, Shalimar, Srinagar, Jammu and Kashmir (Table 12.1) (Bhat *et al.*, 2021).

12.2.2   *Alternaria* Leaf Blotch

Alternaria leaf blotch of apple has been associated with the epidemic-like condition throughout the world. The infection begins in the late spring to early summer when conidia are released from the over-wintering structure (mycelium) of the fungus. The major predisposing factors for the disease development are temperature ranging between 25°C–30°C in combination with heavy rains. The infection can be aggravated by the severe mite infection. The causal organism is *Alternaria mali* Robert.

Conidiophores arise from the short, septate, unbranched, dark-olive colored mycelium and are multi-septate, golden brown producing conidia from the apical cells, rarely from intercalary with conidial scars. Conidia produced as black tiny masses on blotched surfaces in the summer are obclavate or ovate, muriform, 1–6 septate, very short beaked, dark-olive to golden brown in color and measure 13.8–36.8 × 9.2–15.2 μm in size (Filajdic and Sutton, 1991).

12.2.2.1   Symptoms

The infection first appears as small, round, purplish or blackish spots on the leaves in the late spring or early summer. Later the lesions coalesce or enlarge, become irregular and acquire the 'frog eye'

TABLE 12.1

Recommended Fungicide Spray Schedule of SKUAST-Kashmir (2021) for the Management of Fungal Foliar Diseases of Apples

Serial No.	Phenological stage of the apple tree	Fungicide
1	Green tip	Mancozeb 75 WP @ 0.3% or Propineb 70 WP @ 0.3% or Zineb 75 WP @ 0.3% or Captan 50 WP @ 0.3%
2	Pink bud	Dodine 65 WP @ 0.06% or Mancozeb 75 WP @ 0.3% or Propineb 70 WP @ 0.3% or Zineb 75 WP @ 0.3% or Captan 50 WP @ 0.3%
3	Petal fall	Fenarimol 12 EC @ 0.05 % or Hexaconazole 5 EC @ 0.05% or Flusilazole 40 EC @ 0.02% or Difenaconazole 25 EC @ 0.03%
4	Fruit let	Mancozeb 75 WP @ 0.3% or Propineb 70 WP @ 0.3% or Zineb 75 WP @ 0.3% or Captan 50 WP @ 0.3%
5	Fruit development stage I	Dodine 65 WP @ 0.06% or Pyraclostrobin + Metiram 60 WG @ 0.1% or Kresoxim methyl 44.3 SC @ 0.04%
6	Fruit development stage II	Hexaconazole 5 EC @ 0.05% or Flusilazole 40 EC @ 0.02% or Difenaconazole 25 EC @ 0.03%
7	Fruit development stage III	Dodine 65 WP @ 0.06% or Fenarimol 12 EC @ 0.05 % or Hexaconazole 5 EC @ 0.05% or Flusilazole 40 EC @ 0.02% or Difenaconazole 25 EC @ 0.03%
8	Fruit development stage IV	Mancozeb 75 WP @ 0.3% or Propineb 70 WP @ 0.3% or Zineb 75 WP @ 0.3% or Captan 50 WP @ 0.3% Hexaconazole 5 EC @ 0.05% or Flusilazole 40 EC @ 0.02%

appearance. Lesions also occur on leaf petioles that result in a yellowing of leaves. In some cultivars, fruit infection occurs as a small pimple-like spots on the fruit surface (Madhu *et al.*, 2020).

12.2.2.2 Disease Cycle

The pathogen *Alternaria mali* over-winters as mycelium on the dead leaves and twig. The infection is initiated approximately 1 month after petal fall. The severe infection occurs at higher temperatures followed by heavy rains. The fungus produces alternaric acid as a chemical toxin that increases the disease severity (Madhu *et al.*, 2020).

12.2.2.3 Management

The disease can be managed through the integration of both cultural and chemical practices.

12.2.2.3.1 Cultural

* Ensure collection and destruction of fallen leaves and pruned twigs.
* Keep orchard area clean from weeds, bushes and wild plants to avoid excessive humid conditions during the rainy season.
* Ensure proper irrigation during hot and dry periods to avoid stress.
* Remove crowded branches for proper aeration and sunlight to avoid high humidity conditions in the microclimate of the fruit trees.
* Maintain a low mite population, particularly in the spring to early summer, in apple orchards to avoid *Alternaria* blotch (*A. mali*) outbreaks.

12.2.2.3.2 Chemical

The disease can be effectively managed with three to four fungicidal sprays starting from the pink bud stage at intervals of 12–15 days depending upon disease severity. Among the protectant fungicides, Dodine 65 WP @ 0.06 %, Mancozeb 75 WP @ 0.3%, Propineb 70 WP @ 0.3%, Zineb 75 WP @ 0.3% and Captan 50 WP@ 0.3% have been found effective in managing the disease, while among the systemic fungicides, Fenarimol 12 EC @ 0.05%, hexaconazole 5 EC @ 0.05%, flusilazole 40 EC @ 0.02% and difenaconazole 25 EC @ 0.03% have been found effective at different phonological stages of apple (Table 12.1). However, protectant and systemic fungicides should be sprayed alternatively to reduce the chance of resistance development.

12.2.3 MARSSONINA LEAF BLOTCH

Marssonina leaf blotch is one of the major diseases of apple crop causing premature leaf fall. The disease has been reported in United States since 1900s and is found in almost all of the apple growing regions worldwide. Major predisposing factors for *Marssonina* leaf blotch of apple are temperature and moist conditions. The average temperature for the disease development is 25°C with 22 mm of rainfall. The causal organism is Anamorph-*Marssonina coronaria* (Ell. and J. J. Davis) J. J. Davis Telomorph-*Diplocarpon mali* Harada and Sawamura.

The fungus produces asexual fruiting bodies as acervuli on the infected leaves and fruits. Acervuli measuring 100–200 μm in diameter are subcuticular structures and become erumpent at maturity. They are dark brown to black, globose to subglobose and botryose. Conidia borne on small clavate conidiophores are hyaline, bi-celled, guttulated, constricted at the septum and measure 20.0–24.0 × 6.5–8.5 μm in size. Spermatia produced both in cultures and on the hosts in the same stromata solely or mixed with conidia are hyaline, rod to elliptical shaped and measure 4.0–6.0 × 1.0–2.0 μm in size. Apothecia are also reportedly produced on the diseased leaves (Bohr *et al.*, 2018).

12.2.3.1 Symptoms

The symptoms first appear on the upper surface of leaves as grayish brown to dark spots with tinged purple margins. Later the lesions coalesce and cover both the leaf surfaces. Older leaves are more prone to disease compared to younger leaves. Under the favorable humid conditions, the disease becomes severe and results in defoliation within a few weeks, and only fruits are seen hanging on the naked branches. Repeated premature defoliation results in the failure of the crop in the following season. Acervulli (dark colored pin point specks) are visible in and around the diseased area on the leaves (Bohr *et al.*, 2018).

12.2.3.2 Disease Cycle
The fungus over-winters as apothecia on the infected fallen leaves and fruits. The primary infection has been reported to be initiated by the ascospores produced in apothecia on over-wintered leaves. Primary infections occur in late summer, and the fungal spores are released after the rainfall up to 3–4 weeks. The secondary infection occurs more frequently by the conidia. The optimum temperature range of 20°C –25°C and an incubation period of 8 days is necessary for the disease development (Sharma *et al.*, 2004).

12.2.3.3 Management
12.2.3.3.1 Cultural
Disease control can be achieved through the orchard sanitation and adequate pruning to facilitate proper aeration and light penetration around the tree canopy. As the pathogen over-winters on the fallen diseased leaves in the orchard floor, removal of over-wintering leaf litter will reduce the primary inoculum level. The orchard area should also be kept free from weeds and bushes to avoid excessive humid conditions during the rainy season.

12.2.3.3.2 Chemical
Protective spray of fungicides like mancozeb 75 WP @ 0.3% or propineb 70 WP @ 0.3% or dodine 65 WP @ 0.06% or captan 50 WP @ 0.3% when applied at regular intervals during the summer have been found effective in checking the disease. Systemic fungicides like carbendazim 50 WP @ 0.05% or thiophanate methyl @ 0.05% can also be effectively used at selected phonological stages of apple (Table 12.1). Triazoles have been found to be less effective in managing this disease.

12.2.4 POWDERY MILDEW

Powdery mildew of apple is present globally. It is one of the most serious diseases of apple under favorable conditions. The disease is particularly problematic in the western and mid-Atlantic regions of the United States of America. Major predisposing factors for the disease are temperature and dry weather conditions. The average temperature for the disease development ranges from 18°C –27°C with high relative humidity. Powdery mildew is also known as 'dry weather disease' as dry weather favors the disease. The causal organism is Telomorph-*Podosphaera leucotricha* (Ell. & Ev.) Salmon Anamorph-Oidium farinosum.

Powdery mildew is an obligate heterothallic ascomycetous fungus. Conidia, produced in long chains of thin and effused mycelium, are hyaline and ellipsoidal. They also contain hooked fibrosin bodies. Perethecia embedded in external felt-like growth of the fungus are subglobose, dark brown in color and have apical and basal appendages. Ascospores are hyaline, subglobose or oblong and unicellular. Each perithecium contains only one ascus (Marine *et al.*, 2010).

12.2.4.1 Symptoms
The symptoms appear on almost all parts of the plant like the buds, young shoots, leaves, flowers and occasionally on the fruits. The infection first occurs in spring as grayish patches on the underside of leaves. These patches are covered by fungal mycelia and spore mass (conidia). As the disease progresses, entire leaf is covered by the mycelial mass. Severely infected leaves become

longitudinally folded, hard and brittle. Young shoots also start showing the typical disease symptoms. Internodal shortening occurs in severely infected shoots. Grayish powdery mass covers the heavily infected twigs which show die-back-like symptoms with a silvery appearance. Diseased flowers fail to open and rarely set the fruit. Infected areas of fruit become russeted as a very closely interwoven network of the fine lines (Marine *et al.*, 2010).

12.2.4.2   Disease Cycle

The fungus grows superficially on the host surface but withdraws water and nutrition by the haustorium. The fungus produces barrel shaped conidia that are carried to different parts by the air currents. Primary infection occurs in the spring season after their emergence. Conidia initiate the new infection followed by the secondary phase of infection which continues several times during the growing season (Marine *et al.*, 2010).

12.2.4.3   Disease Management

Powdery mildew disease can be managed by the integration of both cultural and chemical practices. Pruning of the dormant shoot tip and silvered twigs reduces the primary source of infection. Sanitation plays a significant role in checking the disease development and subsequent spread. Three to four sprays of fungicides like wettable sulfur 80 WP @ 0.3% or dinocap 48 EC @ 0.05% or triadimefon 25 WP @ 0.05% or flusilazole 40 EC @ 0.02% or tebuconazole 250 EW @ 0.05% or pyraclostrobin + boscalid 38 WG @ 0.03% or hexaconazole 5 EC @ 0.05% at pink bud, petal fall and 2 weeks later are effective in controlling this disease (Table 12.1).

12.2.5   Sooty Blotch and Flyspeck

Sooty blotch and flyspeck are the two different diseases that occur in most of the moist and temperate regions of the world. These diseases do not affect the fruit yield, but they do reduce the market value of the fruit. The major predisposing factors for the diseases are temperature, moisture and humidity. These diseases are favored by high humidity and temperature from 18°C – 27°C. These diseases are further aggravated when cool, rainy weather in spring is coupled by frequent rains in summer. The causal organism for sooty blotch is *Gloeodes pomigena* (Schweinitz) Colby, and the causal organism for flyspeck is *Zygophiala jamaicensis* (Mason).

Gloedoes pomigena, the causal agent of sooty blotch, produces dark brown pycnidia which are scattered or aggregated. The conidia released through a crack in the top of the pycnidium are of variable length, 1–5 septate, generally cigar shaped with slight constriction at the point of septation and are cream to pink in color. *Zygophiala jamaicensis*, the cause of flyspeck, produces conidia on unique conidiophores. The conidiophores consist of four distinct parts. Conidia are bi-celled, elliptical to obovate in shape and often shed in pairs. Pseudothecial production reported elsewhere cracks from the center to release oval to bitunicate asci containing eight hyaline bi-celled ascospores with the upper cell shorter, wider and less pointed than the lower cell (Sharon *et al.*, 2000).

12.2.5.1   Symptoms

Both the diseases are superficial and appear simultaneously on the fruit surfaces, therefore they can be easily mistaken as a single disease. The severity of both diseases increase in late summer with the onset of rains. In case of sooty blotch, black to grayish spots with indefinite outlines occur on the fruit surface (Sharon *et al.*, 2000).

In flyspecks, definite black and often glistened circular spots resembling flyspecks appear on the fruit surface. Both these diseases are common on the fruit on the shady and crowded branches of trees. Economic losses by these diseases are caused primarily through the lowering fruit quality (Sharon *et al.*, 2000).

12.2.5.2   Disease Cycle
12.2.5.2.1   Sooty Blotch
In spring, a large number of spores are produced and released from pycnidia. These spores then spread to healthy plant parts and fruits by air currents and water splashes. Initially, the fungus attack twigs and then secondary colonies are formed on the fruit surface (Brown and Sutton, 1993).

12.2.5.2.2   Flyspeck
Ascocarps are the fruiting bodies of the fungus. In the late spring, spores are produced on the wild host and are then carried in the orchard by air currents. When these spores come in contact with the fruit surface under favorable conditions, they germinate and cause infection. After infection, symptoms develop within 15 days under favorable weather conditions (Brown and Sutton, 1993).

12.2.5.3   Management
12.2.5.3.1   Cultural
Adoption of proper training, pruning and thinning to avoid the overcrowding of branches not only directly reduces the primary source of infection but improves the microclimate of the orchards thereby helping in disease management. Fruits in clusters provide favorable microclimate for the disease spread, therefore thinning of such fruits reduces the disease incidence. Destruction of wild hosts helps in managing the disease.

12.2.5.3.2   Chemical
Spraying dithiocarbamate fungicides like ziram 80 WP @ 0.2% or ziram 27 w/v @ 0.6% or zineb 75 WP @ 0.3% or captan 50 WP @ 0.3% or mancozeb 75 WP @ 0.3% immediately at disease appearance are effective in managing the diseases. Carbendazim 50 WP @ 0.05% is also effective. Triazoles are less effective fungicides against the disease. Dipping of the harvested fruits in bleaching powder (5%) or sodium chlorate (3%) for 10 minutes and wiping with a coarse cloth before packing is also effective for disease management.

12.2.6   Canker

Apple canker is a diseased area on the woody portion of trees, which is usually well defined and often results in the death of the affected bark through girdling, cracking of bark vertically and/or horizontally and sometimes culminating in the death of the tree. On large tree trunks, it usually remains superficial and causes a roughening of the bark. In addition, canker diseases also cause losses due to fruit rotting and premature defoliation. These cankers are usually caused by many fungal and bacterial pathogens (Gupta and Agarwal, 1973).

12.2.6.1   Predisposing Factors
The tree bark is composed of layers of tissue that surround the woody core of trunk or branch. Outer bark serves as a physical and protective barrier against the entry of most canker causing pathogens. To infect the bark, the pathogens gain entry through wounds, which include pruning cuts, branch stubs, broken twigs or branches, insect injury, sun burn, hail damage, leaf scars, hydathodes, crotch angles and/or bark lenticels. The development of cankers in varying proportions may be attributed to unusual harsh summers and winters, faulty pruning and training practices, cultivation of apples or pears on water scarce uplands and rocky foothills which are deficient in water retention capacities and nutrients. Other factors affecting the prevalence of cankers are tree age, cultivar, site and altitude of orchard, etc. (Khan *et al.*, 2010).

12.2.7 SMOKY OR BLACK ROT CANKER

The smoky canker disease is also known as black rot canker or New York apple tree canker. It is worldwide in occurrence, and is found particularly in European counties, Southern Africa, New Zealand, Western Australia, Tasmania, Japan and New Guinea and India.

12.2.7.1 Symptoms

The disease is of major significance on pome fruits especially apple crops. It occurs in three phases: leaf spot, fruit rot and canker. The canker is the most predominant and destructive phase.

12.2.7 LEAF SPOT OR FROG EYE LEAF SPOT

It is usually known as frog eye leaf spot and the symptoms appear as small purple specks on the leaves at around the petal fall stage of the apple tree which later on enlarges to form circular spots. Light brown spots with grayish centers surrounded by the purple margins develop on the leaf giving a frog eye appearance. Severely infected leaves become chlorotic and defoliate prematurely (Khan, 2010).

12.2.7 FRUIT ROT

This disease is also known as black rot. Light brown spots usually start appearing on the fruit at the calyx end. A series of alternating brown and black colored concentric rings develop on the affected fruit surfaces and the rotten fruit finally turns black. Numerous black pimple-like asexual fruiting bodies (pycnidia) of the fungus are produced on the affected fruit surface. Completely rotten fruits dry up and shrivel into a wrinkled black mummy and remain attached to the tree up to several months (Khan *et al.*, 2010).

12.2.8 SMOKY CANKER

This phase is also known as smoky blight and considered to be the most destructive on tree trunks. On the bark of tree trunk and branches, the initial symptoms appear as small, sunken, reddish brown well-demarcated lesions or areas which enlarge lengthwise more rapidly and become elliptical in shape which finally results in the development of a series of concentric rings. The affected area turns smoky and results in complete girdling of the trunks or scaffold branches. However, superficial roughening of the bark is a common feature of the disease on large tree trunks and major limbs. Small branches and twigs show initial symptoms like browning of the bark with simultaneous yellowing of leaves. The disease causes cracking of the bark exposing reddish brown stained wood with the advancement of the infection. The spurs on the twigs and branches above the cankered portion are killed without exhibiting any conspicuous canker-symptom (Shah, 2007; Shah *et al.*, 2007; Khan *et al.*, 2010). The causal organism is *Botryosphaeria obtuse* (Schew.) Shoemarker (Anamorph: *Diplodia seriata* de Not.).

12.2.8.1 Disease Cycle

The causal pathogen usually over-winters asexually as dormant mycelium and pycnidia and sexually as pseudothecia on the cankered wood underneath the bark beside mummified fruits. Conidia and ascospores serve as a primary source of inoculum during the bloom stage in the spring season and their discharge is facilitated by the rains. A relative humidity of more than 75 percent and temperature 21 ± 1°C favors the disease development (Shah, 2007; Shah *et al.*, 2007; Khan, 2010).

12.2.8.2 Management of Canker Diseases

The integrated disease management for canker diseases of apples included the integration of good horticulture practices promoting tree health that will enable the trees to maintain their defense system. Some of the important strategies for optimizing tree growth include the application of balanced doses of fertilizers, maintenance of soil moisture by organic mulches, irrigation particularly during prolonged hot and dry weather and judicious pruning for harvesting moderate crops every year rather than a heavy crop alternatively.

12.2.8.2.1 Cultural

Cultural practices are prerequisite for the effective management of many important canker diseases. White washing/painting the tree trunks and major limbs facing south-west prevent sun burn injury, especially on the young trees that are too small and don't have enough wood/bark which allow them to quickly fall victim to the canker diseases. Adoption of proper training and pruning operations without leaving any pruning stubs minimizes the chances of infection. Avoid injuries caused by faulty agronomical, horticultural, mechanical, cultural practices or chemical applications. Avoid pruning during wet weather conditions. These precautions will restrict the conditions favorable for occurrence and subsequent development of various canker causing pathogens.

The removal and destruction of dead and diseased wood, pruned snags and mummified fruits from the orchard floor eliminates the maximum source of over-wintering or over-summering inocula of the canker causing pathogens. The removal of diseased branches and limbs by pruning at least 6 to 10 inches beyond the point of apparent infection or at least 2 to 3 inches beyond the canker margin followed by the scarification of the existing cankered portions is recommended prior to chemical treatment. Even then, scarification of cankered areas with a sharp edged knife to remove the pathogen(s) from the infected site followed by disinfection with 70% ethanol or spirit for wound dressing has been a routine practice throughout the world.

12.2.8.2.2 Wound Dressing

Pruned ends/cuts, wounds or other mechanical injuries should be dressed immediately with a wound dressing paint or paste that has no adverse effect on the callus formation. However, existing cankered areas need to be scarified with a sharp edged knife to remove the dead and dried wood for effective application of wound dressers. Some of the pastes/paints that have been found effective in healing the canker wounds of smoky canker, European canker, etc. in apple crops are:

- Admixture of carbendazim plus triadimefon (5.0 + 2.5% w/v) or carbendazim plus captan (4.0 + 10% w/v) in gel formulation
- Admixture of copper oxychloride or carbendazim with paint-based cow dung, clay soil and linseed oil (1:1:1)
- Paints containing admixture of mercuric oxide or thiram (8%) and zinc diethyl dithiocarbamte (0.2%)
- Admixture (1:2:5) of carbendazim and copper oxychloride in linseed oil
- Chaubattia paint (copper carbonate, lead oxide and linseed oil in the ratio of 1:1:1.25)
- Bordeaux paint (copper sulfate, lime and linseed oil in the ratio of 1:2:3).

12.2.8.2.3 Fungicide Sprays

Chemical sprays are not usually effective for controlling the existing or established cankers but could be helpful to reduce the fungal inoculum on the tree surface. However, various fungicides have been tested effective against different cankers caused by fungal pathogens. The concentration and time of application of some of fungicides proved to be effective elsewhere are:

- Captan (0.2%) and ziram (0.1%) for management of smoky canker, its leaf spot and fruit rot phase
- Copper oxychloride at leaf fall followed by a summer spray program (against apple scab) comprising of two to three sprays of carbendazim or thiophonate methyl
- Constant application of benomyl, kerosoxim-methyl and trifloxystrobin for reducing the incidence of apple cankers caused by *Botryosphaeria obtuse*.

12.3 SOIL-BORNE DISEASES OF APPLE AND THEIR MANAGEMENT

12.3.1 WHITE ROOT ROT

White root rot of apple is one of the most serious diseases that occurs in almost all of the temperate regions of the world. Major predisposing factors for the disease are temperature, moisture and pH. Favorable temperature for the disease is 25°C and is further aggravated by the wet conditions. The causal organism is Teleomorph-*Rosellina necatrix* (Prillieux) Anamorph-Dematophora necatrix (Hartig).

12.3.1.1 Symptoms

Symptoms occur on both below and above ground parts of the tree. Initially, white mycelial growth occurs on the root surface. Later, the fungus penetrates and rots the root tissue. The above symptoms are responsible for the early decline of the growth and vigor of the tree which later results in a bronzing of leaves (Sharma *et al.*, 2017).

12.3.1.2 Disease Cycle

The entire cycle of *Rosellina necatrix* occurs in the soil. The fungus over-winters as mycelium on the rotten tissue of roots. The fungus spreads either by root-to-root contact or by growth of mycelium in the soil. The mycelium later colonizes the root surface and penetrates to cause infection. Penetration occurs through the wounds or lenticels. The fungus produces pectolytic and cellulolytic enzymes on the host tissues in advance and also toxins which help kill the tree bark. No spore formation seems to play any role in dissemination of the pathogen. The disease is most serious in water-logged acidic soils at pH 6.1 to 6.5. High soil moisture and temperature (20°C –25°C) favor the disease severity (Duan *et al.*, 1990).

12.3.2 COLLAR/CROWN ROT

Collar rot of apple is one of the most serious soil-borne diseases that occur globally. The disease affects the scion portion of the trees. Temperature and high soil moisture are the major predisposing factors for the disease development. The causal organism is *Phytopthora cactorum* (Lebert & Cohn) Schroter.

12.3.2.1 Symptoms

The first symptoms that appear in spring are delayed bud break, leaf discoloration and twig die back. As the infection progresses, the leaves turn yellow to bronze. Branches on the infected tree remain stunted with poor terminal growth. Fruits remain smaller than the normal size and colors prematurely. The most conspicuous symptom on the affected trees is partial or complete girdling of the trunk. The affected tree declines progressively over several seasons and eventually dies (DuPont *et al.*, 2019).

12.3.2.2 Disease Cycle

The fungus *Phytopthora cactorum* over-winters in the form of oospore in plant decay. Cool and wet conditions favor the disease. The infection occurs in spring and during the onset of dormancy. The over-wintering structure, oospores germinate to produce sporangia. Numerous unicellular sporangia are liberated from these sporangia into the soil water. The spores, known as zoospores, swim short distances in the soil pore spaces or can be transported to longer distances by run off or irrigation water. When zoospores come in contact with the roots of susceptible hosts, they germinate and establish new infections. Zoospores are liberated only when soil is saturated. Prolonged flooding favors the production of sporangia and dissemination of zoospores and may also predispose the host to infection. Optimum temperature for the disease development is 20°C–30°C (DuPont *et al.*, 2019).

12.3.2.3   Management of Soil-Borne Diseases

Although, the management of soil-borne diseases is very difficult due to various factors, the integration of various approaches/practices has been found useful in keeping these diseases under check. These practices include cultural, biological, chemical and host resistance.

12.3.2.3.1   Cultural

The cultural practices recommended for the management of soil-borne diseases include improving poorly drained soils by using a central drainage system. Channels 60 cm deep and 90 cm wide should be prepared to separate the infected zone from the healthy one. The acidic soils should be amended by applying hydrated lime. Soil amendment with pine or deodar needles or cruciferous trash or neem cake has been found effective in reducing the disease incidence. Tree grafting should be done about 30 cm from the soil level to avoid entry of the pathogen through graft union. The affected trees should be approach-grafted with apple seedling, preplanted during dormancy around the damaged roots for better anchorage. Apple seedling nurseries should be rotated every 4–5 years and planted with a non-host crop like maize for some years before selecting the site again for nursery raising. Rotten roots should be cut, and the cut ends painted with disinfectant paste. The known infested pits should be treated with 3% formaldehyde 3 weeks before a new plantation.

12.3.2.3.2   Biological Control

Various fungal and bacterial antagonists like *Trichoderma viride*, *Trichoderma harzianum* and *Enterobacterae rogenes* have been found effective to protect the plants from the root rot pathogen *Dematophora necatrix*. Repeated application of these antagonists increased their efficacy against the pathogen.

12.3.2.3.3   Resistant Rootstocks

Use of resistant rootstocks against the soil-borne diseases has been found the most effective and desirable approach of disease management.

12.3.2.3.4   Chemical

Drenching the soil under the tree canopy with fungicides like carbendazim 50 WP (0.1%) or mancozeb + carbendazim 75 WP (0.5%) have been found effective for the control of white root rot disease. Prior to the application of fungicides, the tree basin should be brought to a moisture level of 30–40 percent and the fungicide should be applied through deep holes (15–20 cm) made in the basin of trees at 30 cm distance from each other. The treatment should be repeated three to four times at the interval of 12–15 days during the rainy season. For managing the existing collar rot cankers, the cankered or affected area should be first scarified to healthy tissue and disinfected with methylated sprit followed by an application of Bordeaux or copper or metalaxyl or Chaubattia paint. Drenching of the tree basin either with mancozeb + carbendazim 75 WP @ 0.5 % or with mancozeb + metalaxyl 72 MZ @ 0.5% in 30 cm radius around the tree trunk has also been found effective in controlling the disease.

12.4   DISEASES CAUSED BY BACTERIA

12.4.1   Crown Gall

Crown gall is one of the important bacterial diseases of apples. This disease is most common in apple nurseries and can also occur in orchards. This disease is caused by a gram negative soil-borne bacterium. The causal organism is *Agrobacterium tumefaciens*.

12.4.1.1   Symptoms

In this particular disease, overgrowths appear as galls on the roots and base of the woody plants such as apples. This bacterium can survive in soil with good aeration such as sandy loam soils. It can also survive on the plant root surfaces (Kado, 2002).

12.4.1.2 Disease Cycle

Agrobacterium tumefaciens survive on the infected roots in soil. This bacterium is disseminated by irrigation water and also by infected planting material originating from the uncertified sources. Secondary spread occurs through the pruning and cultivation equipment, particularly when galls are removed manually with same cutting tools used in pruning. The bacterium enters through the wounds and cause infection. Severe and large galls will be seen at the infection site easily with the naked eye (Kado, 2002).

12.4.1.3 Management

12.4.1.3.1 Selection of Site

Fields that have been grown cereal crops for a longer period are favored as crown gall-free sites rather than fields previously used for fruit or nut crops.

12.4.1.3.2 Planting Stock

The selection of disease-free planting stock is important.

12.4.1.3.3 Crop Rotation

Crop rotation of cereal crops with green manuring crop helps reduce the bacterium population.

12.4.1.3.4 Chemical

Copper-based fungicides such as copper oxychloride, copper hydroxide, Boudreaux mixture, etc. and strong oxidants such as sodium hypochlorite are transitorily effective. Antibiotics such as tetracycline, streptomycin, etc. are also recommended for disease management. Generally, chemicals are rarely used for the crown gall.

12.4.1.3.5 Biological Control

Agrobacterium tumafaciens are sensitive to the antibiotic agrocin produced by *A. radiobacter* strain K84 genetically modified as strain K1026. Therefore, the *A. radiobacter* strain K1026 has been recommended for biological control of this disease.

12.4.2 Fire Blight of Apple

Fire blight of apple is a very contagious and destructive disease caused by the bacterium *Erwinia amylovora* affecting many plants of the *rosaceae* family. Under favorable conditions it can destroy a whole orchard in a single growing season. The causal organism is *Erwinia amylovora*.

12.4.2.1 Symptoms

The symptoms of fire blight disease are observed on above ground plant parts like blossoms, leaves, fruits and twigs. The blossom blight symptom appears soon after bloom, i.e., 1–2 weeks after petal fall. Initial symptoms appear as water-soaked lesions on blossoms which eventually shrivel and ultimately changes color from dark green to a dull black or brown. The shoot symptoms appear after one to several weeks of petal fall. First, discoloration of the young succulent tissues occurs followed by the shoot tip curling and wilting to resemble a shepherd's crook. Under warm conditions, bacterial ooze appears on the affected shoots. Blackening is often observed on the leaves of diseased shoots along midrib and veins. As the shoot infection progress downwards, the bark of larger branches and limbs become water soaked, and later, cracks develop resulting in sunken cankers which serve as source of inoculum. Fruit infection occurs through wounds caused by insects, wind and rain or hail damage. Initially small, gray or water-soaked lesions appear on immature fruit which later shrivels and turns brown or black and ultimately becomes mummified. As the disease advances the pathogen migrates toward the rootstock. The bark of the infected rootstock develops water-soaked purplish to black discoloration, cracking, oozing and ultimately results in the death of the entire tree.

12.4.2.2 Disease Cycle

The cankers formed on the infected shoots and branches aid the pathogen during over-wintering. The infection occurs only when the three basic parameters of availability of the pathogen (bacterium), a susceptible host and favorable weather conditions coincide with each other. The disease is favored by the hot and wet weather conditions, and a temperature of 25°C –28°C is optimal for bacterial growth and multiplication. The fire blight disease occurs soon after bloom and spreads through bees, flies, ants, etc. The ooze coming out of the cankered portions gets displaced either through wind or rain to healthy flowers thus initiating infection. After infecting the open blossom, the bacterium colonizes and moves internally to other plant parts and ultimately results in the death of whole plant.

12.4.2.3 Management

- Use of resistant cultivars and rootstocks.
- Removal and destruction of diseased/infected cankered tissues which aid in over-wintering of the pathogen.
- Insects such as leaf hoppers, bees and aphids should be controlled as they act as carriers of the pathogen.
- Spray of copper-based fungicides at bud break and green tip stage.
- Application of streptomycin at trauma blight stage is also recommended.

12.5 VIRUSES INFECTING APPLE

Apple has been infected by 21 viruses from 20 genera and nine families. Apple mosaic virus (ApMV), apple chlorotic leaf spot virus (ACLSV), apple stem grooving virus (ASGV) and apple stem pitting virus (ASPV) are the most destructive viruses infecting apple and other stone fruits throughout the world (Umer et al., 2019).

ApMV infects a wide range of woody hosts and is commonly found in infections with other apple-infecting viruses such as ACLSV, ASPV, ASGV and others. For the first time, the virus was discovered in apple (Bradford and Joly, 1933). On the spring leaves of infected apple trees, light yellow to bright cream sporadic patches or bands appear around main veins. After being exposed to the sun and heat in the summer, the spots on badly infected leaves can become necrotic. The symptomatic leaves fall off too soon. The distribution of symptomatic leaves in the tree may be irregular or restricted to a single branch (Digiaro et al., 1992).

ACLSV was isolated for the first time from apple trees in the United States after transmission to *Malus platycarpa* by Mink and Shay in 1959 (Burnt et al., 1996). It infects most of the fruit trees of *Rosaceae* family including apple and other pome and stone fruits. In most of the commercial apple cultivars, the infection is latent and may be associated with severe fruit deformations, yield reduction, graft incompatibility and bud necrosis (Myrta et al., 2011). ACLSV is commonly found in apple trees in coinfection with other viruses like ApMV.

ASGV is a member of the *Capillovirus* genus in the *Betaflexi viridae* family. This virus was discovered in *Malus sylvestris* cv. 'Virginia Crab' from the United States in the 1960s and is latent. When an infected cultivar is grafted on a susceptible rootstock like *Malus pumila* cultivars, it causes stem grooving, brown line and graft union anomalies (Massart et al., 2011), and the infected plants remain smaller and weaker and spread through contaminated propagative material and grafting.

ASPV is a member of the *Foveavirus* genus and the *Betaflexiviridae* family. The incompatibility between some apple cultivars and the rootstock *M. sylvestris* 'Virginia crab' was first described in the United States in the 1940s. ASPV's wide spread in commercial apple cultivars occurred, but the virus was mainly latent. It induces various symptoms in vulnerable cultivars of apple and other indicator plants. Leaf symptoms appear in many commercial pear cultivars only within the first several years of flowering after which the trees become symptomless. It causes diffused leaf spots and extreme fruit deformation in quince. ASPV is spread by grafting and contaminated propagative material (Jelkmann and Paunovic, 2011).

12.5.1 MANAGEMENT

- Removal and destruction of affected planting material.
- Certified and disease-free planting material should be used for propagation of apple and other crops because the virus is transmitted through grafting.
- Tagging and marking of infected trees, especially bud and graft wood donor plants, should be followed to avoid use of bud wood and/or graft wood from such trees.
- Sterilization of equipment while following intercultural operations, especially training and pruning, is recommended to avoid the spread of infection.
- Viral indexing of planting material for certification should be followed to trace the infected plants and their rouging and destruction.

REFERENCES

Bhat, M.S., Lone, F.A. and Rather, J.A. 2021. Evaluation of long term trends in apple cultivation and its productivity in Jammu and Kashmir from 1975 to 2015. *GeoJournal* 86(3):1193–1202.

Bohr, A., Buchleither, S., Hechinger, M. and Mayr, U. 2018. Symptom occurrence and disease management of Marssonina blotch. *18th International Conference on Organic Fruit-Growing: Proceedings of the Conference*, Hohenheim, Germany, pp. 36–42.

Bradford, F.C. and Joly, L. 1933. Infectious variegation in the apple. *Journal of Agricultural Research* 46:901–908.

Brown, E.M. and Sutton, T.B. 1993. Time if infection of *Gloeodes pomigena* and *Schizothyrium pomi* on apple in North Carolina and potential control by an eradicant spray program. *Plant Disease* 77:451–455.

Burnt, A.A., Crabtree, K., Dallwitz, M.J., Gibbs, A.J. and Watson, L. 1996. *Viruses of Plants Descriptions and Lists from the VIDE Database* (p. 100). Wallingford: CAB International.

Digiaro, M., Savino, V. and Di Terlizzi, B. 1992. Ilarvirus in apricot and plum pollen. *Acta Horticulturae* 309:93–98.

Duan, C.H., Tsai, W.H. and Tu, C. 1990. Dissemination of white root rot disease of loquat and its control. *Journal of Agricultural Research of China* 39:47–54.

DuPont, S.T., Hewavitharana, S. and Mazzola, M. 2019. *Phytopthora Crown, Collar and Root Rot of Apple and Cherry WSU Tree Fruit IPM Startegies* (pp. 1–6). Washington State University.

Filajdic, N. and Sutton, T.B. 1991. Identification of *Alternaria mali* on apples in North Carolina and susceptibility of different apple varieties to *Alternaria* blotch. *Plant Disease* 75:1045–1048.

Gupta, G.K. and Agarwal, R.K. 1973. Canker diseases of apple trees in Himachal Pradesh. *Indian Journal of Mycology and Plant Pathology* 3:189–192.

Jelkmann, W. and Paunovic, S. 2011. Apple stem pitting virus. In A. Hadidi, M. Barba, T. Candresse, and W. Jelkmann (Eds.), *Virus and Virus-like Diseases of Pome and Stone Fruits* (pp. 35–40). St. Paul, MN: APS Press.

Kado, C.I. 2002. Crown gall. *The Plant Health Instructor.* https://doi.org/10.1094/PHI-I-2002-1118-01

Khan, N.A. 2010. *Status and Etiology of Canker Disease of Apple in Kashmir* (Doctoral dissertation, SKUAST Kashmir).

Khan, N.A., Ahmad, M. and Ghani, M.Y. 2010. Botyrosphaeria dothidea associated with White Rot and Stem Bark Canker of Apple in Jammu & Kashmir. *Applied Biological Research* 12(2):69–73.

MacHardy, W.E. 1996. *Apple Scab Biology, Epidemiology and Management.* The American Phytopathological Society. Press St. Paul, 545 pp.

Madhu, G., Nabi, S.U., Raja, W.H., Sharma, O.C., Sheikh, M.A. and Singh, D.B. 2020. Alternaria leaf and fruit spot in apple: Symptoms, cause and management. *Journal of Bioscience and Biotechnology* 8:24–26.

Marine, S.C., Yoder, K.S. and Baudoin, A. 2010. Powdery mildew of apple. *The American Phtopathological Society* 10:1021.

Massart, S., Jijakli, M.H. and Kummert, J. 2011. Apple stem grooving virus. In A. Hadidi, M. Barba, T. Candresse, and W. Jelkmann (Eds.), *Virus and Virus-like Diseases of Pome and Stone Fruits* (pp. 85–90). St. Paul, MN: APS Press.

Myrta, A., Matic, S., Malinowski, T., Pasquini, G. and Candresse, T. 2011. Apple chlorotic leaf spot virus in stone fruits. In A. Hadidi, M. Barba, T. Candresse, and W. Jelkmann (Eds.), *Virus and Virus-like Diseases of Pome and Stone Fruits* (pp. 85–90). St. Paul, MN: APS Press.

Reddy, P.P. (2014). *Biointensive Integrated Pest Management in Horticultural Ecosystems.* Springer Science & Business Media, Germany.

Shah, M.D. 2007. *Characterization and Management of Botryodiplodia Theobromae Pat. Causing Die-back and Bark Canker of Pear* (Doctoral dissertation, Plant Pathology Department, Punjab Agricultural University, Ludhiana).

Shah, M.D., Verma, K.S. and Dhillon, W.S. 2007. Severity and management of die-back and bark canker (*Botryodiplodia theobromae*) of pear in Punjab. *Journal of Research PAU* 44:216–218.

Sharma, J.N., Sharma, A. and Sharma, P. 2004. Outbreak of Marssonina blotch in warmer climates causing premature leaf fall problem of apple and its management. *Acta Horticulture* 662:405–409.

Sharma, Y.P., Pramanick, K.K., Thakur, J.S., Watpade, S. and Kumar, S. 2017. Technique for screening of apple and pear germplasm against white root rot (*Dematophora necatrix*). *Journal of Applied Horticulture* 19:75–77.

Sharon, M.W. and Sutton, T.B. 2000. Sooty blotch and flyspeck of apple: Etiology, biology and control. *The American Phytopathological Society* 84:714–724.

Umer, M., Liu, J., You, H., Xu, C., Dong, K., Luo, N., Kong, L., Li, X., Hong, N., Wang, G. and Fan, X. 2019. Genomic, morphological and biological traits of the viruses infecting major fruit trees. *Viruses* 11:515.

13 Major Insect Pests and Physiological Disorders and Their Management

Deelak Amin, Rovidha S. Rasool, Sumaira-Hamid, Asha Nabi, Rafiya Mushtaq, Nazir Ahmad Ganai and Mehraj D. Shah*

CONTENTS

DOI: 10.1201/9781003239925-13

13.1   INTRODUCTION

Apple is widely cultivated in almost all the temperate regions of the world, and India ranks seventh in the world in apple production. India also ranks second in area under apple cultivation (Ahmad and Sajad, 2017). The major apple growing states of India are Uttarakhand, Jammu and Kashmir and Himachal Pradesh. Jammu and Kashmir contributes around 80% of the overall apple production in the country. The apple crop has grown over a number of years under the same habitat which serves as a permanent home for a multitude of diverse pests (Sherwani *et al.*, 2016; Moinina et al., 2019). It is the main backbone of the Jammu and Kashmir economy and also serves as a main cash crop for farmers. However, the productivity and quality are very low when compared to advanced countries because of various diseases, insect pests and physiological disorders. Insect pests are one of the major causes of losses in production and quality of apple crops. They are responsible for both direct and indirect losses. Direct loses include fruit injuries caused by insect feeding, reduction in fruit quality and quantity while indirect losses are expenses incurred on the management of these insect pests (Hussain *et al.*, 2018). Proper pest management strategies are to be followed to increase production (Beigh *et al.*, 2015). Physiological disorders are fruit abnormalities which are not associated with any insect pests or diseases (Ferguson *et al.*, 1999; Martins *et al.*, 2013). These disorders can appear both in the growing season, i.e. during the preharvest and postharvest under the storage leading to significant losses in the market value of the fruit. Important insect pests and physiological disorders of apples are discussed in this chapter.

13.2 INSECT PESTS OF APPLE

13.2.1 SAN JOSE SCALE

San Jose scale (*Quadraspidiotus perniciosus*) belongs to family *Diaspididae* order Hemiptera. The name San Jose scale has been derived from San Jose, California, where it was discovered for the first time by Comstock in 1881. It is a sucking pest which injects toxins into the plant tissues while feeding and causes localized discoloration.

13.2.1.1 Nature of Damage

The pest sucks the sap in both stages as nymphs and adults from the stem, twigs, leaves and fruits (Pinero *et al.*, 2020). The plant parts and sometimes the whole bark of the tree can be seen covered with ashy gray scales which can be easily scraped off to expose the pigmented area underneath. This pest is mostly found infesting large and older trees which are densely branched resulting in poor spray coverage (Hix *et al.*, 1999). On fruits, pink pigmented areas can be seen around the scale which decrease its market value. When infestation is less, scales on fruit are confined to the calyx area. Infested trees result in loss of vigor, low production, lower fruit quality and eventually the death of the trees (Sofi and Hussain, 2008; Hussain *et al.*, 2018).

13.2.1.2 Life Cycle

The scale completes its two generations in a year, and third generation is partially incomplete which over-winters from November to March on trees. The population of all stages enters into hibernation and only nymphs of first instar survive the cold temperature. The female reproduces ovoviviparously, i.e., the eggs develop inside the female and are born as nymphs. Each female produces 200–400 nymphs in a period of 6 weeks (Pinero *et al.*, 2020). The first instar nymphs (crawlers) crawl randomly for almost 12–24 hours before settling down at a suitable place on the host tree and starting to feed by sucking sap (Hix *et al.*, 1999). Later, they secrete waxy scale coat over themselves which gives them the name scale insects. They become fully grown scales in 3–4 days and after 10–14 days, the females again start giving birth to young ones. The pest develops first generation crawlers in early and mid-June with white and black cap stages developing approximately over the next month. A second generation of scales appear between July and early September. However, third generation crawlers of San Jose Scales occur during late October which is when they enter into hibernation (Sofi and Hussain, 2008; Sherwani *et al.*, 2017).

13.2.1.3 Management

13.2.1.3.1 Cultural

- Orchard sanitation is the first step to prevent the infestation.
- Removal of alternate hosts from the orchard like willow, poplar, etc. has been found effective.
- Pruned heavily infested twigs and branches must be removed from the orchards and burned.

13.2.1.3.2　Biological Control
- Release of biological control predatory beetle namely *Chilocorus infernalis* and parasit-oids such as *Encarsia perniciocus* and *Aphytis proclia*.

13.2.1.3.3　Oil Spray
- Spraying of delayed dormant oils @ 2 lit/100 liter of water.

13.2.1.3.4　Chemical
- When the population exceeds economic threshold level (ETL) systemic insecticides can be sprayed.

13.2.2　Woolly Apple Aphid

The woolly apple aphid (*Eriosoma lanigerum*) belongs to the family *Aphididae*, order Hemiptera. This pest was first identified in the United States in 1842. However, this pest is now present in all the apple growing areas of the world. The aphid colonies are covered with long white cottony filaments which give them a woolly appearance and justify the name woolly apple aphid. This pest sucks the limbs and roots of the apple trees (Sherwani *et al.*, 2016).

13.2.2.1　Nature of Damage
The woolly apple aphid infestation is observed on both above and below ground parts of the trees. Colonies are mostly formed at the wounded sites on trunks, limbs and twigs and also on the water sprouts in the center of the trees. Galls or swellings are formed at the feeding sites of the aphid colony which may seriously disfigure the young trees and nursery plants. These galls may serve as entry sites for various pathogens like canker (Lal and Singh, 1947). Infestation on roots by woolly apple aphid causes more damage to the trees, especially young trees. Pale foliage can be one of the signs of root infestation. Infested roots often become fibrous, which predisposes the tree to being easily uprooted. Galls on roots may also interfere with the nutrient uptake affecting growth and vigor of the tree (Hussain *et al.*, 2018).

13.2.2.2　Life Cycle
Woolly apple aphids over-winter as naked nymphs either on roots or under the loose bark and other protected places on apple trunk and branches. In spring, these aphids resume activity and migrate to their preferred feeding sites. In summer, they are mostly found in leaf axils on tender shoots. There are many generations in a single year. Like other aphids, both winged (oviparious) and wingless (viviparous) adults are produced, but eggs produced by winged adults fail to hatch and population is wholly dependent on the viviparous females (Lal and Singh, 1947).

13.2.2.3　Management

13.2.2.3.1　Cultural
All the cracks, crevices and wounded areas on the tree (like pruning wounds) must be covered with Chaubatia paste. Summer pruning must be done to remove excessive water sprouts, and planting resistant rootstocks like the MM series is also recommended for management.

13.2.2.3.2　Biological Control
The release of an endoparasitoid such as *Aphelinus mali* has been found effective against the pests.

13.2.2.3.3　Oil Spray
Application of delayed dormant spray oil @ 2lit/100 lit of water also helps in its management.

13.2.2.3.4 Chemical

For adult infestations, when the pest population is above ETL, spray systemic insecticide (Table 13.1). For root infestation, soil drenching in tree canopy area with chloropyriphos 20 EC @ 0.3% in water during December or March can be done or carbofuran 3 CG @ 100 g/tree can be hoed in the soil around the tree canopy.

13.2.3 European Red Mite (ERM)

The European red mite (*Panonychus ulmi*) belongs to family *Tetranychidae: Trombidiformes*. ERM is native to Europe and was first identified in the United States in 1911. This pest causes discoloration to leaves and russeting to fruits.

TABLE 13.1

Insecticide Spray Schedule for the Management of Different Insect Pests of Apples

Spray	Phenological stage	Insecticides/acaricides/oils per 100 liters of water
1	Delayed Dormancy	Horticultural Mineral Oil (2 Lit.)
2	Pink bud	**Need Based for Insects:**
		If HMO spray is missed: Dimethoate 30 EC (100 ml)
		Need Based for Blossom Thrips:
		Apply when two or more thrips per flower are observed
		Thiacloprid 21.7 SC (40 ml)
3	Petal fall	**Need Based for Insects:**
		Dimethoate 30 EC (100 ml) or Quinalphos 25 EC (100 ml)
		Need Based for Mites:
		Apply acaricide when 4–5 mites per leaf are observed Hexythiazox 5.45 EC (40 ml) or Spiromesifen 22.9 SC (40 ml) or Fenazaquin 10 EC (40 ml) or Cyenopyrafen 30 SC (30 ml)
4	Fruitlet (pea size)	**Need Based for Insects:**
		Chlorpyriphos 20 EC (100 ml) or Dimethoate 30 EC (100 ml)
		Need Based for Mites:
		Apply acaricide when more than five mites per leaf are observed Hexythiazox 5.45 EC (40 ml) or Spiromesifen 22.9 SC (40 ml)
5	Fruit Development-I	**Need Based for Insects:**
		Chlorpyriphos 20 EC (100 ml) or Quinalphos 25 EC (100 ml)
		Need Based for Mites:
		When population is more than ten mites per leaf Hexythiazox 5.45 EC (40 ml) or Spiromesifen 22.9 SC (40 ml)
6	Fruit Development-II	**Need Based for Insects:**
		Dimethoate 30 EC (100 ml) or Chlorpyriphos 20 EC (100 ml)
		Need Based for Mites:
		Hexythiazox 5.45 EC (40 ml) or Fenazaquin 10 EC (40 ml) or Spiromesifen 22.9 SC (40 ml) of summer spray oil (750 ml)
7	Fruit Development-III	**Need Based for Insects:**
		Chlorpyriphos 20 EC (100 ml) or Dimethoate 30 EC (100 ml)
		Need Based for Mites:
		If Population is more than ten mites per leaf Fenazaquin 10 EC (40 ml) or Spiromesifen 22.9 SC (40 ml) or Cyenopyrafen 30 SC (30 ml)

Note: As recommended by the SKUAST-K for the year 2020–2021

13.2.3.1   Nature of Damage

The ERM causes injury by sucking sap from the leaves. It sucks out the chlorophyll from the leaf tissue which results in increasing respiration. Affected leaves appear as speckled or in severe cases may become bronze in color (Faruque and Dan, 2013). Heavily infested orchards give a burned appearance. When a population of this pest is very high, it may result in defoliation, stunted shoot growth, reduced fruit quality and a decreased number of fruit buds in the following year (Sherwani *et al.*, 2017).

13.2.3.2   Life Cycle

Bright red colored over-wintering eggs are laid in groups under the cracks and crevices, on roughened bark and around the bud scales on twigs and branches. These eggs begin to hatch on the arrival of spring usually at the pre-pink bud stage of apples. Newly emerged young mites move toward tender leaves where they feed, become adults and reproduce (Faruque and Dan, 2013). Depending upon the temperature, the pest completes 6–8 generations in a year. The first generation develops in approximately 3 weeks, the summer generations are overlapping and all the stages (eggs, nymphs and adults) are present at the same time. Mite development is completely temperature dependent; therefore, hot and dry weather favors development of mites while cool and wet weather delays mite activities. A gravid female lays approximately 35 eggs during her average life span of 18 days (Sherwani *et al.*, 2016).

13.2.3.3   Management

13.2.3.3.1   Cultural

In order to keep the trees healthy, applications of balanced fertilizer doses and proper irrigation are ensured. Pruned infested branches need to be removed from the orchards and burned. Proper orchard sanitation should be followed and is a prior requirement for its management.

13.2.3.3.2   Biological

Biological control can be achieved by the release of lady bird beetle *Stethorus punctum*, a predatory mite like *Amblyseius fallacis* and all stages of an anthocorid bug namely *Blaptostethus pallences.*

13.2.3.3.3   Oil Spray

Delayed dormant spray of HMO @ 2 lit/100 liters of water is effective.

13.2.3.3.4   Chemical

If the population is above ETL, Acaricides (Table 13.1) can be sprayed.

13.2.4   Blossom Thrips

Blossom thrips (*Frankliniella occidentalis*) belongs to family *Thysanoptera: Thripidae*. These are minute insects about 1–1.5 mm long and pale yellow, brown or black in color.

13.2.4.1   Nature of Damage

Both adults and nymphs attack floral parts like flowers and buds of apple tree. They lacerate the floral tissues and feed on sap oozing out from the injured areas. As a result of feeding on stamen, the style and petals of the injured flower will develop brownish patches. A severe attack may result in a significant reduction in fruit set. The main damage to fruit is caused by egg laying on juvenile fruit before petal fall. The egg laying site results in yellow patching or russeting called as pansy spots (Hussain *et al.*, 2018).

13.2.4.2 Life Cycle

Thrips have multiple overlapping generations is a year which survive on apple flowers from March to late April and later migrate to various flowering weeds. In the month of November, the adult thrips move to the cracks, crevices, weed debris and other protected places for hibernation (Faruque and Dan, 2013). They emerge during late March or early April when the apples are ready to bloom. Later, females will lay eggs in pistils, stamens and juvenile fruits (Sherwani *et al.*, 2016).

13.2.4.3 Management
13.2.4.3.1 Cultural

Weeds growing around the orchard harbor blossom thrips which later transfer to apple trees; therefore, the weeds need to be removed before the apple blooming. Some weeds flowering on apple orchard floors bloom in synchronization with apples and must be present as a trap crops.

13.2.4.3.2 Chemical

If population is above ETL, spraying of systemic insecticide is recommended (Table 13.1).

13.2.5 Chafer Beetle or White Grub

Chafer beetle or white grub (*Adoretus simplex* and *Holotricha* spp.) belongs to family *Scarabaeidae*, order Coleoptera. This pest feeds on a variety of crops ranging from weeds, vegetables, fruits, etc.

13.2.5.1 Nature of Damage

These beetles are also called June beetles or defoliating beetles. Adults are nocturnal feeders and attack leaves, buds, flowers and fruitlets of apple trees while the larvae (white grubs) only feed on the roots (Sofi, 2017). These are voracious feeders and live gregariously. In cases of large populations, the complete defoliation of trees may occur and infested trees may become weak and eventually dry up (Altaf *et al.*, 2019).

13.2.5.2 Life Cycle

Chafer beetles complete one generation in a year. They over-winter as larvae under the soil surface. After emerging from hibernation during March and April, they start feeding on the roots, and later the adults emerge in the month of June to feed, mate and reproduce (Altaf *et al.*, 2019).

13.2.5.3 Management
13.2.5.3.1 Cultural

Adults can be collected from non-bearing trees by shaking them at dusk during mass emergence periods. Rotten farmyard manure (FYM) must be applied to the tree as fresh FYM can attract the pests and therefore should be avoided.

13.2.5.3.2 Mechanical
Light traps can be installed to trap and kill the adults to reduce the infestation.

13.2.5.3.3 Chemical

For adults, spray contact insecticide (Table 13.1) in the evening hours during the mass emergence period. For grubs, soil drenching in tree canopy areas with Chlorpyriphos 20 EC @ 0.3% of water can be done or Carbofuran 3% CG @ 100 g/tree can be hoed into soil around the tree canopy.

13.2.6 HAIRY CATERPILLAR

This pest is also known as the Indian gypsy moth (*Lymantria obfuscate*) and belongs to family *Lymantriidae*, order Lepidoptera. It is an important defoliator of a wide range of crops and forests (Sofi, 2017). However, this pest is considered a minor pest of apple fruit in India.

13.2.6.1 Nature of Damage

The larvae or caterpillar is the most damaging stage of this pest. They are voracious night feeders and eat in gregarious manners. In cases of severe infestation, they can completely defoliate the trees leading to poor produce and low quality fruit production (Altaf *et al.*, 2019).

13.2.6.2 Life Cycle

The Indian gypsy moth completes only one generation in a year. They enter into hibernation as eggs during the fall season and later revive activity on the onset of spring in mid to late March. The larval or host damaging stage lasts for approximately 2–3 months and they later pupate and transform into adults whose sole purpose is to mate and reproduce (Khan, 1941).

13.2.6.3 Management

13.2.6.3.1 Cultural

We can culturally control this pest to a significant level by destroying the egg masses which are accessible for almost 9 months on branches and trunk of the host trees like willow, poplar, etc. and sometimes on walls, stones, soil clods etc. Keeping trees healthy by applying balanced doses of fertilizers reduce the chances of a pest infestation. As larvae are nocturnal in nature and hide during the daytime, the gunny bag burlap (dipped in Chlorpyriphos 20 EC @ 0.1% in water) can attract the larvae and the chemical will kill them.

13.2.6.3.2 Chemical

At fruit-let stage, Chlorpyriphos 20 EC @ 0.1% in water can be sprayed if the larval population is above ETL.

13.2.7 GREEN APPLE APHID

Green apple aphid (*Aphis pomi*) belongs to the family *Aphididae*, order Hemiptera. This pest feeds on apple trees especially leaves and/or tender shoot tips and also attacks other members of *rosaceae* family, e.g., pears, quinces, etc (Gupta and Tara, 2015).

13.2.7.1 Nature of Damage

The insect pest sucks the sap in both stages as adults and nymphs from young leaves and tender terminal shoots of the trees. This pest can be found feeding on the underside of leaves turning them yellow to brown in color. Feeding may result in severe distortion and curling of leaves (Gupta and Tara, 2015. Aphids secrete honey dew which, in case of large aphid populations, trickle down to leaves and fruits, and may result in the development of black mold. Infested trees remain stunted and weak (Altaf *et al.*, 2019).

13.2.7.2 Life Cycle

Several overlapping generations of green apple aphids can be seen throughout the year. This pest over-winters as eggs laid under a protected placed on the host tree. In the onset of spring, the viviparous females are hatched from these over-wintering eggs which contain later generations of both apetrous and winged individuals (Gupta and Tara, 2015). Later towards fall, oviparous females and males mate, and over-wintering eggs are deposited which will survive through the fall season to start the life cycle again in the next season (Biswas and Thakur, 2020).

13.2.7.3 Management

13.2.7.3.1 Biological

Natural enemies that are predacious on aphids are brown lacewings, green lacewings, lady beetles and syrphid fly larvae.

13.2.7.3.2 Oil Spray

Delayed dormant oil sprays of HMO @ 2% are found effective against hibernating eggs.

13.2.7.3.3 Chemical

When the population is above ETL, Dimethoate 30 EC @ 0.1% of water or Imidacloprid 17.8 SL @ 0.28% in water can be sprayed.

13.2.8 Bark Beetle/Shot Hole Borer/Pin Hole Borer

Bark beetle (*Scolytus nitidus*) belongs to family *curculionidae*, order Coleoptera. This pest attacks a wide range of forest and fruit trees.

13.2.8.1 Nature of Damage

Adults and grubs bore into the sapwood making tunnels inside. Pin poles can be seen on the trunk with frass. Leaves of infested trees may wilt and turn yellow. Unhealthy and young trees are usually attacked where females lay their eggs and cause injury (Sofi, 2017). The diseases, drought and unsuitable soil conditions may favor the infestation. A severe infestation may result in the death of the trees.

13.2.8.2 Life Cycle

There are two complete generations and a partially incomplete generation in a year whose larvae enter into hibernation. They over-winter as larvae inside the feeding tunnels under the bark of host trees. Females lay eggs at the entrance hole, and the newly emerged grubs bore into the bark and feed inside these feeding tunnels (Sofi, 2017). They pupate and adults emerge by chewing exit holes out of the tree to attack new trees, and a new cycle starts again.

13.2.8.3 Management

13.2.8.3.1 Cultural

Infested branches need to be pruned and destroyed. Keep trees healthy by applying balanced doses of fertilizers to reduce the chances of pest infestation. Piles of old tree trunks or pruned branches from the previous year should be kept in the orchard to trap egg-laying females; burn these branches to kill the females and eggs.

13.2.8.3.2 Chemical

Holes in the tree trunk and branches can be plastered with Chloropyriphos @ 1.5%WP and soil mixed in 1:1 ratio. At the time of adult emergence, Dimethoate 30 EC @ 0.1% in water can be sprayed to manage the insect pest.

13.2.9 Apple Stem Borer

Apple stem borer (*Aeolesthes sarta*) belongs to the family *Cerambycidae*, order Coleoptera.

13.2.9.1   Nature of Damage

Larvae of the apple stem borer bore holes into the apple stem which may range from 0.5–4 inch in diameter. Larvae form typical zig zag galleries under the bark and later enter the wood. Frass and sawdust can be seen coming out of live holes. In young trees, resin may ooze out from egg laying scars and larval tunnels (Gupta, 2014). After leaving the stem through exit holes, the adults also feed on leaves and bark. Both healthy and stressed trees are attacked; however, stressed trees are more likely to be infested. Infested trees become weak and eventually die (Biswas and Thakur, 2020).

13.2.9.2   Life Cycle

One generation is completed in 2 years; in other words, it partially completes one generation a year. Female beetle lays eggs on the stem inside the oviposition scars which are crescent shaped and are easily visible in the months of July-August. After emergence, the larvae bore into the stem and start feeding inside the feeding tunnels. At the bottom of this gallery, larvae enter into hibernation protected by a double plug made through boring (Gupta, 2014). Larvae later resume feeding at the onset of spring and tunnel deep into the wood. They pupate inside and after 2 weeks adults emerge. Adults stay in these pupal cells until next spring (Biswas and Thakur, 2020).

13.2.9.3   Management

13.2.9.3.1   Chemical

Live holes are plugged with cotton which is impregnated with Dichlorvos 76 EC @ 0.3 % in water, petrol or formalin 4% and then these holes are plastered with mud. Also Naphthalene balls @ 1 ball per each hole can be placed into holes and later plastered with mud. At the time of adult emergence, Chloropyriphos @ 0.1% in water can be sprayed.

13.2.10   Codling Moth

Codling moth (*Cydia pomonella*) belongs to family *Tortricidae,* order Lepidoptera. This pest originated from Asia Minor. This pest is commonly found in the Ladakh area of Jammu and Kashmir, India and in other European countries (Zaki, 1999).

13.2.10.1   Nature of Damage

Larvae of this pest initially feeds on the leaves, but fruit is the major source of food for them. Larvae enters fruit either through the stem end, sides or calyx end (usually calyx end is preferred by the pest) and bore its way into the pulp for feeding. Later, the core area is attacked causing extensive damage to the crop (Zaki, 1999). Frass plugged holes on apple fruit are the characteristic damage of codling moth. Later when larvae are fully grown and about to pupate, they will eat their way out of the apple leaving the large exit holes. Infested fruit will fall prematurely; so in severe cases production will be reduced drastically (Hussain *et al.*, 2018).

13.2.10.2   Life Cycle

There are two generations in a year. This pest hibernates as larvae inside the silken cocoon. In the late spring, adults emerge from the hibernating larvae and start laying eggs in developing fruits. Eggs hatch in approximately 6–20 days, and then the larvae will emerge and bore into fruit for feeding. Larvae will pupate, and a second generation of adults will be born. Adults will lay eggs and then the more damaging second generation of larvae arises. Later, they will enter into hibernation under the loose bark of the tree to survive the harsh winters (Hussain *et al.*, 2018).

13.2.10.3 Management

13.2.10.3.1 Cultural

Remove the loose bark of the tree trunk so the over-wintering larvae cannot take refuge there. Burlap the trees with gunny bags or cardboard during the months of August-September which will attract the over-wintering larvae so they can be easily killed.

13.2.10.3.2 Mechanical

Pheromone traps are used for mass trapping of the adults during 1st and 2nd emergence period @ 4 traps/orchard.

13.2.10.3.3 Biological

For biological control, parasitoids *Trichogramma cacocciae* or *T. embryophagum* @ 2500–5000/ tree can be released twice, once during the emergence of the first generation and once during the second generation.

13.2.10.3.4 Chemical

Apply contact insecticides (Table 13.1) prior to codling moth emergence and repeat the spray 3 weeks after first spray.

13.3 PHYSIOLOGICAL DISORDERS OF APPLE

13.3.1 BITTER PIT

13.3.1.1 Cause

Bitter pit in apple is caused by low levels of calcium in the fruit and is favored by hot and dry summers (Rosenberger *et al.*, 2004).

13.3.1.2 Symptoms

The symptoms appear as minute brownish lesions of 2–10 mm in diameter (depending up on the cultivars) on the surface of the fruit but most commonly on the calyx end. The flesh under the skin becomes dark and corky. After a certain period in storage, the fruit skin develops water soaked and sunken spots (Rosenberger *et al.*, 2004).

13.3.1.3 Management

The incidence of bitter pit can be managed by calcium sprays prior to harvest and calcium dips before storage. The trees must be sprayed 45 days prior to harvest followed by a repeated spray at an interval of 15 days. The calcium dip for 1–2 minutes must be given before fruit storage.

Pruning of apple fruit trees to reduce the leaf area will help control the vigor of the trees and redirect calcium to foliage as well as fruits. Use a balanced dose of fertilizers and avoid excessive usage of nitrogenous and potassium-rich fertilizers.

13.3.2 WATER CORE

13.3.2.1 Cause

Water core in apple occurs due to a high leaf to fruit ratio, high levels of boron and nitrogen in the fruits, high temperatures, low levels of calcium and excessive thinning (Singh *et al.*, 2019).

13.3.2.2 Symptoms

Water core occurs before harvesting the fruits and is characterized as hard, water soaked translucent regions on the fruit surfaces. These regions present on the outer surface are visible only when the infection is very severe. Later, these severely infected fruits may smell and have a fermented taste (Ferguson *et al.*, 2013).

13.3.2.3   Management
- Avoid delayed harvesting.
- Use balanced fertilizer doses.

13.3.3   BROWN HEART

13.3.3.1   Cause
It occurs in storage when the CO_2 concentration increases above 1%. This disorder is mostly associated with large and over matured fruits (Sandhu and Gill, 2013).

13.3.3.2   Symptoms
The symptoms originate in or near the core as brown discoloration of flesh. Initial symptoms are light brown spots which later turn to a dark browning of flesh with a margin of healthy white flesh remaining just below the skin. The symptoms develop in storage and may increase in severity with storage time (Sandhu and Gill, 2013).

13.3.3.3   Management
- Avoid delayed harvesting.
- CO_2 concentrations in controlled atmosphere should be < 1% to reduce the development of brown heart disorder incidence.

13.3.4   SUN SCALD

Incidence and severity of sun scald is favored by hot and dry weather before harvesting, immature fruit at harvest and low calcium and high nitrogen concentrations in the fruit. Inadequate ventilation in cold storage rooms also promotes this disorder (Sandhu and Gill, 2013).

13.3.4.1   Symptoms
Initially, white, tan and yellowish patches appear on the area exposed to direct solar radiations. Later, the patches turn brown and sunken.

13.3.4.2   Management
- Avoid delayed harvesting.
- Proper ventilation in cold storage helps decrease the sun scald incidence.
- Application of an antioxidant like diphenylamine is the most common method used to control scald. The application should be done immediately after harvesting.

13.3.5   SUN BURN

13.3.5.1   Cause
Sun burn occurs due to intense heat of the sun. Drought conditions favor the incidence of sun burn (Saquet *et al.*, 2000).

13.3.5.2   Symptoms
Initially white, tan or yellow patches appear on the fruits exposed to direct solar radiations. The patches turn blackish brown and cracks develop on the affected area under severe conditions (Saquet *et al.*, 2000).

13.3.5.3 Management
- Training and pruning should be done properly to avoid heat and sun exposure.
- Pruned orchards should be irrigated regularly to avoid heat stress.
- Application of balanced nutrient doses for maintenance of proper foliage to cover the fruits directly facing the sun is recommended.

13.3.6 RUSSETING

13.3.6.1 Cause
In apple crop immediately after petal fall, russeting occurs. Cultivars with thin cuticles are more susceptible to russeting (Bhat and Khan, 2010). This disorder is commonly noticed on exposed fruits rather than on shaded fruits. Russeting can be a sign of a serious problems like phototoxicity, frost damage, fungal and bacterial growth (Singh *et al.*, 2019).

13.3.6.2 Symptoms
A brown layer of suberized cells forms in the lower epidermal region. As corky cells develop in this area, they push outward and become exposed to the surface as the fruit matures (Singh *et al.*, 2019).

13.3.6.3 Management
- Proper pruning should be adopted for good aeration and to prevent humid conditions.
- Adequate manuring, irrigation and effective pest management can reduce the russeting.

13.3.7 CORK SPOT

13.3.7.1 Cause
It is caused due to a deficiency of boron and calcium.

13.3.7.2 Symptoms
Initial symptoms are localized green to brown spots on the surface of fruit. The cork spots later turn brown and corky and are more prominent toward the calyx half of apple. Under severe conditions, fruit cracking is also observed (Saquet *et al.*, 2000).

13.3.7.3 Management
- Proper management of nutrients, especially boron and calcium, help prevent cork spots of the apples.
- Calcium chloride @ 1.5 tablespoons/gallon of water should be sprayed in four sprays beginning 2 weeks after full bloom and continuing at 12 to 15 day intervals until harvest.

13.4 CONCLUSION

Apple is a temperate fruit crop cultivated throughout the world for table purposes due to their high nutritious value which contains almost all the important nutrients. It is infested with various insect pests and physiological disorders responsible for huge yield losses. Insect pests such as the San Jose scale, woolly apple aphid, European red mite (ERM), blossom thrips, hairy caterpillar, green apple aphid, bark beetle/shot hole borer/pin hole borer, apple stem borer, codling moth, etc. are the major constraints in quality apple production. Major physiological disorders such as bitter pit, water core, brown heart, sun scald, sun burn, russeting, cork spot, etc. also play a significant role in quality reduction in apples, therefore reducing the marketable value of the produce. Management of these insect pests and physiological disorders are essential to increase the production as well as the quality of apples. Insect pest and physiological disorder management in an integrated and eco-friendly manner helps to improve the quality assurance of the apple produce.

REFERENCES

Ahmad, M. and Sajad, S. 2017. Major Insect pests of Apple grown under temperate conditions of Kashmir and their Integrated Pest Management. In: *Insect Pests of Apple.* www.researchgate.net.

Altaf, S., Ahad, I., Pathania, S. S., Lone, G. M., Peer, F. A. and Maqbool, S. 2019. Insect pest complex of apple nurseries in North Kashmir. *Journal of Entomology and Zoology Studies* 7(3):697–700.

Beigh. M. A., Peer, Q. J. A., Kher, S. K. and Ganai, N. A. 2015. Disease and pest management in apple: Farmers perception and adoption in J&K state. *Journal of Applied and Natural Science* 7(1):293–297.

Bhat, S. A. and Khan, F. A. 2010. Physiological disorders of apple and pear. In *Training Compendium-Advanced Technologies for Post Harvest Management of Food Crops*, ed. M. Y. Ghani, F. A. Khan and S. A. Bhat, 25–27.

Biswas, D. and Thakur, S. A. 2020. Seasonal incidence of major insect pests of apple in mid hills of megha-laya. *Indian Journal of Entomlogy.* http://doi.org/10.5958/0974-8172.2020.00204.7.

Faruque, Z. and Dan, G. 2013. European red mite management in tree fruit orchards. *Agricultural News Magazine* 97:9–10.

Ferguson, I., Volz, R. and Woolf, A. 1999. Pre-harvest factors affecting physiological disorders of fruit Postharvest. *Biology and Technology* 15:255–262.

Gupta, R. 2014. Management of apple tree borer. *Aeolesthes Holosericea* Fabricius on apple trees (*Malus domestica* Borkh.) In Jammu Province, Jammu and Kashmir State, India. *Journal of Entomology and Zoology Studies* 2:96–98.

Gupta, R. and Tara, J. S. 2015. Life history of *Aphis pomi* De Geer (Green apple aphid) on apple plantations in Jammu province, J&K, India. *Munis Entomology & Zoology Journal* 10(2):388–391.

Hix, R. L., Pless, C. D., Deyton, D. E. and Sams, C. E. 1999. Management of San Jose Scale on apple with soybean-oil dormant sprays. *HortScience* 34(1):106–108.

Hussain, B., Buhroo, A. A., War, A. R. and Sheerwani, A. 2018. Insect-pest complex and integrated pest management of apple in Jammu and Kashmir, India. In *Apple: Production and Value Chain Analysis*, ed. N. Ahmed, S. A. Wani, and W. A. Wani, 16–35. New Delhi: Daya Publishing House.

Khan, R. A. 1941. Occurrence of the gypsy moth, *Lymantria obfuscate* Walk. In Shimla hills. *Indian Journal of Entomology* 31(2):338.

Lal, K. B. and Singh, R. N. 1947. Seasonal history and field ecology of the wooly aphid in the Kamaon hills. *Indian Journal of Agricultural Sciences* 17(4):211–218.

Martins, C. R., Hoffmann, A., Rombaldi, C. V., Farias, R. D. M. and Teodoro, A. V. 2013. Apple biological and physiological disorders in the orchard and in postharvest according to production system. *Revista Braileira de Fruticultura* 35(1):1–8.

Moinina, A., Lahlila, R. and Baulif, M. 2019. Important pest, diseases and weather conditions affecting apple production: Current state and perspectives. *Revue Marocaine des Sciences Agronomiques et Veterinaires* 7(1):71–87.

Pinero, J., Garofalo, E. and Schloemann, S. 2020. *Apple IPM – San Jose Scale.* UMass Extension. http://ag.umass.edu/fruit (accessed September 23, 2021).

Rosenberger, D. A., Schupp, J. R., Hoying, S. A., Chang, L. and Wakins, C. B. 2004. Controlling bitter pit in 'Honeycrisp' apples. *Horticulture Technology* 14:3.

Sandhu, S. and Gill, B. K. 2013. *Physiological Disorders of Fruit Crops.* New Delhi: New India Publishing Agency.

Saquet, A. A., Streif, J. and Bangerth, F. 2000. Changes in ATP, ADP and pyridine nucleotide levels related to the incidence of physiological disorders in conference peas and Jona gold apples during controlled atmosphere storage. *Journal of Horticultural Science and Biotechnology* 75:243–249.

Sherwani, A., Mukhtar, M., Sofi, M. A. and Maqsood, S. 2017. Field evaluation of new ovicidal acaricides (Tafethion) against two major insect pests of apple in Kashmir, India. *International Journal of Current Microbiology and Applied Sciences* 6(7):3905–3916.

Sherwani, A., Mukhtar, M. and Wani, A. A. 2016. Insect pests of apple and their management. In *Insect Pests Management of Fruit Crops*, ed. A. K. Pandey and P. Mall, 295–306. New Delhi: Biotech Books.

Singh, N., Sharma, D. D., Singh, G., Thakur, K. K. and Kumari, S. 2019. Physiological disorders and their management in apple and pear fruit production. *Advanced Botany* 1:39–66.

Sofi, M. A. 2017. *Major Insect Pests of Apple Grown under Temperate Conditions of Kashmir and Their Integrated Pest Management.* www.researchgate.net/publication/315118003 (accessed September 24, 2021).

Sofi, M. A. and Hussain, B. 2008. Pest management strategies against black cap stage of San Jose scale in apple orchards of Kashmir valley. *Indian Journal of Entomology* 70(40):398–399.

Zaki, F. A. 1999. Incidence and biology of codling moth, *Cydia pomonella* L., in Ladakh (Jammu and Kashmir). *Applied Biological Research* 1:75–78.

14 Recent Developments in Harvest and Postharvest Management

Nusrat Jan, Gousia Gani, Haamiyah Sidiq, Sajad Mohd Wani, and Syed Zameer Hussain

CONTENTS

14.1 INTRODUCTION

The operations used for harvesting and postharvesting like sizing, cleaning, sorting, packaging and transportation mostly depend on the physical characteristics of fruits (Tabatabaeefar and Rajabipour, 2005). The projected area, volume and mass of an object are vital physical properties central in determining sizing systems (Khodabandehloo, 1999). It is therefore important to decrease the product loss by considering all the criteria necessary for the determination of these physical properties. Harvesting the fruits at an appropriate time is the principal component of determining the shelf life and quality of fruits. When harvested at an advanced maturity, fruits have a shorter shelf life and are susceptible to mechanical injury, physiological disorders, and attacks by pathogens (Juan et al., 1999). Apart from this, immaturity and over-maturity of fruit result in other serious causes of postharvest losses which are characterized by its careless harvesting (Ingle et al., 2000).

Apple (*Malus domestica* L.) belongs to the genus *Malus* and family *Rosaceae*, and thousands of cultivars are grown all over the world. From an economic point of view, this fruit is a significant species for the union territory of Jammu and Kashmir. Apples are beneficial due to the presence

DOI: 10.1201/9781003239925-14

of a good amount of bioactive compounds. Flavonol, quercetin and their derivatives are found in abundance in apples which belong to the major dietary sources of flavonoids (Herranz et al., 2019), and several studies have demonstrated that these bioactives have antioxidant potential (Wang et al., 2020a), antimicrobial (Tian et al., 2021) anti-inflammatory properties (Yong et al., 2020) anti-depressive (Dimpfel, 2009) and anti-carcinogenic effects (Imran et al., 2020). These compounds also protect against diabetes (Ebrahimpour et al., 2020), arteriosclerosis (Ishizawa et al., 2011), oral diseases (Wang et al., 2020b) neurodegenerative (Sahoo et al., 2020) and cardiovascular diseases (DiNicolantonio and Mccarty, 2020). The beneficial effects of anthocyanins, flavones, and flavonols (Chang et al., 2018); phenolic acids (Veeriah et al., 2008); triterpenoids (He et al., 2012); pectin and pectic oligosaccharides (Chung et al., 2017) and polysaccharides (Li et al., 2020a) found in apples on intestinal inflammation and colorectal cancer (Bars-Cortina et al., 2020) have been studied by various researchers. The type and concentration of bioactive molecules in apples differ markedly according to the cultivar and species under study (Herranz et al., 2019; Williams et al., 2013).

Apples are consumed extensively all around the world in all countries, and they are popular because of their nutritional content, texture, color, juiciness and appreciated taste. It is a healthy food available at relatively low prices in markets throughout the year (Bars-Cortina et al., 2020; Mohebbi et al., 2020). Marketable cultivation of apples in India is limited to Jammu and Kashmir, Himachal Pradesh and Uttrakhand, which account for about 2.5% of the world production. The productivity of apples in the union territory of Jammu and Kashmir per hectare is 11.43 tons, which is comparable to the world average of 10.82 tons per hectare and is much higher than the national average (Hassan et al., 2020). Some of the districts yield higher than the production in European countries such as the district of Baramulla with an average yield of 21 tons per hectare (Naqash, 2015). Moreover, modern technologies can be a boon toward increasing apple yield. For high density plantations, the lands with a slight gradient or the flatlands are of great potential. The supply of apples lasts for more than half a year in the union territory of Jammu and Kashmir which is due to the harvest season extending from July to November. The apple fruit still does not fetch the price of their counterparts; however, the apples of the union territory of Jammu and Kashmir are much tastier than their counterparts (Bhat and Choure, 2014).

14.2 HARVESTING OF APPLES

Apples are climacteric in nature and continuously ripen after harvest, so they can be harvested before full ripeness and stored to obtain a good price in the markets. Early harvesting can make the fruit susceptible to bitter rot, internal breakdown and scald, and the smaller-sized fruits have less color and flavor. In these apples, the outer surface, which is waxy, is not fully formed and there is more reduction in mass due to the loss of water (Juan et al., 1999; Zerbini et al., 1999). In early harvested apples, the structure of the cuticle is not fully developed, which is another reason for more intensive evaporation. The cuticle acts as a shield for the entry of pathogens (Ihabi et al., 1998). Even under optimal conditions, late-harvested apples are over-mature which makes their storage complicated (Braun et al., 1995; Ingle et al., 2000). The proneness to mechanical injures, water core, sensitiveness to low-temperature breakdown and more rot is mostly found in apples that are harvested too late (Hribar et al., 1996). The apples picked at an optimal harvest time endure a storage period of more than 6 months (Casals et al., 2006). Furthermore, apple fruit gradually loses its quality during storage due to fungal diseases and metabolic, oxidative or enzymatic changes. In developing countries, the losses at postharvest range from 20% to 50%, and in South Africa it ranges from 25% to 50% (Wand et al., 2006; Kitinoja and Kader, 2015; Den Breeyen et al., 2020). The important thing to consider with the storage of apples is fungal attack. *Botrytis cinerea* is the more common reason for the formation of blue, gray and black mold (Moscetti et al., 2013) (Table 14.1). Thus, postharvest management of apples is necessary in order to increase their shelf life.

TABLE 14.1
Characterization of Postharvest Fungal Diseases

	Disease name	Causal agent	References
Lenticel rot	Bitter rot	*Colletotrichum gloeosporioides*	Jones and Aldwinckle (1990)
	Bull's eye rot	*Cryptosporiopsis curvispora*	Jones and Aldwinckle (1990)
	Gloeosporium rot	*Trichoseptoria fructigena*	Edney (1983)
Core rot	Dry core rot and moldy core rot	*Alternaria spp* *Botrytis cinerea*	Jones and Aldwinckle (1990)
Wound pathogens	Blue mold	*Penicillium spp*	Bondoux (1992)
	Gray mold	*Botrytis cinerea*	Bondoux (1992)
	Brown rot	*Monilinia fructicola* and *M. fructigena*	Lichou et al., (2002)
Other alterations	Mucor rot	*Mucor piriformis*	Sanderson (2000)
	Phytophtora rot	*Phytophthora syringae* and *P. cactorum*	Creemers (1998)

14.3 POSTHARVEST MANAGEMENT OF APPLES

Processes done immediately after harvesting the produce are called postharvest management and include cooling, sorting, cleaning and packing. When an apple is harvested from its parent plant or removed from the ground, it starts to deteriorate. Therefore, handling the apples once harvested is absolutely essential for quality and shelf life of produce. Produce perishability can be minimized by using efficient postharvest management which maintains its quality and increases its shelf life, thereby helping remuneration among farmers and leading to a better productivity of fruit.

14.3.1 Postharvest Treatments

14.3.1.1 Sorting, Grading and Packing

Postharvest losses are usually the result of improper sorting, grading and packaging practices; inadequate sanitation and cleaning; damage from long fingernails; over-packing crates and throwing the fruits from a distance into crates (IALC, 2006). Organic matter (soil or irrelevant plant parts) acts as a substrate for fruit decay. Therefore, all areas where the product is packed should be kept under sanitary conditions. The primary purpose of package is to contain and protect the fruit. There are two primary types of packaging for fresh fruits and vegetables. In the first one, the products are offered to the consumers in a single pack. In the second one, the pack contains the consumer packs for transportation to the retail shops. For market or customer needs, the consumer packs are shaped in terms of the quantity which determines the size of the package. In some conditions, the overall package sizes are decided according to the person's logical pick up (Thompson, 2003). For packing of horticultural commodities, different materials of various sizes and shapes can be used. Marketing and distribution of horticultural commodities involve packages of suitable size (Thompson and Mitchel, 2002). For product presentation, a feature of quality or grading is a vitally essential practice because it is often judged by standardization and/or uniformity. Handling (e.g. unitization) and marketing of a standard product needs a good presentation which can be attained by important terms like standardization and/or uniformity. The features like weight, size, defects, composition, color and/or a combination of these are usually used to grade the products (Wills et al., 2007).

14.3.2   Pre-Cooling

After harvesting the produce, reduce the temperature of the commodity as soon as possible to enhance its shelf life and quality by reducing the respiration rate which also decreases the rates of moisture loss and decay by reducing volume losses. Pre-cooling is the first step since it is done as early as possible after harvest and is completed before the placement of product into marine containers, cold storage or trucks used for refrigeration for shipment to market (Lisa and Thompson, 2010). Wijewardane and Guleria (2009) compared the pre-cooled apples and control samples and reported that there was relatively less loss in the weight of pre-cooled apple fruits when equated with the control sample. There is less loss of moisture from the fruits which are pre-cooled due to an elevated relative humidity in the storage atmosphere. A temperature reduction can help effortlessly attain such conditions. Commercially there are various hydro-cooler designs in operation. The overall process efficiencies and cooling rates are different for different hydrocooling methods. The way in which the product is placed or moved in the cooler and the method of cooling is unique for an individual technique. The different types of hydro-coolers include batch type, conventional (flood) type and immersion type (Brosnan and Sun, 2003).

14.3.2.1   Room Cooling

An old and well-established technique for pre-cooled apples is to place them in a pre-cooling chamber or cold storage. The produce is exposed to cold air when it is placed into a cold room using boxes (plastic, fiberboard or wooden) or other containers. In cold storage, the cooled air is produced by using induced draft or forced coolers. Consequently, proper packaging (fully vented) or containers, as well as stacking patterns, should be used to attain efficient and fast cooling (Ryall and Pentzer, 1982). For good storage management, more space is required to get the best cooling rates and to utilize the storage space completely. Some rearranging is therefore done necessarily after the cooling of the product. However, room cooling results in a variable and slow temperature reduction because most of the cooling occurs through conduction (Gibbon, 2009).

14.3.2.2   Hydrocooling

The reduction of the product temperature in smaller or bulk containers using chilled or cold water before further packing is known as hydrocooling. The overall process efficiencies and cooling rates are different for different hydrocooling techniques, and a variety of hydrocooler designs are commercially available. The way in which the produce is positioned or moved in the cooler and the type of cooling reveal the differences between the various techniques. There are several varieties of hydrocoolers including immersion types, conventional (flood) types and batch types (Brosnan and Sun, 2003). The decay threat linked with recirculated water is a problem associated with most hydrocoolers as it can result in contamination of the cooled produce and may accumulate decay-producing organisms in the system (Reid, 2000). Approved phenol compounds or chlorine components at the concentration of 100 ppm act as a mild disinfectant and are used to prevent the accumulation of microorganisms and produce-cooled air to the chamber that is not influenced by the use of these chemicals. It is relatively quick in comparison to other available pre-cooling techniques, which is another advantage of this process.

14.3.2.3   Forced Air Cooling

The modification involves cooling the room and exposing produce packaging to elevated air pressure on one side compared to the other side is often known as pressure cooling or forced air cooling. The cooling air is driven through the individual containers through the use of stack bagging and stacking patterns which are usually done in this technique. A number of forced air-cooling arrangements can be used to cool produce. These include (a) high-velocity air in refrigerated rooms, (b) forcing air through spaces in bulk products as it passes through a cooling tunnel on continuous conveyors and (c) using the pressure differential technique to encourage forced airflow through packed produce. Each of these technologies is commercially viable and suitable for specific commodities when appropriately applied (Nunes et al., 2005).

14.3.2.4 Vacuum Cooling

The evaporation of free water from any porous material is achieved by a rapid evaporative cooling technique known as vacuum. It is a competent cooling method accessible and is expansively used for cooling some food and agricultural produce (Ozturk et al., 2017. Some advantages of vacuum cooling involve an improvement of product quality and safety, extension of product shelf life and short processing time. During vacuum cooling, evaporation is slow, cooling is minimal and produce temperatures are constant until saturation pressure is reached. The flashpoint occurs at approximately this pressure. It is the point where evaporation results in quick cooling and moisture loss of the produce (Deng et al., 2011).

14.3.3 CONTROLLED ATMOSPHERE STORAGE

Controlled atmosphere (CA) storage remains one of the most common methods for storing apples which are among the most cultivated and consumed fruits in the world. The most essential parameters for extending the storage period in CA storage technology are lessening the oxygen (O_2) concentration and raising the carbon dioxide (CO_2) concentration in the storage system. As a result, the excellence of the fruit is well-maintained for extended periods of time, and the losses incurred during postharvest are minimized during storage in a CA. The determination of the possibility of the extended storage stretch in a CA is governed by (a) harvest time fruit maturity, (b) composition of the CA atmosphere throughout storage and (c) the kind of fruit (Thompson, 2010). For the maintenance of quality, it is of paramount importance to store fruits that are climacteric like apple in ideal settings (Bertone et al., 2012). Storage in a CA inhibits the ripening of fruit by downsizing the O_2 levels, respiration rates and ethylene biosynthesis (Veltman et al., 2003). When placed in a CA, fruits deescalate the synthesis of ethylene, and the rate of respiration also diminishes (Wright et al., 2015). Metabolic activity can be immensely decreased after harvest by a decrement in the level of O_2, and therefore, this technique is of utmost significance (Tuna Gunes and Horzum, 2017).

14.3.4 DYNAMIC CONTROLLED ATMOSPHERE STORAGE

The pome industry utilizes various postharvest treatments and a few of the nonchemical treatments are controlled atmospheres (CA) and initial low oxygen stress (ILOS) (Sabban-Amin et al., 2011; Zanella, 2003). These techniques are notable for lowering the degree of ethylene production and the rate of respiration which is the basic biochemical processes happening during the storage of fruits (Wright et al., 2015). Fruit maintained in low oxygen environments has a lower occurrence of greasiness and superficial scald (DeLong et al., 2004). Low oxygen storages (LOS) have a number of advantages, but there are also some drawbacks associated with this technology including the damage done due to the utilization of lower O_2 levels and an occurrence of off flavors (Wright et al., 2015; Imahori et al., 2005). To enhance the effectiveness of LOS, dynamic controlled atmosphere (DCA) technology is employed (Weber et al., 2015; Wright et al., 2015). DCA is one of the recent and widely used techniques in the apple industry (Weber et al., 2015; Wright et al., 2015). For prevention of the formation of scald and retaining the apple quality, diphenylamine (DPA) was used. But now DCA storage is harnessed which uses fluorescence-based technology to detect fruit responses to lower concentrations of O_2 and enables the prevention of disorders (DeEll et al., 1999; Prange et al., 2003; Zanella et al., 2005; DeLong et al., 2007). In this way, tolerable low levels of O_2 can be set while preventing the formation of anaerobic conditions (Prange et al., 2003; Gasser et al., 2008). The growing and increasing demands for organic production of apples can be met by the application of this innovative technology (Prange et al., 2006).

14.3.5 Heat Treatments

Heat treatments have been found to produce changes in the ripening of the fruit, for instance by inhibiting the formation of ethylene and the performance of the enzymes that are accountable for degrading the cell wall due to the alterations in the formation of proteins and a changed gene expression (Paull and Chen, 2000). There is a noticeable outcome of heat treatment on the physiology, quality and life of apples, and the principal outcome of the heat treatment is the reduced synthesis of ethylene (Wang et al., 1997). The growth of mold can be reduced at a temperature of 38°C. Hot water treatments are also employed to control postharvest disorders as an alternative method (Margonsan et al., 1997). Because ripe fruits are very prone to disease infection, various techniques of disease control are required. To increase the quality, dipping apples in hot water at harvest or following the opening of CA storerooms can be utilized (Maxin et al., 2012).

14.3.6 1-Methylcyclopropene (1-MCP) Treatment

An inhibitor by nature, 1-MCP (Smart Fresh™) gas slows down the progression of senescence and the ripening of numerous products, and it has a major role to play in the apple industry (Watkins, 2006; Watkins, 2008; Huber, 2008). It reduces ethylene generation, softening, and the ripening of apples during its storage in air and CA (Bai et al., 2005; Rupasinghe et al., 2000). There are abundant aspects that contribute to the regulation of fruit ripening by this gas, for instance the variety of fruit, internal ethylene concentration (IEC), the duration and temperature of the treatment employed and the time interval between harvest and 1-MCP treatment which is generally a 3–7 day interval depending upon the fruit variety and maturity (Argenta et al., 2005; Amarante et al., 2010; Tatsuki et al., 2011; AgroFresh, 2011; Watkins and Nock, 2005). Inhibition of physiological diseases such as superficial scald (Lurie and Watkins, 2012) and senescent problems (Fan et al., 1999) have also been linked to the introduction of 1-MCP. However, 1-MCP can accelerate other problems including injury due to carbon dioxide (Fawbush et al., 2008; Argenta et al., 2010) and browning of flesh (Watkins, 2008; Jung and Watkins, 2011).

14.3.7 Irradiation

Gamma radiation is commonly used in combination with cold storage of fruits to delay ripening, prevent weight loss and extend shelf life. A low dose of 1 kGy on apples can cause softness in the fruit through a wrinkling of cell membranes and a dissolution of the middle lamella (Kovacs et al., 1988). This can be explained possibly by a decrease in the total pectin and protopectin or the change of insoluble pectin to a more soluble form or by degradation of pectin and cellulose (Al-Bachir, 1999). If higher doses are used, fruit tissue will be injured and damaged (Shi et al., 1993). The probable multiple benefits of low doses of irradiation (0.3 and 0.6 KGy) coupled with CA results in considerable firmness, total soluble solids, weight loss, antioxidant activity and phenolic contents retention. Mostafavi et al., (2012) found that gamma rays are an effective strategy for reducing apple quality losses and improving postharvest preservation.

14.3.8 Waxing and Coating

Nowadays, waxing and coating is a typical postharvest technique employed to lengthen the shelf stability of fruits. It is generally employed for 'Red Delicious' apples since it conceals the bruises, makes them glossy and improves firmness and shelf stability. The aim of such treatments is the suppression of the biochemical processes which are responsible for aging, and, as a result, a shorter storage life (Bai et al., 2002; Ganai et al., 2015; Alleyne and Hagenmaier, 2000).

14.4 PROCESSED APPLE PRODUCTS

Depending on the processing technology used, a variety of products can be developed from apples (Li et al., 2020b) beside fresh consumption. Some of these apple products include juices (Pruksasri et al., 2020; Sauceda-Galvez et al., 2021), canned products (Li et al., 2020b; Dobias et al., 2006), dehydrated products (Kidon and Grabowska, 2021; Guine et al., 2014) or purees (Lan et al., 2021). In addition, other apple products such as ciders (Sousa et al., 2020) and probiotic fermented apple juices (Peng et al., 2020) are also developed using the fermentation processes.

14.5 CONCLUSION

In conclusion, it has been demonstrated that postharvest treatments are quite effective in maintaining apple quality and controlling physiological disorders. A suitable method of storage for different varieties of apples can therefore be selected while taking into consideration the consequences of various treatments on the quality of each variety.

REFERENCES

AgroFresh, 2011. *SmartFreshSM Quality System, Apple Use Recommendations.* AgroFresh Inc., PA.

Al-Bachir, M. 1999. Effect of gamma irradiation on storability of apples (*Malus domestica* L.). *Plant Foods for Human Nutrition*, 54:1–11.

Alleyne, V. and R. D. Hagenmaier, 2000. Candelilla-shellac-an alternatie formulation for coating apples (*Malus domestica* Borkh). *Horticultural Science*, 35:691–693.

Amarante, C. V. T., L. C. Do-Argenta, M. J. Vieira, and C. A. Steffens, 2010. Changes of 1-MCP efficiency by delaying its postharvest application on 'Fuji Suprema' apples. *Revista Brasileira de Fruticultura*, 32:984–992.

Argenta, L. C., X. F. Fan, and J. P. Mattheis, 2005. Factors affecting efficacy of 1-MCP to maintain quality of apple fruit after storage. *Acta Horticulturae*, 682:1249–1255.

Argenta, L. C., J. P. Mattheis, and X. Fan, 2010. Interactive effects of CA storage 1- methylcyclopropene and methyl jasmonate on quality of apple fruit. *Acta Horticulturae*, 857:259–266.

Bai, J. H., E. A. Baldwin, K. L. Goodner, J. P. Mattheis, and J. K. Brecht, 2005. Response of four apple cultivars to 1-methylcyclopropene treatment and controlled atmosphere storage. *HortScience*, 40:1534–1538.

Bai, J. H., E. A. Baldwin, and R. D. Hagenmaier, 2002. Alternatives to shellac coatings provide comparable benefits in terms of gloss, internal gases modification, and quality for "Delicious" apple fruit. *Horticultural Science*, 37:559–563.

Bars-Cortina, D., A. Martinez-Bardaji, A. Macia, M.J. Motilva, and C. Pinol-Felis, 2020. Consumption evaluation of one apple flesh a day in the initial phases prior to adenoma/adenocarcinoma in an azoxymethane rat colon carcinogenesis model. *Journal of Nutritional Biochemistry*, 83:108418.

Bertone, E., A. Venturello, R. Leardi, and F. Geobaldo, 2012. Prediction of the optimum harvest time of 'Scarlet' apples using DR-UV – Vis and NIR spectroscopy. *Postharvest Biology and Technology*, 69:15–23.

Bhat, T. A. and T. Choure, 2014. Status and strength of apple industry in Jammu and Kashmir. *International Journal of Research*, 1(4):277–283.

Bondoux, P. (ed.), 1992. *Maladies de conservation des fruits a pepins, pommes et poires.* INRA and PHM (revue horticole), Paris, France, 173 p.

Braun, H., B. Brosh, P. Ecker, and K. Krumbock, 1995. Changes in quality off apples before, during and after CA-cold storage. *Obstau Und Fruchteverwertung*, 45(5–6):143–206.

Brosnan, T. and D. Sun, 2003. Influence of modulated vacuum cooling on the cooling rate, mass loss and vase life of cut lily flowers. *Biosystems Engineering*, 86(1):45–49.

Casals, M., J. Bonany, J. Carbo, S. Alegre, I. Iglesias, D. Molina, T. Casero, and I. Recasens, 2006. Establishment of a criterion to determine the optimal harvest date of 'Gala' apples based on consumer preferences. *Journal of Fruit and Ornamental Plant Research*, 14:53–63.

Chang, H., L. Lei, Y. Zhou, F. Ye, and G. Zhao, 2018. Dietary flavonoids and the risk of colorectal cancer: An updated meta-analysis of epidemiological studies. *Nutrients*, 10:950.

Chung, W. S. F., M. Meijerink, B. Zeuner, J. Holck, P. Louis, A. S. Meyer, J. M. Wells, H. J. Flint, and S. H. Duncan, 2017. Prebiotic potential of pectin and pectic oligosaccharides to promote anti-inflammatory commensal bacteria in the human colon. *FEMS Microbiology Ecology*, 93.

Creemers, P. 1998. Lutte contre les maladies de conservation: Situation actuelle et perspectives nouvelles. *Le Fruit Belge*, 472:37–49.

DeEll, J. R., O. van Kooten, K. R. Prange, and D. P. Murr, 1999. Applications of chlorophyll fluorescence techniques in postharvest physiology. *Horticutural Reviews*, 23:69–107.

DeLong, J. M., R. K. Prange, and P. A. Harrison, 2004. The influence of 1-methylcyclopropene on 'Cortland' and 'McIntosh' apple quality following long-term storage. *Horticultural Science*, 39(5):1062–1065.

DeLong, J. M., R. K. Prange, and P. A. Harrison, 2007. Chlorophyll fluorescence-based low O2 storage of organic 'Cortland' and 'Delicious' apples. *Acta Horticulturae*, 737:31–37.

Den Breeyen, A., J. Rochefort, A. Russouw, J. Meitz-Hopkins, and C. L. Lennox, 2020. Preharvest detection and postharvest incidence of phlyctema vagabunda on 'cripps Pink' apples in South Africa. *Plant Diseases*, 104(3):841–846.

Deng, Y., X. Song, and Y. Li, 2011. Impact of pressure reduction rate on the quality of Steamed Stuffed Bun. *Journal of Agriculture, Science and Technology*, 13:377–386.

Dimpfel, W. 2009. Rat electropharmacograms of the flavonoids rutin and quercetin in comparison to those of moclobemide and clinically used reference drugs suggest antidepressive and/or neuroprotective action. *Phytomedicine*, 16:287–294.

DiNicolantonio, J. J. and M. F. Mccarty, 2020. Targeting Casein kinase 2 with quercetin or enzymatically modified isoquercitrin as a strategy for boosting the type 1 interferon response to viruses and promoting cardiovascular health. *Medical Hypotheses*, 142:109800.

Dobias, J., M. Voldrich, and D. Curda, 2006. Heating of canned fruits and vegetables: Deaeration and texture changes. *Journal of Food Engineering*, 77:421–425.

Ebrahimpour, S., M. Zakeri, and A. Esmaeili, 2020. Crosstalk between obesity, diabetes, and alzheimer's disease: Introducing quercetin as an effective triple herbal medicine. *Ageing Research Reviews*, 62:101095.

Edney, K. L. 1983. Top fruit. In C. Dennis (Ed.), *Post-harvest pathology of fruits and vegetable*. Academic Press Inc., London, pp. 43–71.

Fan, X. T., J. P. Mattheis, and S. M. Blankenship, 1999. Development of apple superficial scald, soft scald, core flush, and greasiness is reduced by MCP. *Journal of Agricultural and Food Chemistry*, 47:3063–3068.

Fawbush, F., J. F. Nock, and C. B. Watkins, 2008. External carbon dioxide injury and 1-methylcyclopropene. *Postharvest Biology and Technology*, 48:92–98.

Ganai, S. A., H. Ahsan, I. A. Wani, A. A. Lone, S. A. Mir, and S. M. Wani, 2015. Colour changes during storage of apple cv. Red delicious-influence of harvest dates, precooling, calcium chloride and waxing. *International Food Research Journal*, 22(1):196–201.

Gasser, F., T. Eppler, W. Naunheim, S. Gabioud, and E. Hoehn, 2008. Control of the critical oxygen level during dynamic CA storage of apples by monitoring respiration as well as chlorophyll fluorescence. *Acta Horticulturae*, 796:69–76.

Gibbon, T. 2009. Some observations of temperatures and cooling rates of vegetables in commercial cold stores. *Journal of Agricultural Engineering Research*, 17(4):332–337.

Guine, R. P., A. C. Cruz, and M. Mendes, 2014. Convective drying of apples: Kinetic study, evaluation of mass transfer properties and data analysis using artificial neural networks. *International Journal of Food Engineering*, 10:281–299.

Hassan, B., M. Bhattacharjee, and S. A. Wani, 2020. Economic analysis of high-density apple plantation scheme in Jammu and Kashmir. *Asian Journal of Agriculture and rural Development*, 10(1):379–391.

He, X., Y. Wang, H. Hu, and Z. Zhang, 2012. In vitro and in vivo antimammary tumor activities and mechanisms of the apple total triterpenoids. *Journal of Agricultural and Food Chemistry*, 60:9430–9436.

Herranz, B., I. Fernandez-Jalao, M. D. Alvarez, A. Quiles, C. Sánchez-Moreno, I. Hernando, and B. De Ancos, 2019. Phenolic compounds, microstructure and viscosity of onion and apple products subjected to in vitro gastrointestinal digestion. *Innovative Food Science and Emerging Technology*, 51:114–125.

Hribar, J., A. Plestenjak, M. Simcic, R. Vidrih, and D. Patako, 1996. Influence of ecological conditions on optimum harvest date in Slovenia. In: A. de Jager, D. Johanson, E. Hohn (eds.), The Postharvest Treatment of Fruit and Vegetables. Determination and Prediction of Optimum Harvest Date of Apples and Pears. COST 94. *European Commission. Luxembourg*: 49–51.

Huber, D. J. 2008. Suppression of ethylene responses through application of 1-methylcyclopropene: a powerful tool for elucidating ripening and senescence mechanisms in climacteric and nonclimacteric fruits and vegetables. *HortScience*, 43:106–111.

IALC. 2006. Post-harvest management of fruits and vegetables. *Training manual, Volume-V-E. University of Illinois at Urbana-Champaign, USA*. International Arid Lands Consortium and NWFP Agriculture University Peshawar.

Ihabi, M., C. Rafin, E. Veighie, and M. Sancholle, 1998. Storage diseases of apples: Orchard or in storage. In *First transnational workshop on biological, integrated, and rational control*. Lille, France 21–23 January 1998. Service Regional de la Protection des Vegetaux, Nord Pas-de Calais, 91–92.

Imahori, Y., K. Uemura, I. Kishioka, H. Fujiwara, A. Tulio, Y. Ueda, and K. Chachin, 2005. Relationship between low-oxygen injury and ethanol metabolism in various fruits and vegetables. *Acta Horticulturae*, 682:1103–1108.

Imran, M., M. K. Iqubal, K. Imtiyaz, S. Saleem, S. Mittal, M. M. A. Rizvi, J. Ali, and S. Baboota, 2020. Topical nanostructured lipid carrier gel of quercetin and resveratrol: Formulation, optimization, in vitro and ex vivo study for the treatment of skin cancer. *International Journal of Pharmaceutics*, 587:119705.

Ingle, M., M. C. D'Souza, and E. C. Townsend, 2000. Fruit characteristics of York apples during development and after storage. *Horticultural Science*, 35:95–98.

Ishizawa, K., M. Yoshizumi, Y. Kawai, J. Terao, Y. Kihira, Y. Ikeda, S. Tomita, K. Minakuchi, K. Tsuchiya, and T. Tamaki, 2011. Pharmacology in health food: Metabolism of Quercetin in vivo and its protective effect against arteriosclerosis. *Journal of Pharmacological Sciences*, 115:466–470.

Jones, A. L. and H. S. Aldwinckle, (eds.) 1990. *Compendium of apple and pear diseases*, 100. APS Press, Washington, DC.

Juan, J. L., J. Frances, E. Montesinos, F. Camps, and J. Bonany 1999. Effect of harvest date on quality and decay losses after cold storage of "Golden Delicious" apple in Girona, Spain. *Acta Horticulturae*, 485:195–202.

Jung, S. K. and C. B. Watkins, 2011. Involvement of ethylene in browning development of controlled atmosphere-stored 'Empire' apple fruit. *Postharvest Biology and Technology*, 59:219–226.

Khodabandehloo, H. 1999. *Physical properties of Iranian export apples*. M.S. Thesis Tehran University, Karaj, Iran, 1–102.

Kidon, M. and J. Grabowska, 2021. Bioactive compounds, antioxidant activity, and sensory qualities of red-fleshed apples dried by different methods. *LWT*, 136:110302,

Kitinoja, L. and A. A. Kader, 2015. Measuring postharvest losses of fresh fruits and vegetables in developing countries. *PEF White Paper*, 15:26.

Kovacs, E., A. Keresztes, and J. Kovacs, 1988. The effects of gamma irradiation and calcium treatment on the ultrastructure of apples and pears. *Food Microstructure*, 7:29.

Lan, W., S. Bureau, S. Chen, A. Leca, C. M. Renard, and B. Jaillais, 2021. Visible, near- and mid-infrared spectroscopy coupled with an innovative chemometric strategy to control apple puree quality. *Food Control*, 120:107546.

Li, Y., S. Wang, Y. Sun, W. Xu, H. Zheng, Y. Wang, Y. Tang, X. Gao, C. Song, and Y. Long, 2020a. Apple polysaccharide protects ICR mice against colitis associated colorectal cancer through the regulation of microbial dysbiosis. *Carbohydrate Polymers*, 230:115726.

Li, Y., X. Zhang, J. Nie, S. A. S. Bacha, Z. Yan, and G. Gao, 2020b. Occurrence and co-occurrence of mycotoxins in apple and apple products from China. *Food Control*, 118:107354.

Lichou, L., F. Mandrin, J. D. Breniaux, V. Mercier, P. Giauque, D. Desbrus, P. Blanc, and E. Belluau, 2002. Une nouvelle moniliose: Monilia fructicola s'attaque aux abres fruitiers a noyaux. *Phytoma*, 547:22–25.

Lisa, K. and J. F. Thompson, 2010. Pre-cooling systems for small-scale producers. *Stewart Postharvest Review*, 6(2):1–14. http://doi.org/10.2212/spr 2010.2.2.

Lurie, S. and C. B. Watkins, 2012. Superficial scald, its etiology and control. *Postharvest Biology and Technology*, 65:44–60.

Margonsan, D. A., J. L. Smilanick, G. F. Simmons, and D. J. Henson, 1997. Combination of hot water and ethanol to control postharvest decay of peaches and nectarines. *Plant Diseases*, 81:1405–1409.

Maxin, P., R. W. S. Weber, L. H. Pedersen, and M. Williams, 2012. Hot-water dipping of apples to control *Penicillium expansum*, *Neonectria galligena* and *Botrytis cinerea*: effects of temperature on spore germination and fruit rots. *European Journal of Horticultural Science*, 77(1):1–9.

Mohebbi, S., M. Babalar, Z. Zamani, and M. A. Askari, 2020. Influence of early season boron spraying and postharvest calcium dip treatment on cell-wall degrading enzymes and fruit firmness in 'Starking Delicious' apple during storage. *Scientia Horticulturae*, 259:108822.

Moscetti, R., L. Carletti, D. Monarca, M. Cecchini, E. Stella, and R. Massantini, 2013. Effect of alternative postharvest control treatments on the storability of 'Golden Delicious' apples. *Journal of the Science of Food and Agriculture*, 93(11):2691–2697.

Mostafavi, H. A., S. M. Mirmajlessi, M. M. Seyed, H. Fathollahi, and H. Askari, 2012. Gamma radiation effects on physico-chemical parameters of apple fruit during commercial post-harvest preservation. *Radiation Physics, and Chemistry*, 81:666–671.

Naqash, F. A. 2015. *Value chain analysis of apple in Jammu and Kashmir.* Doctoral dissertation, SKUAST, Kashmir.

Nunes, M. C. N., A. M. M. B. Morais, J. K. Brecht, S. A. Sargent and J. A. Bartz, 2005. Prompt cooling reduces incidence and severity of decay caused by Botrytis cinerea and Rhizopus stolonifer in strawberry. *Hort Technology*, 15(1):153–156.

Ozturk, H. M., H. K. Ozturk and G. Koçar, 2017. Microbial analysis of meatballs cooled with vacuum and conventional cooling. *Journal of Food Science and Technology*, 54(9): 2825–2832.

Paull, R. and N. Chen, 2000. Heat treatments and fruits ripening postharvest. *Postharvest Biology and Technology*, 21:21–37.

Peng, W., D. Meng, T. Yue, Z. Wang, and Z. Gao, 2020. Effect of the apple cultivar on cloudy apple juice fermented by a mixture of Lactobacillus acidophilus, Lactobacillus plantarum, and Lactobacillus fermentum. *Food Chemistry*, 340:127922.

Prange, R. K., J. M. DeLong, and P. A. Harrison, 2003. Oxygen concentration affects chlorophyll fluorescence in chlorophyll-containing fruit and vegetables. *Journal of the American Society for Horticultural Science*, 128:603–607.

Prange, R. K., A. A. Ramin, B. J. Daniels-Lake, J. M. DeLong, and P. G. Braun, 2006. Perspectives on postharvest biopesticides and storage technologies for organic produce. *HortScience*, 41:301–303.

Pruksasri, S., B. Lanner, and S. Novalin, 2020. Nanofiltration as a potential process for the reduction of sugar in apple juices on an industrial scale. *LWT*, 133:110118,

Reid, M. S. 2000. Handling of cut flowers for air transport. IATA perishable cargo manual –*flowers*: 1–24.

Rupasinghe, H. P. V., D. P. Murr, G. Paliyath, and L. Skog, 2000. Inhibitory effect of 1- MCP on ripening and superficial scald development in 'McIntosh' and 'Delicious' apples. *The Journal of Horticultural Science and Biotechnology*, 75:271–276.

Ryall A. L. and W. T. Pentzer, 1982. Handling, transportation, and storage of fruits and vegetables. In *Fruits and tree nuts* (2nd ed.). AVI Pub. Co., Westport, CT.

Sabban-Amin, R., O. Feygenberg, E. Belausov, and E. Pesis, 2011. Low oxygen and 1- MCP pretreatments delay superficial scald development by reducing reactive oxygen species (ROS) accumulation in stored 'Granny Smith' apples. *Postharvest Biology and Technology*, 62(3):295–304.

Sahoo, P. K., L. K. Pradhan, S. Aparna, K. Agarwal, A. Banerjee, and S. K. Das, 2020. Quercetin abrogates bisphenol: An induced altered neurobehavioral response and oxidative stress in zebrafish by modulating brain antioxidant defence system. *Environmental Toxicology and Pharmacology*, 80:103483.

Sanderson, P. G. 2000. Management of decay around the world and at home. *16th Annual Postharvest Conference*, Yakima, WA. http://postharvest.tfrec.wsu.edu/pull.php?&article=PC2000Z&position=4

Sauceda-Galvez, J., I. Codina-Torrella, M. Martinez-Garcia, M. Hernández-Herrero, R. Gervilla, and A. Roig-Sagués, 2021. Combined effects of ultra-high pressure homogenization and short-wave ultraviolet radiation on the properties of cloudy apple juice. *LWT*, 136:110286.

Shi, J. X., R. Y. Znahg, and Y. X. Li-Wang, 1993. The influence of post-harvest gamma irradiation on the activity of enzymes in peach fruits. *Journal of Southwest Agricultural University*, 2:157–167.

Sousa, A., J. Vareda, R. Pereira, C. Silva, J. S. Camara, and R. Perestrelo, 2020. Geographical differentiation of apple ciders based on volatile fingerprint. *Food Research International*, 137:109550.

Tabatabaeefar, A. and A. Rajabipour, 2005. Modeling the mass of apples by geometrical attributes. *Scientia Horticulturae*, 105:373–382.

Tatsuki, M., H. Hayama, H. Yoshioka, and Y. Nakamura, 2011. Cold pre-treatment is effective for 1-MCP efficacy in 'Tsugaru' apple fruit. *Postharvest Biology and Technology*, 62:282–287.

Thompson, A. K. 2003. *Fruit and vegetables harvesting, handling and storage* (2nd ed.). Blackwell Publishing Ltd., NJ.

Thompson, A. K. 2010. *Controlled atmosphere storage of fruits and vegetables*. CABI Publishing, London.

Thompson, J. F. and F. G. Mitchel, 2002. *Postharvest technology of horticultural crops* (3rd ed.). Chapter 10, Packages for horticultural crops. University of California, Agriculture and Natural Resources, Davis, CA.

Thompson, J. F., E. J. Mitcham, and F. G. Mitchell, 2002. *Postharvest technology of horticultural crops* (3rd ed.). Chapter 8, Preparation for fresh market. University of California, Agriculture and Natural Resources, Davis, CA.

Tian, C., X. Liu, Y. Chang, R. Wang, T. Lv, C. Cui, and M. Liu, 2021. Investigation of the anti-inflammatory and antioxidant activities of luteolin, kaempferol, apigenin and quercetin. *South African Journal of Botany*, 137:257–264.

Tuna Guneş, N., and O. Horzum, 2017. Physiological events in horticultural crops. In: R. Türk, N. Tuna Güneş, M. Erkan & M. A. Koyuncu (Eds.), *Storage and market preparation of horticultural products*, Somtad Publications Textbook No: 1, Antalya/Türkiye, pp. 61–83 (In Turkish).

Veeriah, S., C. Miene, N. Habermann, T. Hofmann, S. Klenow, J. Sauer, F. Bohmer, S. Wolfl, and B. L. Pool-Zobel, 2008. Apple polyphenols modulate expression of selected genes related to toxicological defence and stress response in human colon adenoma cells. *International Journal of Cancer*, 122:2647–2655.

Veltman, R., J. Verschoor, and J. R. Van Dugteren, 2003. Dynamic control system (DCS) for apples (Malus domestica Borkh. cv 'Elstar'): Optimal quality through storage based on product response. *Postharvest Biology and Technology*, 27(1):79–86.

Wand, S. J., K. I. Theron, J. Ackerman, and S. J. Marais, 2006. Harvest and post-harvest apple fruit quality following applications of kaolin particle film in South African orchards. *Scientia Horticulturae*, 107(3):271–276.

Wang, D., Y. Jiang, D. X. Sun-Waterhouse, H. Zhai, H. Guan, X. Rong, F. Li, J. C. Yu, and D. Li, 2020a. MicroRNA-based regulatory mechanisms underlying the synergistic antioxidant action of quercetin and catechin in H_2O_2-stimulated HepG2 cells: Roles of BACH1 in Nrf2-dependent pathways. *Free Radical Biology & Medicine*, 153:122–131.

Wang, F. M., L. J. Rui, J. A. Mei, F. W. Ma, J. R. Li, and A. M. Ji, 1997. Effect of postharvest heat treatment on fruit storability and physiology. *International Journal of Fruit Science*, 14(Supplement):79–82.

Wang, Y., B. Tao, Y. Wan, Y. Sun, L. Wangab, J. Sun, and C. Li, 2020b. Drug delivery based pharmacological enhancement and current insights of quercetin with therapeutic potential against oral diseases. *Biomedicine & Pharmacotherapy*, 128:110372.

Watkins, C. B. 2006. The use of 1-methyl cyclopropane (1-MCP) on fruits and vegetables. *Biotechnology Advances*, 24:389–409.

Watkins, C. B. 2008. Overview of 1-methylcyclopropene trials and uses for edible horticultural crops. *HortScience*, 43:86–94.

Watkins, C. B. and J. F. Nock, 2005. Effects of delays between harvest and 1- methylcyclopropene treatment, and temperature during treatment, on ripening of air-stored and controlled-atmosphere-stored apples. *HortScience*, 40:2096–2101.

Weber, A., A. Brackmann, V. Both, E. P. Pavanello, R. D. O. Anese, and F. R. Thewes, 2015. Respiratory quotient: Innovative method for monitoring Royal Gala apple storage in a dynamic controlled atmosphere. *Scientia Agricola*, 72(1):28–33.

Wijewardane, R. M., and S. P. Guleria, 2009. Combined effects of pre-cooling, application of natural extracts and packaging on the storage quality of apple (*Malus Domestica*) cv. Royal Delicious. *Tropical Agricultural Research*, 21(1):10–20.

Williams, D. J., D. Edwards, I. Hamernig, L. Jian, A. P. James, S. K. Johnson, and L. C. Tapsell, 2013. Vegetables containing phytochemicals with potential anti-obesity properties: A review. *Food Research International*, 52:323–333.

Wills, R. B. H., W. B. McGlasson, D. Graham, and D. C. Joyce, 2007. *Post-harvest. An introduction to the physiology and handling of fruit, vegetables and ornamentals* (5th ed.). UNSW Press, Sydney.

Wright, A., J. Delong, J. Arul, and R. Prange, 2015. The trend toward lower oxygen levels during apple (*Malus domestica* Borkh) storage. *The Journal of Horticultural Science and Biotechnology*, 90(1):1–13.

Yong, H., R. Bai, F. Bi, J. Liu, Y. Qin, and J. Liu, 2020. Synthesis, characterization, antioxidant and antimicrobial activities of starch aldehyde-quercetin conjugate. *International Journal of Biological Macromolecules*, 156:462–470.

Zanella, A. 2003. Control of apple superficial scald and ripening – a comparison between 1-methylcyclopro-pene and diphenylamine postharvest treatments, initial low oxygen stress and ultra-low oxygen storage. *Postharvest Biology and Technology*, 27(1):69–78.

Zanella, A., P. Cazzanelli, A. Panarese, M. Coser, M. Cecchinel, and O. Rossi, 2005. Fruit fluorescence response to low oxygen stress: modern storage technologies compared to 1-MCP treatment of apple. *Acta Horticultruae*, 682:1535–1542.

Zerbini, P. E., A. Pianezzola, and M. Grassi, 1999. Post storage sensory profiles of fruit of apple cultivars harvested at different maturity stages. *Journal of Food Quality*, 22(1):1–17.

15 Emerging Packaging and Storage Technologies

Aftab Ahmed, Farhan Saeed, Muhammad Afzaal,
Shinawar Waseem Ali, Ali Imran and
Muhammad Awais Saleem

CONTENTS

15.1   TEMPERATURE MANAGEMENT

The fruit cooling process is a dynamic process involves field heat removal, a stepwise temperature lowering process until the fruit reaches the final storage temperature and maintaining the final storage temperature throughout the storage process. For example, in the case of 'Pink Lady™', the temperature is first lowered to 4°C soon after harvesting, andthen the temperature is lowered by 1°C each week for 2 weeks (stepwise) until the final storage temperature of 2°C is reached (Avilova et al., 2021). Stepwise cooling is mild on the fruit, and the fruit becomes dormant. This also reduces stresses on fruit that can result in skin damage in some varieties. Stepwise cooling is also economical as the cooling system operates for fewer hours at a lower intensity. During stepwise cooling less moisture is removed from fruit (Behdani et al., 2019).

15.2   COOLING CONDITIONS

The pace at which apple fruit is cooled impacts its ability to preserve quality, although its efficacy depending on the fruit's nutritional condition, storage history, variety and harvest ripeness. It is imperative to cool apple varieties that attain maturity earlier in the harvest season (summer varieties)

DOI: 10.1201/9781003239925-15

TABLE 15.1
Rapid Cooling Effects on Firmness of CA Stored 'Empire' Apples

Days to cool to 0°C	Days from harvest to 3% O_2	Flesh firmness (N) after removal from CA
1	4	63
7	4	58
14	4	52

as they rapidly become softer than varieties that mature late in the harvest season (Whitehead et al., 2021). Usually at an early maturity period apples of the same variety tend to soften rapidly during later maturity stages than during early maturity periods. Apples soften very quickly at an advanced levels of maturity rather than at initial levels of maturity within a cultivar. As the storage period increases, the effects of slow cooling are magnified. Therefore, insufficient utilization of resources for rapid cooling at the time of harvest is evident during the later phase of storage phases while fruit cannot rally the least decisiveness criteria for selling. For instance, a 7–10 days loss in storage life is seen in the 'McIntosh' variety after a 1-day delay at 21°C (70°F). Table 15.1 presents the effects of rapid cooling in CA storage of 'Empire' apples (Wassermann et al., 2019).

Hydrocooling, forced air cooling or room cooling are the most common methods of chilling apple fruits. Although forced air cooling and hydrocooling systems may effectively decrease fruit temperatures, they are not frequently utilized for apples. The room cooling method is the most common method adopted to reduce field heat by employing normal air circulation within the storage chamber. But air flows around the fruit instead of flowing through bins of fruit; therefore this technique is considered inefficient and slow (Shen et al., 2018). A swift chilling process is often hard to accomplish if the storage chambers are rapidly filled and the refrigeration facility was not appropriately intended for a huge fruit weight. This issue can be resolved in two ways. First, the fruit can be sorted and kept in pre-cooling chambers before being send to a long-term storage facility. The second option can be to hold fruit in a refrigeration facility as per system capacity (Indiarto et al., 2020).

If the refrigeration capacity is limited, no more than two piles of fruit bins should be placed transverse the width of the luggage compartment. If the room temperature is not lower than 0°C (32°F) it should be reduced to one stack by the next morning. Containers in the evaporator's downward outflow, combined with panel rollers oriented in the same path as the air movement, result in an effective and fast chilling operation (Shen et al., 2018). The rate of cooing process directly depends upon refrigeration capacity and the layout of the storage chamber. The development and the initiation of the cooling process must be supervised by a qualified heating, ventilation and air conditioning (HVAC) engineer (Mditshwa et al., 2018).

Fruit freshness must be maintained at a constant temperature not just after collection and during storage but even during packaging, transportation and marketing and sales. The term 'cold chain' refers to the interconnectedness of all of these processes and stresses the need to preserve the link from farm to table (Onursal and Koyuncu, 2021).

Apple fruit reacts strongly to changes in temperature and environment as apple growers have discovered through extensive understanding of apple preservation. Rapid temperature decreases and rigorous preservation of lower temperatures near to the apple variety's freezing threshold can increase the shelf life of apples for approximately to 4–7 months. However, modern commercial stores integrate temperature monitoring with CA storing for protracted apple storing (Bessemans et al., 2020).

15.3   REGULAR AIR STORAGE

Temperature and moisture guidelines for industrial apple storage are -1°C to 4°C (30°F to 39°F) and 90 to 95 percent relative humidity (RH), accordingly, and depends upon the cultivar. Table 15.2

TABLE 15.2

Storage Behaviors of Different Apple Varieties

Variety	Potential months of storage		Superficial scald susceptibility	Comments
	0 °C Air	CA*		
Braeburn	3–4	8–10	Minor	CO_2 sensitive
Cortland	2–3	4–6	Very high	Sensitive to temperature; McIntosh conditions preferred; scald inhibitor
Delicious	3	12	Medium to very high	Intolerant to CO_2 > 2 percent; scald suppressants are advised
Empire	2–3	5–10	Minor	Late harvest must be avoided; sensitive to temperature; sensitive to CO_2
Fuji	4	12	Minor	CO_2 sensitivity exists in late harvest fruit
Gala	2–3	5–6	Minor	During prolonged storage the flavor is lost
Golden Delicious	3–4	8–10	Minor	Vulnerable to skin shriveling
Granny Smith	3–4	10–11	Very high	CO_2 sensitive
Idared	3–4	7–9	Minor	Sensitive to temperature; resistant to field freezing damage
Jonagold	2	5–7	Medium	Late fruit picking is restricted; chance of scald development
Jonamac	2	3	Medium	During prolonged storage the flavor is lost
Law Rome	3–4	7–9	Very high	Scald suppressants are advised
Macoun	3	5–7	Minor	McIntosh and Macoun can be collectively stored
McIntosh	2–3	5–7	Medium	Sensitive to CO_2; due to excessive softening, the storage period is reduced
Mutsu	3–4 6–8	6–8	Minor	Eating quality is less in green apples
Spartan	3–4 6–8	6–8	Minor	Vulnerable to high CO_2; vulnerable to skin shrinkage at 2.2°C to 3.3°C (36°F to 38°F)
Stayman	2–3	5–7	High	Tolerable to CO_2 up to 5 percent, however generally kept in 2 percent to 3 percent CO_2; scald inhibitor essential; vulnerable to skin shrinkage

shows normal storage times for a variety of airborne species (Ludwig et al., 2020). The appropriate time for air storing has gotten shorter as the market requirements have raised. Furthermore, as the accessibility duration for selling air-stored goods has shortened, short-term CA storage is becoming popular (Brizzolara et al., 2017).

The temperature parameters for air kept fruit are directly influenced by the variety's vulnerability to cold temperature diseases. Cold temperatures typically result in harder green fruit, however when stored at temperatures below 3°C (37°F), some cultivars, such as 'McIntosh', experience center darkening, mild blistering, and interior darkening. Because these difficulties only occur in fruits that have been preserved over many months, fruits that have been stored for a short period of time (2 to 3 months) are not in danger of low temperature damage. Another aspect to consider when choosing a temperature for low temperature storage is the influence of temperature on dampness (She et al., 2018). At 1°C (34°F), for instance, it is easier to maintain >90 percent RH than at 0°C (32°F). Ultimately selections must be based on personal experiences with the apple cultivars as well as guidance from agricultural specialists. The majority of apple cultivars are unaffected by cooling and may be kept at temperatures as low as 0°C (32°F). Low temperature sensitive cultivars, on the other hand, should be kept at 2°C to 3°C (36°F to 37°F). Fruit stored in low-oxygen CA is susceptible to low-oxygen damage, which can be mitigated by raising the storage room temperature. Temperatures in storerooms should be constantly reviewed using temperature sensors placed throughout the room throughout the curing time (DeEll and Lum, 2017).

Temperatures within stacks and throughout the room could be lower or higher than reported by temperature detecting equipment; therefore, it is not a good idea to rely on a single thermometer at the entrance. Increased temperatures may cause fruit to ripen quickly, necessitating the use of the refrigeration equipment more often (Table 15.3). High temperature buildup during packaging and shipping may result in lower-quality fruit at the consumer end. Fruit temperature can rise during packing, resulting in a loss of firmness throughout transportation if not removed (do Amarante et al., 2020).

15.4   CONTROLLED ATMOSPHERE (CA) CONSIDERATIONS

Apples are the specific fruit stored under CA conditions, but storage conditions like the mixture of gases and temperature of storage chamber vary with growing region, variety and complexity of the equipment employed for monitoring and controlling the atmospheres. There exists a relation between temperature, CO_2, and O_2. For example, fruit is more susceptible to low-O_2 injury at low storage temperatures. Moreover, CO_2 levels must be reduced to protect fruit from CO_2 damage, especially when low O_2 levels are used during CA storage (Al Shoffe et al., 2020).

Fruit quality resulting from these circumstances progressively becomes unacceptable in the marketplace. In many apple processing industries rapid CA has now become standard practice. Now O_2 in CA rooms can efficiently be lessened below 5 percent within a day or two through sophisticated nitrogen flushing equipment. However, placing fruit first in a storage chamber in CA conditions 5 to 8 days from the harvest is considered 'rapid CA'. The fruit core temperature must be dropped to prearranged thresholds for some varieties before storing it in CA conditions (Feng et al., 2014). Flushing with N_2 is commonly practiced to lower O_2 concentration, even when rooms are filled for prolonged periods, and it has become general practice to flush rooms with N_2 that are opened for short periods of time to remove fruit required for marketing. N_2 utilized for this purpose is either procured in cylinders or produced on site (Gabioud et al., 2010).

The recommended RH for apple storage to prevent shrivel is 90 percent to 95 percent. The main reason for fruit dehydration is less coil exposed area and recurrent defrosting. The HVAC experts

TABLE 15.3
Heat Dissipation Rate of Varieties at Different Temperatures

| | *BTU ton⁻¹ day⁻¹ | | | | |
| | Temperature | | | | |
Cultivar	1°C	0°C	2.2°C	3.3°C	4.4°C
Delicious	690	760	910	1,010	1,110
Golden Delicious	730	800	970	1,070	1,180
Jonathan	800	880	1,060	1,170	1,290
McIntosh	730	800	970	1,070	1,180
Northern Spy	820	900	1,090	1,200	1,320
Rome Beauty	530	580	700	780	850
Stayman Winesap	820	910	1,100	1,210	1,330
Winesap	530	590	710	780	860
Yellow Newton	510	570	690	760	840
York Imperial	610	670	810	900	990
Mean	680	750	900	1,000	1,100

* Conversion factor for BTU ton⁻¹ day⁻¹ to kJ ton⁻¹ day⁻¹ (multiply by 1.055)

demand a possible increased coil surface area when designing CA rooms. To minimize dehydration, the number of freezing cycles is lowered (Köpcke et al., 2015). To avoid the need to defrost, the amount of O_2 is decreased to the least tolerable range and then the temperature is increased to 1°C to 2°C (34°F to 36°F). In some storage facilities at around 0°C (32°F) a high pressure water vapor system is installed to maintain moisture in the storage chamber. The air circulation system should be efficient enough to avoid water condensation on the fruit to prevent decay. In the 'Golden Delicious' variety, shriveling is also minimized by the use of plastic in place of wooden crates (Kafle et al., 2016).

CA storage regimes fall into one of three types after fruit has been chilled and CA conditions have been achieved, depending on the amount of materials and machinery used.

15.4.1 STANDARD CA STORAGE

Standard CA implies traditional storage with the least amount of danger of fuel harm (Table 15.4). The control parameters in this environment can be regulated physically or with the help of computer-assisted technology. Changes in gas percentages in manually managed storage conditions must not cause fruit damage since the safety factor is big enough (Kittemann et al., 2015).

TABLE 15.4
Atmospheric and Temperature Requirements for Standard CA Storage of Apples

Variety	CO_2 (%)	O_2 (%)	Temperature (°C)	Low-O_2*
Braeburn	0.5	1.5–2	1	+
Cortland	2–3	2–3	0	-
	2–3 for 1 month then 5	2–3	2	
Delicious	2	0.7–2	0	+
Empire	2–3[†]	2	2	+
Fuji	0.5[†]	1.5–2	0–1	+
Gala	2–3	1–2	0–1	+
Golden Delicious	2–3	1–2	0–1	+
Granny Smith	0.5	1.5–2	1	+
Idared	2–3	2	1	+
Jonagold	2–3	2–3	0	+
Jonamac	2–3	2–3	0	-
	2–3 for 1 month then 5	2–3	2	
Law Rome	2–3	2	0	+
Macoun	5	2–3	2	-
Marshall McIntosh	2–3 for 1 month then 5	4–4.5	2	-
		3	2	–
	2–3 for 1 month then 5			
	2–3 for 1 month then 5	2	3	
Mutsu	2–3	2	0	+
Spartan	2–3	2–3	0	+
Stayman	2–5	2–3	0	+

* Only use low-O_2 CA storage on apples picked within the first few days of their harvest window.
† Sensitive to CO_2: maintain CO_2 lower than O_2 concentration. If not diphenylamine treated, 1.5 percent to 2 percent CO_2 is advised for the first 30 days.

15.4.2 Low-O_2 CA Storage

Fruit in limited O_2 CA storage is maintained at a concentration of less than 2 percent O_2 yet just above the amount where fermentation begins. Occasionally nondescriptive words like 'ultralow' are being used, although they should be abandoned in favor of expressing the specificO_2 percent. The acceptable O_2 incidence ranges with cultivar (Table 15.4) and geographic location. The proper safety threshold for 'Delicious' apples from British Columbia, Canada, is 0.6 percent (Kweon et al., 2013), which allows for the prevention of superficial scald without the application of diphenylamine (DPA). Fruit from different growing locations of the same cultivar may be damaged if kept at these low O_2 levels. The susceptibility of different strains within the same variety differs as well. In the Marshall variant of 'McIntosh,' O_2 levels beneath 3 percent to 4 percent are not regarded acceptable, although 3 percent to 4 percent O_2 is deemed safe in other 'McIntosh' variants (Lee et al., 2012). During low-O_2 CA storage, it is usually suggested that the storage temperature be raised. For the reliable success of protracted CA storage, a variety of protocols were devised. The danger of low-O_2 injury can be reduced by adopting the subsequent measures:

1. Low-O_2 CA storage can only be used after apples are harvested. Only use low-O_2 CA storage on apples picked within the first few days of their harvest window. Fruit that is overripe should not be stored in a low-oxygen environment since it will cause fruit damage.
2. Avoid apples from farm units with an average seed count of less than five per apple. With certain cultivars, a low seed quantity can be an issue.
3. Within 3 days of fruit harvest, the temperature of the fruit should be lowered to 4°C (particularly for 'McIntosh' and 'Empire').
4. With the exception of 'Fuji' and 'Braeburn,' the O_2 content should be lowered to 5 percent within 8 days of farm work.
5. The storing temperature must be adjusted from 0°C–1°C to reduce the danger of fruit deterioration.
6. To eliminate the chances of O_2 fluctuations, computer-assisted automatic gas concentration controlling equipment should be installed.
7. If possible, the application of DPA should be avoided as it results in low-O_2 injury, for example, for cultivars with low-O_2 blister threat.

15.4.3 Low-Ethylene CA Storage

Apples are a climacteric fruit is best suited for storage in low-ethylene CA storage, autocatalytic ethylene making often occurs secure to harvest. The rate of gas manufacture varies among cultivars. The CA storage results in the inhibition of ethylene production and its effects by lowering O_2 concentration or increasing CO_2 (Ilyas et al., 2007).

To reduce the likelihood of fruit softening plus extra forms of senescence low-ethylene and minimize the risk of superficial scald, CA storage was tested as a best suitable method. 'Empire' apples are a low-ethylene producing variety and were effectively preserved by using low-ethylene CA storage (below 1 μL L^{-1}), but this was later replaced by low-O_2 storage (Lu et al., 2019).

Short-term pressure levels of a little O_2 or elevated CO_2 have been successfully utilized for maintaining the apple fruit quality in conjunction with CA storage. Superficial scald has been controlled in cultivars with 'Granny Smith,' 'Delicious' and 'Law Rome' by storing the apple fruit at 0.25 percent to 0.5 percent O_2 concentration for up to 2 weeks. High-CO_2 concentration (15 percent to 20 percent) is not recommended for 'Golden Delicious' due to fruit damage (Lin et al., 1999).

In commercial CA storage settings fruits should be sampled on a monthly basis to spot any storage issue to reduce chances of major fruit losses. The delegate samples of fruit should be kept close to a case harbor in the CA room door and sampling should be carried out from representative samples. To avoid false positive readings for scale the samples must be kept in mesh bags (Yang et al., 2006).

15.5 CA AND APPLE VARIETIES

The variety and, in certain circumstances, the breed of a specific variety and where it was produced all influence the choice of CA environments and temperatures. According to research, there are two types of cultivars: those that are high CO_2 tolerant and those that are not. 'Gala' and 'Golden Delicious' are CO_2-tolerant cultivars which also profit from fast atmospheric decreases (Soto-Silva et al., 2017).

Quick CA is recommended for these cultivars as it aids in preserving fruit insistence and sourness better than traditional CA. The 'Gala' and 'Golden' varieties are most suited to store at O_2 concentration as low as 1.0 percent and at CO_2 levels of 2.5 percent at 11°C (34°F). The O_2 level is raised if the temperature goes below this point. Usually 0°C (32°F) temperature is employed in regular storage conditions (Yahia, 2009).

The varieties that fall under the CO_2 intolerant category are 'Fuji', 'Braeburn' and 'Granny Smith'. The air exchange within these varieties is reduced due to the densely packed cell structure. Before the reduction of O_2 levels during the storage of these varieties, the flesh temperature must be maintained close to the storage temperature. Due to their natural disposition there exists a tendency to develop internal browning in these varieties. During storage temperatures should be kept slightly elevated while CO_2 should remain well below the O_2 level throughout the storage period. For example, appropriately matured fruits stored at 1.5 percent O_2 are stored with CO_2 below 0.5 percent at 1°C (34°F). The wax-coated fruits are not recommended to store under CA conditions as it can obstruct the air exchange within the fruit (Xiaoyu et al., 2007).

'Red Delicious' exhibits CO_2 tolerance as well as CA tolerance. Unfortunately, unlike 'Golden Delicious' and 'Gala', growers have not observed the significant beneficial impacts of fast CA on 'Delicious'. CA must not be postponed after harvest since 'Delicious' fruit softens more quickly within a container than on the tree. At 0°C to 1°C (32°F to 34°F), 1.5 percent O_2 and up to 2 percent CO_2 are typical CA environments for non-water cored 'Delicious'. The proper operation of CA storage as well as the usage of the authorized concept of a 'controlled environment' for stored apples is covered by state legislation (Jinjin, 2013). The rate at which CA conditions are established, the highest amount of O_2 allowed and the duration fruit spends in CA is all regulated. Because laws differ by state, owners should consult their state's department of agriculture for further information (Ghafir et al., 2009).

To ensure the safety and security of CA storage rooms, several steps must be followed. Workers must be mindful of the dangers of operating with oxygen levels that are less than what is required for breathing. When dealing with CA generators, extra measures need to be taken to minimize implosion or blast dangers. In a nutshell, personnel must be well trained (Pason et al., 1990).

1-MCP treatment delays or inhibits softening, staining, failure of titratable acidity and occasionally a drop in soluble solids contents (SSC) and even prevents the expansion of various diseases. Cultivar and fruit ripeness might influence a fruit's responses to 1-MCP. Apples' volatile production is likewise reduced by 1-MCP which supports the theory that ethylene regulates volatile synthesis. To guarantee that the flavor of 1-MCP-treated fruit is not affected, consumer tests on acceptance are necessary (Kittemann et al., 2015).

15.6 DISORDERS THAT DEVELOP DURING STORAGE

There are three types of disorders: chilling disorders, senescent breakdown diseases and disorders linked with improper storage environments. The occurrence of senescent breakdown is linked to the harvest of overripe or calcium-deficient fruit. This could be worsened by keeping fruit at conditions beyond the recommended range (Lu and Wang, 2004). Fruit from sensitive cultivars is frequently soaked in calcium prior storage, although the frequency of senescent collapse could be decreased by picking fruit at a younger phase and chilling quickly which would minimize storage time. Surface injury, spongy blister, coldness breakdown, brown central part, interior browning, low-O_2 damage and high-CO_2 harm are the most frequent diseases related with temperature and atmospheres (Pandey et al., 2006).

15.7   SPECIFIC DISORDERS

15.7.1   SUPERFICIAL SCALD

The physical condition known as superficial scald (storage scald) is linked to prolonged storage. Until the development of postharvest DPA remedies, this was the leading cause of apple fruit waste. Fruit vulnerability to the disease is affected by cultivar, climate and harvest timing; therefore intervention with DPA must be decided after consulting with a regional agricultural professional (Yang and Ren, 2016). DPA is typically used in conjunction with a fungicide to minimize deterioration; furthermore calcium salts could be added at the same time to reduce unpleasant depression or senescent breakdown. Clean DPA at tag charge should avoid DPA-induced fruit destruction while also exceeding residual limits. As DPA is not eliminated as dirt develops in the solution, the danger of DPA harm to fruit escalates (Sharma et al., 2009).

In certain countries, both the usage of DPA and the presence of DPA remnants on fruit are illegal. Ethoxyquin, another antioxidant, is no longer allowed to be used on apples. Small amounts of O_2 in CA storerooms minimize the danger of injury and might allow less DPA quantities to be used. Different solutions to prevent surface scald are now being researched, and reservoir owners are limiting their usage of DPA to the extent feasible. The use of low-O_2 and low-ethylene CA storage further reduces the risk of scalding. In British Columbia, Canada, 0.7 percent O_2 storage is utilized as a DPA treatment alternative. Although fruit cultivated in different locations may be sensitive to low-O_2 damage or the danger of blister due to atmosphere or diversity, this approach cannot be utilized globally (Dunn and Garafola, 1986).

15.7.2   SOFT SCALD

Soft scald is marked by random but precise patches of fragile, brown tissue that might spread into the cortex. Fruit vulnerability to squashy blister varies by diversity and environment, but the property of harvest ripeness is unclear (Sharma and Singh, 2010). Overripeness is nearly usually a component in the development of 'Golden Delicious'. The condition can occasionally be controlled by storing fruit at 2°C rather than at cooler temperatures; DPA, which is used to treat surface scald, may also minimize the possibility of soft scald. Soft scald on 'Golden Delicious' can be reduced by storing it at a cooler temperature once it has been quickly cooled (Indiarto et al., 2020).

15.7.3   CHILLING RELATED DISORDERS

Coldness breakdown, russet center and interior browning are all chill-related diseases. Variety vulnerability to cold temperatures is a factor in many diseases which tend to develop in frequency and severity as storage time increases. Fruit susceptibility to diseases is affected by climate with more issues developing during colder rainy growing seasons. Russet vascular bundles, browning of soft tissue and a visible circle of light of undamaged tissue beneath the covering are all signs of low-temperature breakdown (Onursal and Koyuncu, 2021). The afflicted tissues, in contradiction to senescent disintegration, are expected to be stiffer, moister and darker in color. Russet center (center soft tissue) is a browning of the tissue that occurs first in the central part and then in the cortex, making it complicated to differentiate from coolness disintegration. Interior browning is a grizzling of the fleshy tissue that is visible after apples are sliced, rather than a disintegration of the flesh. Intrinsic browning and core flush are frequently linked to elevated CO_2 since both can develop in CA when CO_2 exceeds O_2 (DeEll and Lum, 2017).

15.7.4 LOW O_2 INJURY

Fruit is harmed by low oxygen levels in a variety of ways. The loss of taste is the first sign of damage accompanied by fermentation-related smells. If storage issues are detected early enough and no significant harm has occurred, these scents may go away. Browning of the covering in a red-colored variety, formation of russet tender spots like spongy blister and unusual softening and breaking of fruit are all signs of injury. As previously stated, cultivars respond differently to low O_2 levels with the vulnerability to damage determined by a variety of preharvest and postharvest variables.

15.7.5 CO_2 INJURY

The effects of CO_2 might be extrinsic or intrinsic. Crumpled, depressed, dull or bright patches limited to the covering surface generally lying on the greener surface of the fruit characterize the exterior shape. Interior forms manifest as a brown heart and/or flesh cavities. Both exterior and internal CO_2 damage can be reduced by DPA (Mditshwa et al., 2018).

15.8 TRANSPORT

This is used to get fruits to marketplaces. This form of packing is typically made up of fiberboard, hardwood or plastic containers/crates/boxes or bags ranging between 4 kg and 20 kg. Box or pallet packing are heavier bundles that are harder to manage and require sophisticated instruments to transport. Transport packing should be simple to handle, foldable by one individual and the right size to load into cargo trucks (highway cargo trucks, air shipment containers and transport containers) (Dunn and Garafola, 1986; Lu and Wang, 2004). Biodegradable, noncontaminating and recyclable materials are ideal. To save space on the return voyage, repeat-use packaging should be simple to clean and disassemble. It should be able to withstand usage and the environmental circumstances in which the product is promoted and eventually sold. Packaging must also fulfill quantity and weight requirements without becoming overstuffed and damaging the produce. Inserts, dividers and immobilizers are used to keep the fruit from being destroyed. These immobilizers may also act as a form of reinforcement for the container. The produce can be immobilized in the container using wood, wool and shredded paper (Yahia, 2009). Pallets seem to be the most common unit load packing technique across the world. Dimensions are consistent with marine and air transport containers, vehicles, loaders and storage facilities, among other things (Yang et al., 2006). Handling stages in the supply chain are reduced when pallets are utilized. A screen tension cover, side hooks and vertical and lateral belting maintain pallet unit loads. Apples cannot grown in huge solo unit boxes which are utilized for further fruit crops (Lin et al., 1999).

15.9 TYPES OF PACKAGING MATERIAL

The majority of crates or barrels are made with a collapsing shape, and the producer or shipper assembles them. Fitting the boxes with paste or pins is required to build the boxes in some situations. These boxes are available in a variety of sizes, styles and capacities and are hygienic and incredibly lightweight. Waxing the carton/box adds even more strength by preventing dampness from weakening the fireboard. Waxed crates or barrels are especially beneficial in the tropics where damp amalgamation is prevalent due to substantial humidity and temperature differences across storage units (Sharma and Singh, 2010).

Xitex cardboard is used to make the majority of contemporary trays. The Xitex board is made up of two flutes that are connected at their tips and have liners on both sides. The flute points are perfectly aligned, resulting in increased potency at a similar cost. This technique produces lighter, tougher, smoother, shorter and squarer crates (Yang and Ren, 2016). Usually the following boxes and liners are used for apple fruit storage:

- 15 kg three layer box and lid (482 mm x 313 mm x 250 mm), stackable at 590 boxes per pallet
- P6 box with lid 12 kg (569 mm x 377 mm x 148 mm) has a pallet capacity of 780 boxes

- Lifespan® liners come in a variety of sizes including 11 kg–13 kg and 20 kg carton liners, as well as 3 kg, 5 kg and 10 kg bags
- P12 liners for apple trays are available in sizes 22, 25, 27, 28, 30, 32, 33, 35, 38, 39, 41, 42 A, 42 B, 45, 49, 51 and 54.

15.10 CONCLUSION

Apples are a nutritious and delicious fruit popular around the globe. Extending the shelf stability for extended used and for transportation to remote areas requires specific requirements and storage conditions. The most practiced and suited storage for apple fruit is controlled atmosphere storage. There are different forms of CA storage that are practiced in commercial settings for apple storage depending upon the variety, origin, harvest time and expected shelf life. The apple fruit shelf life varies from 3 to 5 months with optimum quality in terms of taste, aroma and texture. More research and innovation is in progress to further extend the shelf life of this commodity.

REFERENCES

Al Shoffe, Y., Nock, J.F., Baugher, T.A., Marini, R.P. and Watkins, C.B. 2020. Bitter pit and soft scald development during storage of unconditioned and conditioned 'Honeycrisp' apples in relation to mineral contents and harvest indices. *Postharvest Biology and Technology* 160: 111044.

Avilova, S.V., Gryzunov, A.A. and Kornienko, V.N. 2021, February. The use of negative temperatures during storage and transportation of apple fruits. In *IOP Conference Series: Earth and Environmental Science* (Vol. 640, No. 2. 022007). IOP Publishing.

Behdani, B., Fan, Y. and Bloemhof, J.M. 2019. Cool chain and temperature-controlled transport: An overview of concepts, challenges, and technologies. *Sustainable Food Supply Chains*: 167–183.

Bessemans, N., Verboven, P., Verlinden, B.E., Janssens, M., Hertog, M.L.A.T.M. and Nicolaï, B.M. 2020. Apparent respiratory quotient observed in headspace of static respirometers underestimates cellular respiratory quotient of pear fruit. *Postharvest Biology and Technology* 162: 111104.

Brizzolara, S., Santucci, C., Tenori, L., Hertog, M., Nicolai, B., Stürz, S., Zanella, A. and Tonutti, P. 2017. A metabolomics approach to elucidate apple fruit responses to static and dynamic controlled atmosphere storage. *Postharvest Biology and Technology* 127: 76–87.

DeEll, J.R. and Lum, G.B. 2017. Effects of low oxygen and 1-methylcyclopropene on storage disorders of 'Empire' apples. *HortScience* 52(9): 1265–1270.

do Amarante, C.V., Silveira, J.P.G., Steffens, C.A., de Freitas, S.T., Mitcham, E.J. and Miqueloto, A. 2020. Post-bloom and preharvest treatment of 'Braeburn' apple trees with prohexadione-calcium and GA 4+ 7 affects vegetative growth and postharvest incidence of calcium-related physiological disorders and decay in the fruit. *Scientia Horticulturae* 261: 108919.

Dunn, J.W. and Garafola, L.A. 1986. Changes in transportation costs and interregional competition in the US apple industry. *Northeastern Journal of Agricultural and Resource Economics* 15(1): 37–44.

Feng, F., Li, M., Ma, F. and Cheng, L. 2014. Effects of location within the tree canopy on carbohydrates, organic acids, amino acids and phenolic compounds in the fruit peel and flesh from three apple (*Malus×domestica*) cultivars. *Horticulture Research* 1(1): 1–7.

Gabioud, S., Bozzi Nising, A., Gasser, F., Eppler, T. and Naunheim, W. 2010. Dynamic CA storage of apples: monitoring of the critical oxygen concentration and adjustment of optimum conditions during oxygen reduction. *Acta Horticulturae* 876: 39–46.

Ghafir, S.A., Gadalla, S.O., Murajei, B.N. and El-Nady, M.F. 2009. Physiological and anatomical comparison between four different apple cultivars under cold-storage conditions. *African Journal of Plant Science* 3(6): 133–138.

Ilyas, M.B., Ghazanfar, M.U., Khan, M.A., Khan, C.A. and Bhatti, M.A.R. 2007. Post harvest losses in apple and banana during transport and storage. *Pakistan Journal of Agricultural Sciences* 44(3): 534–539.

Indiarto, R., Izzati, A.N. and Djali, M. 2020. Post-harvest handling technologies of tropical fruits: A review. *International Journal* 8(7): 3951–3957.

Jinjin, T.A.N. 2013. A strategic analysis of Apple computer Inc. & Recommendations for the Future direction. *Management Science and Engineering* 7(2): 94–103.

Kafle, G.K., Khot, L.R., Jarolmasjed, S., Yongsheng, S. and Lewis, K. 2016. Robustness of near infrared spectroscopy based spectral features for non-destructive bitter pit detection in honeycrisp apples. *Postharvest Biology and Technology* 120: 188–192.

Kittemann, D., McCormick, R. and Neuwald, D.A. 2015. Effect of high temperature and 1-MCP application or dynamic controlled atmosphere on energy savings during apple storage. *European Journal of Horticultural Science* 80(1): 33–38.

Köpcke, D. 2015. 1-Methylcyclopropene (1-MCP) and dynamic controlled atmosphere (DCA) applications under elevated storage temperatures: Effects on fruit quality of 'Elstar', 'Jonagold' and 'Gloster' apple (*Malus domestica Borkh.*). *European Journal of Horticultural Science* 80(1): 25–32.

Kweon, H.J., Kang, I.K., Kim, M.J., Lee, J., Moon, Y.S., Choi, C., Choi, D.G. and Watkins, C.B. 2013. Fruit maturity, controlled atmosphere delays and storage temperature affect fruit quality and incidence of storage disorders of 'Fuji' apples. *Scientia Horticulturae* 157: 60–64.

Lee, J., Mattheis, J.P. and Rudell, D.R. 2012. Antioxidant treatment alters metabolism associated with internal browning in 'Braeburn' apples during controlled atmosphere storage. *Postharvest Biology and Technology* 68: 32–42.

Lin, S., Jiping, S. and Yunbo, L. 1999. Influence of mechanical stress on the active oxygene metabolism system of apple during transportation. *Journal-China Agricultural University* 4(5): 107–110.

Lu, F., Xu, F., Li, Z., Liu, Y., Wang, J. and Zhang, L. 2019. Effect of vibration on storage quality and ethylene biosynthesis-related enzyme genes expression in harvested apple fruit. *Scientia Horticulturae* 249: 1–6.

Lu, L.X. and Wang, Z.W. 2004. Study of mechanisms of mechanical damage and transport packaging in fruits transportation [J]. *Packaging Engineering* 4.

Ludwig, V., Thewes, F.R., Wendt, L.M., Berghetti, M.R.P., Schultz, E.E., Schmidt, S.F.P. and Brackmann, A. 2020. Extremely low-oxygen storage: Aerobic, anaerobic metabolism and overall quality of apples at two temperatures. *Bragantia* 79: 458–471.

Mditshwa, A., Fawole, O.A. and Opara, U.L. 2018. Recent developments on dynamic controlled atmosphere storage of apples – A review. *Food Packaging and Shelf Life* 16: 59–68.

Onursal, C. and Koyuncu, M.A. 2021. Role of controlled atmosphere, ultra low oxygen or dynamic controlled atmosphere conditions on quality characteristics of 'scarlet spur' apple fruit. *Journal of Agricultural Sciences* 27(3): 267–275.

Pandey, G., Verma, M.K. and Tripathi, A.N. 2006. Studies on storage behaviour of apple cultivars. *Indian Journal of Horticulture* 63(4): 368–371.

Pason, N.S., Timm, E.J., Brown, G.K., Marshall, D.E. and Burton, C.L. 1990. Apple damage assessment during intrastate transportation. *Applied Engineering in Agriculture* 6(6): 753–758.

Sharma, R.R. and Singh, D. 2010. Effect of different packaging materials on shelf-life and quality of apple during storage. *Indian Journal of Horticulture* 67(1): 94–101.

Sharma, R.R., Singh, D. and Singh, R. 2009. Studies on transportation losses and quality parameters in apple packed in different containers. *Indian Journal of Horticulture* 66(2): 245–248.

She, J., Mlsna, D.A., Baird, R.E., Mohottige, C.U. and Mlsna, T.E. 2018. Volatile metabolomics with focus on fungal and plant applications-a review. *Current Metabolomics* 6(3): 157–169.

Shen, Y., Nie, J., Dong, Y., Kuang, L., Li, Y. and Zhang, J. 2018. Compositional shifts in the surface fungal communities of apple fruits during cold storage. *Postharvest Biology and Technology* 144: 55–62.

Soto-Silva, W.E., González-Araya, M.C., Oliva-Fernández, M.A. and Plà-Aragonés, L.M. 2017. Optimizing fresh food logistics for processing: Application for a large Chilean apple supply chain. *Computers and Electronics in Agriculture* 136: 42–57.

Wassermann, B., Kusstatscher, P. and Berg, G. 2019. Microbiome response to hot water treatment and potential synergy with biological control on stored apples. *Frontiers in Microbiology* 10: 2502.

Whitehead, S.R., Wisniewski, M.E., Droby, S., Abdelfattah, A., Freilich, S. and Mazzola, M. 2021. The apple microbiome: Structure, function, and manipulation for improved plant health. *The Apple Genome*: 341–382.

Xiaoyu, L., Xiaofang, W., Wei, W. and Jun, Z. 2007. Estimation of apple storage quality properties based on the mechanical properties with BP neural network. *Transactions of the Chinese Society of Agricultural Engineering* 23(5): 150–153.

Yahia, E.M., ed. 2009. *Modified and Controlled Atmospheres for the Storage, Transportation, and Packaging of Horticultural Commodities.* CRC Press.

Yang, D.S., Luo, J.M., Xu, G.Y., Liu, D. and Wang, M. 2006. Preliminary study on controlling apple diseases during its transportation and storage by the treatment of Natamycin. *China Plant Protection* 10.

Yang, Y. and Ren, X. 2016. Research progress of apple cuticular wax and its effect on storage and transportation quality of postharvest apple fruit. *Storage and Process* 16(6): 1–15.

16 Innovative Processing Technologies

Muhammad Aamir, Muhammad Afzaal,
Farhan Saeed, Ifrah Usman and Iqra Ashfaq

CONTENTS

DOI: 10.1201/9781003239925-16

16.1   INTRODUCTION

Apple is known as pomaceous or garden fruit and is considered the 4th most nutritionally rich fruit worldwide. Utilization of this vital fruit as an appreciable raw material for the development of different food products is now emerging. Apples belongs to *Malus domestica* species and is a member of the *Rosaceae* family (Zarein *et al.*, 2015). According to a food and agriculture organization 2011 evaluation, about 75 million tons of apples are produced worldwide (Kim *et al.*, 2013). Apples are consumed raw and also used processed for juice, cider and puree, etc. A commonly used sentence "An apple a day keeps the doctor away" is highly advertised to make people aware of the health benefits of this important fruit. This saying is highly appreciable because of apples' significant nutritional importance, easy accessibility and high abundance in fruit markets.

Apples are among the most widely consumed fruits worldwide, and their high antioxidant and phenolic content makes it a fruit of major importance. A study observed the phenolic contents of 62 fruits by using the Folin-Ciocalteu reagent, and it ranged from 11.88 for Pyrus communis to 585.52 for Ziziphus jujube Mill mg GAE/100 g wet weight. Among these red, rose-red and green apples exhibited values such as 73.9, 70.5 and 68.2 mg GAE/100 g wet weight respectively (Fu *et al.*, 2011). Iacopini *et al.* (2010) and Minnocci *et al.* (2010) performed studies to compare polyphenolic contents in ancient and commercial cultivars of apples. It was observed that the polyphenolic contents in 'Paniaia' red cultivars resulted in 221 mg GAE/100 g and 56 mg GAE/100 g in case of 'Gala' cultivars. These studies revealed that variations in genetic makeup for polyphenolic contents can be the result of genetic variability of germplasm of apple fruits. As apples contain high amount of moisture, they are more vulnerable to decay and exhibit low storage life. For the past few years a major concern has been to increase storage life by drying or other processing activities. Depending upon varying varieties of apples, post storage technologies have enhanced storage life up to 1 year.

16.2   BY-PRODUCTS OF THE APPLE INDUSTRY

16.2.1   APPLE POMACE AND APPLE PEEL

An overview of by-products (Figure 16.1) generated from the apple processing industry is extensively discussed in previous literature (Rabetafika *et al.*, 2014; Perussello *et al.*, 2017). After harvesting, apples are picked up, washed thoroughly and transporting to processing units for further processing. The juice is squeezed off from the fruit by pressing it. Before pressing the fruit, apples are processed in the required manner depending upon the requirements of end production and the desires of different companies. Production of juice is also linked with the generation of apple pomace and peels as a by-product in apple processing industries. Potential utilization of apple peel is described in Table 16.1

Huge amounts of by-products as solid residues (such as seed, peel, stem, core and soft tissues) are obtained during the production of syrup, juice, vinegar and cider (Kennedy *et al.*, 1999). From all these, apple pomace is the most widely generated by-product which ranges from 25%–30% of apple weight (Gullón *et al.*, 2007). Composition of apple pomace principally comprises flesh, peel, seed, seed core and stem by 54%, 34%, 7%, 4% and 2% (Kolodziejczyk *et al.*, 2007). Apple

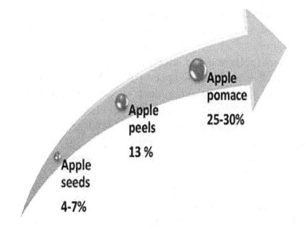

FIGURE 16.1   Apple by-products.

TABLE 16.1
Potential Utilization of Apple Peels

Example as food additive	Application method	Source
• Dietary fibers	Direct use	Roberts *et al.* (2004)
• Fruit juice preparation		Singh and Narang (1992)
• Animal feed		Gassara *et al.* (2011)
• Organic acid	Products gained via fermentation	Attri and Joshi (2005)
• Pigments		Bhushan and Joshi (2006)
• Enzymes		Longo and Sanromán (2006)
• Organic acid		Villas-Bôas *et al.* (2003)
• Enzymes		Hang and Woodams (1984)
• Animal feed		
• Aroma compounds		
• Ethanol		
• Baker's yeast		
• Antioxidant	Bioactive compound extractions	Tian *et al.* (2010)
• Pectin		Wijngaard and Brunton (2009)
• Oil		Figuerola *et al.* (2005)
• Dietary fiber		
• Biogas	Fuel production	Hang (1987)
• Ethanol		
• Oyster mushroom	Substrate for edible mushrooms	Worrall and Yang (1992)
• Functional food	Bulk peel powder via drum drying	Henríquez *et al.* (2010)
• Fiber formulation		Rupasinghe *et al.* (2008)
• Phenolic compounds		
• Functional foods	Antioxidant peel extract via freeze	Wegrzyn *et al.* (2008)
• Nutraceutical	drying	Wolfe *et al.* (2003)

peel is also major constituent of apple pomace, and it accounts for about 13% of the original weight of apples (Tarazona-Díaz and Aguayo, 2013). Their by-products are mainly generated from apple sauce, freshly cut apples, canned and dried apple manufacturing (Tian *et al.*, 2010). In Chile, processing of apple fruit generate about 9000 metrics tons of peels annually (Henríquez *et al.*, 2010).

Inferior quality fruits (not suitable for consumption purposes) generate tremendous quantity of residues especially pulp and peel, seeds as well as stems by 95%, 2%–4% and 1% respectively (Perussello *et al.*, 2017). Pomace of this fruit can be significantly used as a source of essential components, i.e., antioxidant (polyphenol) (García *et al.*, 2009), aroma compounds (Medeiros *et al.*, 2000), fiber (edible) (Masoodi *et al.*, 2002) and pectin (May, 1990). It is also used as an important raw material for the development of various food products such as apple pomace spirits (Moscetti *et al.*, 2013). In addition, apple pomace can also be used for the efficient synthesis of pectolytic enzyme (Berovič and Ostroveršnik, 1997) and protein rich animal feeds (Alibes *et al.*, 1984). Natural aroma compounds can also be prepared by fermentation processing of this by-product (Daigle *et al.*, 1999).

The variety of apple used and the process used to extract juices in different industries determines the nutritional value of pomace. These by-products possess a higher moisture level and carbohydrate contents such as simple sugars (sucrose, glucose and fructose), cellulose, lignin and hemicellulose. It also contains small amounts of vitamins, minerals and proteins as well as a good source of pectic compounds (Villas-Bôas *et al.*, 2002; Jin *et al.*, 2002; Canteri-Schemin *et al.*, 2005). Apple pomace is a significant by-product that can be widely used for fermentation processes, especially the solid-state fermentation, as it possesses vital nutrients required for the growth of fermentative microbes. Baker's yeast was produced by using the extract of apple pomace in an aerobically fed-batch process (Bhushan and Joshi, 2006). The outcome of this study showed incredible results as the ability of the baker's yeast (produced by apple pomace) to raise the dough proved to be the same as the commercial yeast. So it was concluded that the apple pomace can be best used as a substrate to provide carbon for the production of baker's yeast. It was also found that apple pomace can be used as a natural stabilizer to enhance the cohesiveness and uniformity of yogurt fermentation by using *Lactobacillus delbrueckii subsp. bulgaricus* and *Streptococcus thermophilus* (Wang *et al.*, 2019).

Composition of apple pomace mainly comprises carbohydrates (84.7%), total sugars (54.2%) and starch (5.6%) (Dhillon *et al.*, 2013), so it could be used to produce flavorings and beverages (alcoholic). Fermented apple pomace also has the ability to be used as beer flavoring as discussed by (Ricci *et al.*, 2019). This fermented by-product also possess the capability to induce complex aroma profile thus proving to be a significant by-product for enhancing aroma profiles of alcoholic beverages like beer. Li *et al.* (2015) evaluated the ability of apple pomace to be used to impart a fruity flavor to cider (that is typically produced by apple fermentation). It is also observed that fermented apple pomace can be used as a natural flavoring compound in different beverages (Madrera *et al.*, 2015). Fermented apple pomace also has the application of production of pomace spirits (using the required procedure) (Silva *et al.*, 2000).

16.3   BIOACTIVE POLYPHENOLS IN APPLES

Apple pomace is rich in polyphenols, especially in its peel parts (Kennedy *et al.*, 1999). Presence of polyphenolic compounds, its composition and contents can be varied depending on the variety of plant, environmental conditions, postharvest strategies and the part of plant such as pomace, peels, flesh and seeds (Heimler *et al.*, 2017; Oszmiański *et al.*, 2018). McRae *et al.* (1990), Pérez-Ilzarbe *et al.* (1991) and Kalinowska *et al.* (2014) evaluated differences and compared polyphenols of different varieties of apple plants. The studies proved the presence of flavonols as well as their glycosides in flesh and peels parts of apples and confirmed that these compounds are present in good amounts in peels but in trace amounts in the flesh portions of the plants.

16.4   CONVENTIONAL PROCESSING TECHNIQUES

In food industries one of the most commonly used process technologies is thermal processing which helps to ensure and maintain microbiological safety of food products (Pereira and Vicente, 2010). These technologies fundamentally rely on the heat generation outside of the food products by fuel combustion or electric resistive heater. The heat is transferred by conduction or convection

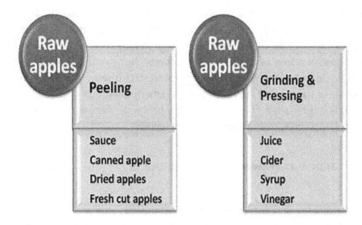

FIGURE 16.2   Conventional processing of apples.

mechanisms inside the food. Highly perishable food products like fruits are thermally processed which assist to preserve and enhance their shelf life. Apart from canned or dehydrated methods, these fruits are processed into nectar, juices, purees, smoothies, etc. (Figure 16.2). These thermal processing techniques affect the bioactive content of these processed products.

16.4.1   BLANCHING

Blanching is processing technique which helps to inactivate enzymes and maintain the nutritional aspects, color textures and other qualities of fruit products. Blanching can be done by using steam, water, hot air, vacuum steam in can, etc. The most widely used method is water blanching 75°C –95°C for 1 to 10 minutes. The running cost of this method is relatively low. For various processing techniques such as drying and freezing, blanching is a pretreatment (Sian and Ishak, 1991). One of the drawbacks of this technique is the loss of heat sensitive compounds; for example, in blanched papaya and pineapple cubes anthocyanins loss occurs. During blanching of papayas and pineapples the major cause of anthocyanins loss was leaching (Patras *et al.*, 2009). It was reported that anthocyanins are one of the heat sensitive phytochemicals that go through decomposition following heat treatment. In exotic fruits blanching may affect various other phytochemicals. Many researchers have reported that ascorbic acid is an extremely thermosensitive compound which undergoes degradation by heat application. It was described by Beyers and Thomas (1979) that ascorbic acid degrades easily as a result of blanching. Furthermore, according to Vieira *et al.* (2000), ascorbic acid reacts through two major paths, the most common one being in the presence of oxygen ("aerobic pathway"), which leads to the formation of dehydroascorbic acid, which can then follow different modes of degradation.

16.4.2   PASTEURIZATION AND STERILIZATION

Pasteurization of purees and juices is done to destroy microbes as well as inactivate pectin methyl esterase (PME). Parallel to various other thermal techniques, it may lead to changes in the bioactive compounds of food products. As a result of severe heat processes, in exotic fruits the pasteurization process may become a cause of reduction in bioactive contents (Elez-Martínez *et al.*, 2006). For various fruit juices such as mulberry fruits, pineapple, cashew and apple the result of pasteurization appear as a decrease in ascorbic acid, anthocyanin, carotenoids, etc. (Aramwit *et al.*, 2010; Chin *et al.*, 2010; Zepka and Mercadante, 2009). Many researchers report that pasteurization causes a reduction in bioactive contents but exceptions exists. According to Hoffmann-Ribani *et al.* (2009)

there is a decrease in quercetin, myricetin and kaemferol levels in various industrial processed fruit juices. Furthermore Beyers and Thomas (1979) describe a reduction in carotene as well as ascorbic acid contents in papaya and litchis, etc. following sterilization. A simple thermal decomposition process would cause a loss of bioactive compounds; however, it depends upon their chemical structure. The molecules containing higher unsaturated structure are more susceptible to a thermal degradation process. In the case of liquid food products dissolved oxygen may enhance the rate of bioactive compounds degradation. To overcome this problem, degassing of the liquid food prior to the pasteurization process may aid in their longer retention.

16.4.3 FERMENTATION AND FERMENTED APPLE PRODUCTS

16.4.3.1 Cider

The term cider is basically defined according to the specific region. Some countries classify and label cider in a well-defined manner, but in other countries like Eastern Europe some adaptations are still occurring. The European Cider and Fruit Association defined cider as an alcoholic drink that is produced directly from the partial or complete fermentation of only fresh apple juice (or concentrated one). The alcoholic content of apple ciders vary from 1.2% to 8.5% by volume (ABV), but it is necessary to retain the characteristics of fermented apple juice. Ciders that have a lower amount of alcohol are also common in the market now including alcohol-free ciders that contain even less than 0.5% ABV. Low-alcohol ciders are also used and the alcoholic contents in these ciders range between 0.5%–1.2% ABV. In the modern world other ciders such as ice ciders (no alcohol, just water and sugar) and flavored ciders (apple flavors combined with flavorings, juices and extracts of other fruits) are also available. Increasing demand and production of apples has increased the interest toward cider production. The development of cider from apples comprises few basic steps. First the apples are washed and sorted before being cut into smaller pieces with a thickness of about 4 mm to 5 mm. The juice is squeezed from the smaller pieces and is clarified and then depectinized. Inoculation of yeast, alcoholic fermentation and the incorporation of some specific nutrients for the growth of lactic acid bacteria (LAB) is employed. The last steps include malolactic fermentation followed by stabilization and maturation as an optional part in cider preparation.

The sensory characteristics of the cider, such as aroma properties, basically depend on the apple cultivars, type or species of microorganisms used, geographical areas and processing techniques. Consecrated cider assortments are typically French and English, therefore the market is varying constantly, and emerging varieties are being produced. Previous studies showed that some unusual and unconventional cultivars of apples including dessert apples are depicted to have better quality attributes and consumer acceptability (Nicolini *et al.*, 2018; Ye *et al.*, 2014). Another product that is mainly obtained from the apple cider by secondary fermentation in bottles is termed sparkling cider (obtained by the Champenoise method) and is of high interest now (Valles *et al.*, 2008; Picinelli *et al.*, 2005; Gomis *et al.*, 2004). Apple seasonality is a cost-effective substitute of apple juice thus concentrated forms of apple juice can be significantly used in cider preparation. If the media for fermentation is enriched with essential nutrients, concentrated apple juice can be preferably used (Rosend *et al.*, 2020). Some pre-fermentation techniques applied to apple juice prior to fermenting apples can give better results and improve the overall quality of the cider (Tarko *et al.*, 2020).

16.4.3.2 Vinegar

In previous times, acetic acid fermentation or simple fermentation was mainly done for preserving food from spoilage. Apples are cultivated in huge areas and result in large production, and thus apple production becomes overfull in some of areas. Apple processing products basically belong to beverage and canning industries (both alcoholic and non-alcoholic). Because of the large production of apple cider, it is sometimes reprocessed to the production of vinegar. European countries often use apple cider vinegar as a functional drink which is also alcohol free. Apple vinegar, also known as cider vinegar, is prepared from concentrated or fresh apple juice by double fermenting the juice. Both the alcoholic as well as acetic acid fermentation is used in this process. It is in high demand because of its therapeutic properties and food flavoring and preservation abilities. Previous studies also showed that alcoholic as well as acetic acid fermentation can alter the nutrient profile and functionality to a more complex one (Budak and Guzel-Seydim, 2010; Mudura *et al.*, 2018). Composition, especially its acidity, is maintained specifically, but the minimum amount is regulated at 4% (weight/volume) acetic acid. Traditionally apple cider vinegar is prepared by crushing fresh fruits and then fermenting and maturing in specific barrels (wooden). The disadvantages of the traditional method are that it is highly time consuming, the fermentation processes may end up uncompleted and it may even result in reduced yield (acetic acid fermentation) (Joshi *et al.*, 2009).

16.4.3.3 Apple Spirit

European Regulation EC 110/2008 defined fruit marc spirit as a specific drink having greater than 37.5% (volume by volume) alcoholic potential in it, and the amount of volatile substances should be more than 200 g/hectoliter of purified form of alcohol (Scotch Whisky Association, 2018). The specific aroma, the quality of the apple spirit and the amount of methanol used all depend on the fruit that is used during fermentation which is mainly affected by specific varieties, their origins, storage and ripening status (Januszek *et al.*, 2020; Hang and Woodams, 2010; Versini *et al.*, 2009). To prepare apple spirit, the juice is squeezed out of the fruit and then used to prepare cider by fermentation followed by the distillation process. It is found that the aging process can enhance the overall quality of drinks. Studies show that aging apple spirits for about 60 days improve the phenolic content and volatile characteristics of the product (Coldea *et al.*, 2020). 'Calvados' is the most famous apple spirit that is prepared by the fermentation of a mixture of specific fruits with bitter, sweet and tart flavors to produce cider. Fermented cider is then distilled (double distillation) followed by aging (Small *et al.*, 2011). Aroma characteristics of the spirits are affected by the cider (Madrera *et al.*, 2010).

16.4.3.4 Probiotic Fermented Apples

Fermentation food technology deals with the texture, flavor and odor of fruits and vegetables to enhance the nutritional profile and preservative ability to minimize the dependence of refrigeration or freezing processes. It is also observed that the fermentation process could enhance antimicrobial abilities and deteriorate toxic substances, thus promising significant properties of the end product (Daeschel, 1989), thus generally termed as probiotics. Lactic acid bacteria have the ability to alter these sensorial attributes and can also affect the lipolytic and proteolytic performance of the product. The quality of probiotic fermented apples is mainly influenced by fruit cultivars which could affect aroma and the taste of the end product (Peng *et al.*, 2021). Quality can be varied due to different cultivars' flavor, taste and aroma attributes and varied chemical compositions, mainly volatile substances, soluble sugars and organic acids (Braga *et al.*, 2013).

16.4.3.5 Properties of Fermented Apple Products

Fruits juices have been identified as a significant substitute for lactic acid (bacterial) fermentation because of their considerable nutritional profile including antioxidant potential, vitamins, dietary fibers and minerals contents. Fruits fermented with lactic acid bacteria can be a better functional food possessing probiotics, improved nutritional contents, sensory characteristics and physiochemical properties (Costa *et al.*, 2013; Pereira *et al.*, 2011). During fermentation process, lactic acid bacteria utilizes sugars such as sucrose, fructose and glucose present in juices and converts them into lactic acid and enhances their growth thus decreasing the amount of sugar, especially fructose content (Verón *et al.*, 2019). Metabolism of carbohydrates depends on the variety or cultivar of the fruit. A study was conducted to evaluate the utilization of sugar contents during fermentation. For this purpose nine cultivars of apple fruit were investigated. It was observed that fructose in the 'Golden

Delicious' cultivar exhibited a significant amount of reduction (Peng *et al.*, 2021). In another study it was estimated that using a strain of commercial lactic acid bacteria, the total content of sugar can be reduced up to 23% in the juice of the 'Fuji' apple cultivar during the process of fermentation (Wu *et al.*, 2020). Different strains of lactic acid bacteria impart various organic acids and aroma substances to stabilize the taste, flavor, color, overall acceptability, storage and chemical stability of the end product (Filannino *et al.*, 2013). Thus strains of lactic acid bacteria can impart distinctive effects on flavors produced during fermentation. Figure 16.3 illustrates the major products of apple fermentation.

16.4.4 THERMAL DRYING

Dehydrated fruits are mainly consumed as standalone food or constituent of other food products. The drying technique assists in shelf life enhancement as well as a decrease in the volume of fruits (Prakash *et al.*, 2004). There are several dehydration techniques such as microwave, cross flow, osmotic air drying, oven air drying, spray drying, freeze drying, drum drying, etc. which are applied on different fruits. Although the drying process extend the shelf life of fruits, it may also affect bioactive compounds like polyphenls, carotenoids and ascorbic acid present in them as these are heat sensitive compounds. It was reported by Dahiya and Dhawan (2003) that the storage of fruits at elevated temperatures may lead to oxidation. In addition, in the samples which were dried by applying fluidized bed drying techniques more ascorbic acid retention was observed in comparison with fruits which were dried by applying hot air or sun drying methods (Murthy and Joshi, 2007). It was also observed that low temperature super-heated steam drying (LPSSD) showed a higher retention of ascorbic acid as compared to vacuum drying. Among bioactive compounds, the carotenoids were more stable as compared to anthocyanins in the drying process (Sian and Ishak, 1991). Various scientists (Saxena *et al.*, 2010; Kong *et al.*, 2010) reported similar results for various other fruits. Contrary to this, Saxena *et al.* (2010) reported that the total hydrolysable tannins and gallic acids are more stable.

In general thermal processing may cause a loss of bioactive content in fruits. In order to minimize this loss and ensure maximum bioactive compounds stability, the processor may need to reevaluate existing thermal processing techniques. This scenario has urged researchers to direct their efforts at

FIGURE 16.3   Fermented apple products.

novel nonthermal processing techniques. Innovative nonthermal processing techniques can ensure the safety of fruits as well as minimize the negative thermal effects on them. Modern nonthermal processes are effective at ambient/sublethal temperatures and could aid in improving nutritional and quality parameters.

16.4.5 VALORIZATION

In the past, apple peels were used for worthless purposes mainly for direct use including fertilizers, compost, animal feed, press aid for the preparation of vinegar and juice and pressed to form a cake. Wolfe and Liu (2003) stated that apple peels have begun to be used as a significant source of fiber and pectin (Miceli-Garcia, 2014).

Łata and Tomala (2007) depict that apple peels are composed of various vital nutrients including dietary fibers, phenolics, minerals and ascorbates. Apple peel contain even higher antioxidant activity and antioxidant contents and thus possess better health-improving characteristics including inhibiting cancerous cells in the body (Wolfe *et al.*, 2003; Drogoudi *et al.*, 2008; Khanizadeh *et al.*, 2008). Gorinstein *et al.* (2001), Leontowicz *et al.* (2003) andHenríquez *et al.* (2010) also observed that mineral content, as well as dietary fibers, are present in much higher amounts in apple peels than in apple flesh. The same results were observed by Burda *et al.* (1990) andJu *et al.* (1996) that apple peels are more nutritious in the case of polyphenols concentration than in its flesh part. Studies also prove that similar to apple flesh, apple peels are composed of catechins, phloridzin, procyanidins, caffeic acid, cholorogenic acid and phloretin glycosides. Unlike apple flesh, apple peel contains more flavonoids like cyaniding glycosides and quercetin (Escarpa and Gonzalez, 1998; van der Sluis *et al.*, 2005). Apple peels can be significantly used as a source of phytochemicals and health-improving substances in different food products.

16.5 INNOVATIVE APPLE PROCESSING TECHNIQUES

16.5.1 DENSE PHASE CARBON DIOXIDE (DPCO$_2$)

This is a collective term for supercritical CO_2 or liquid CO_2 or high pressurized carbon dioxide (HPCD). DPCD is nonthermal as well as a continuous system for foods which make use of CO_2 in combination with a pressure to destroy microbes and to ensure food preservation. This innovative technique is an alternative to thermal techniques such as pasteurization. According to Del Pozo-Insfran *et al.* (2006) in the food industrial sector worldwide this method is gaining interest as it utilize CO_2 gas which is nontoxic, inexpensive, and nonflammable. In addition to this, it has gained the status of generally recognized as safe (GRAS) generally recognized as safe. Super critical CO_2 or dense phase carbon dioxide indicated phases of matter that exist as fluid, however dense, with respect to CO_2 in gaseous form. Fraser (1951) was the first to explore that DPCD can cause the inactivation to bacterial cells. Furthermore in a supercritical state the CO_2 has less viscosity, which allows it to penetrate in complex structure as well as porous material of fruits (Zhang *et al.*, 2006). Initially the dense phase CO_2 application was treatment of the whole fruits to inhibit the growth of mold. But one of the drawbacks of this technique is that even at low pressure it may cause tissue damage to some fruits (Haas *et al.*, 1989). The application of DPCD has been used for different fruit-based juices. For example Gunes *et al.* (2006) reported the application of DPCO$_2$ for apple cider. Furthermore many other researchers reported that orange juice (Gunes *et al.*, 2005), grape juice (Yagiz *et al.*, 2005) and mandarin juice (Garcia-Gonzalez *et al.*, 2007) showed high pressure CO_2 effects on microbe inactivation. The researchers found that ascorbic acid (vitamin C) concentration reduced following DPCO$_2$ processing, and the loss percentage was less than untreated samples. The reason behind this is that CO_2 may lower pH when dissolved in aqueous portions of food, but contrary to this vitamin C has more stability at a lower pH and in the presence of oxygen, it oxidizes easily. They concluded that precise investigation is required to evaluate the parameter needed to optimize bioactive content and quality of fruit juices. $DPCO_2$ has proved to prevent the loss of many other valuable bioactive compounds like beta carotene. Many authors have studied the positive effect of $DPCO_2$ on antioxidant as well as polyphenols of fruits.

16.5.2 MICROWAVE TREATMENT

Microwave treatment is extensively used for processing various food products. It has been reported by many researchers that the application of microwave on different fruits and their products has proved advantageous. In addition to this, microwave treatment application has been gaining attraction to extract valuable bioactive compounds from apple pomace. According to Bai *et al.* (2010) at optimal condition microwave-assisted extraction gives higher TPC extraction (approx. 62.68 mg GAE/100 g) yields as well as less solvent utilization as compared to convectional extraction. After this it was reported by Rezaei *et al.* (2013) that parallel results have been observed by application of microwave in combination with ethanol water solvents. Currently the researchers are in search to apply microwave in combination with other innovative techniques to get desired results for fruits and their products processing. Ajila *et al.* (2011) reported the effects of microwave and ultrasound treatment on bioactive polyphenol extraction from apple pomace.

16.5.3 PULSED ELECTRIC FIELD

Pulsed electric field (PEF) is a technique which has been used recently for its application in fruit processing. By using electro permeabilization mechanism, PEF has proved an efficient technology for safe beverage production. Moreover PEF showed a positive influence on the texture of fruits leading to increased yields of juice as well as metabolites extraction (Zulueta *et al.*, 2010). The principal difference in PEF utilization is that the use of the intensity of fields and higher intensity fields was more efficient toward inactivation of microbes. With the interest of bioactive compound availability in foods, especially fruit-based beverages; PEF has been studied by Soliva-Fortuny *et al.* (2009). The effect of PEF was appreciable over storage for 31 days. 800 μs treatment showed greater retention in comparison with thermal treatment. The results proved that shorter PEF treatment times lead to higher retention of vitamin C. Contrary to this, longer exposure of PEF produced free radicals and caused degradation (Ade-Omowaye *et al.*, 2001; Corrales *et al.*, 2008) for the preservation of fruit juices; favorable electric field strengths showed a higher extraction of metabolites from the fruits in range of 1–10 kV/cm. The application of this technique and the positive outcomes to increase yield extraction of fruits like apples has been reported by many (Corrales *et al.*, 2008; López *et al.*, 2009; El Belghiti and Vorobiev, 2004). Vorobiev and Lebovka (2010) reported that PEF treatment can be employed to extract biomolecules such as proyein, carbohydrates, pigments, flavors, pectin, etc. from apples. Furthermore PEF increase the extraction of polyphenols from apple peels, flesh, juice and pomace (Lohani and Muthukumarappan, 2016; Wiktor *et al.*, 2016; Turk *et al.*, 2012). For high quality apple juice, PEF in combination with pressing could be an excellent technique (Grimi *et al.*, 2011).

16.5.4 OZONE PROCESSING

There is an ever increasing interest to utilize ozone as an efficient technique for the preservation of foods because of its biocidal efficacy as well as wide antimicrobial spectrum. In food industries this innovative technology is utilized to wash and store fruits using gaseous treatment. According to Cullen *et al.* (2009), with FDA approval ozone can be used as direct additive in food products; in liquid food such as fruits the potential of ozonation has started to be exploited. Ozone has a wide application in the food industrial sector as an antimicrobial agent and has proven to more efficient compared to conventional antimicrobial agents such as potassium sorbate, chlorine, etc. The utilization of ozone for storage and disinfection purposes for fruits and their products have been reported by many researchers (Barboni *et al.*, 2010; Akbas and Ozdemir, 2008; Meyvaci Sen and Uthaibutra,

2010; Graham and Tyman, 2002). However in the case of exotic fruits there are limited studies related to microbiological analysis. It was reported by Jay *et al.* (2005) that normally 0.15 to 5.0 concentration of ozone have been observed to be effective against spoilage, bacteria and yeast. The effects of ozone on bioactive compounds of fruits such as apples are more limited than thermal treatment. Furthermore Alothman *et al.* (2009b) studied the effects of ozone applications on vitamin C, total flavonoid and total phenols content of fresh cut fruits such as apples. The application of large doses of ozone may change sensory qualities of foods. In contrast with the positive results, the excessive use of ozone treatments may cause discoloration and undesirable odors (Khadre *et al.*, 2001). According to ozone dose, chemical composition of fruits, application type and time, the ozonation affects differently on quality as well as the physiology of fruits or other foods (Whangchai *et al.*, 2006).

16.5.5 ULTRASOUND PROCESSING

In last decade the use of ultrasound (US) techniques in the food industry has emerged as a novel application and an alternative to conventional thermal approaches such as pasteurization and sterilization on fruits and their products. The advantage of US includes low energy utilization, minimum time consumption and higher outputs (Zenker *et al.*, 2003). In the case of fruit juices ultrasound has the potential to achieve a desired (5 log) food-borne pathogens (Salleh-Mack and Roberts, 2007). However it can negatively affect other food properties like flavor, color and nutritional value. During the processing of fruit juices like apple juice, US enhances storage stability and has a smaller effect on the ascorbic acid content of juices. As compared to thermal processing, US has the positive result of removing oxygen from juices (Knorr *et al.*, 2004). It has been proved by many researchers that US has high potential to extract polyphenols from apples; it is beneficial in terms of process duration as well as yields (Virot *et al.*, 2010). According to Pingret *et al.* (2012), ultrasound-assisted extraction was advantageous to enhance antioxidant activity of the apple pomace extracts. Almost 30% higher total phenolic content was obtained from ultrasound-assisted extraction than conventional methods. For apples, bioactive compounds extraction, ultrasound-assisted extraction (UAE) is an environmental-friendly, time saving technique (Yue *et al.*, 2012). Moreover they studied the polyphenol extraction from unripe apples using ethanol solvent (Pingret *et al.*, 2012; Wang *et al.*, 2018). Ultrasound treatment is favorable selectively on the extraction of polyphenol from apple products.

16.5.6 HIGH HYDROSTATIC PRESSURE PROCESSING (HHP)

One of modern nonthermal processing technique for food processing as well as preservation is High hydrostatic processing (HHP). The HHP technique is obtained from material science. In HHP 100 MPa is applied to food products. HHP has been widely reviewed by Rastogi *et al.* (2007). The benefit of HHP is that the pressure given at the precise position and time is alike in all directions, uniformly transferred and remains independent of geometry (Oey *et al.*, 2008). Many researchers have found that HHP helps preserve the nutritional value of fruits and their products. In addition, this technique has minimal effects on the bioactive compounds of food products. After HHP processing during storage the ascorbic acid degradation can take place but will depend upon conditions. It has been stated by many authors that a different influence has been observed by different HHP combinations and the vitamin C stability differs in fruits according to HHP conditions (Yen and Lin, 1996). According to Suthanthangjai *et al.* (2005) and Zabetakis *et al.* (2000), during storage in processed juice the concentration of oxygen and the enzyme's residual activity can cause anthocyanin degradation.

16.5.7 RADIATION PROCESSING

The exposure of fruits or their products to radiation (ionizing or nonionizing) for processing or preservation is called radiation processing. Ionizing radiation (IR) can be gamma rays, Cobalt

TABLE 16.2
Advantage of Modern Techniques During Bioactive Compounds Extraction from Apples

Modern techniques	Advantages	References
Microwave	Reduction in extraction time and solvent utilization	Hemwimon et al. (2007)
Pressurized fluid	Less cost with chemicals; lower solvent utilization; less ecological impacts	Grigoras et al. (2013)
Ultrasound	Utilize less fossil energy and solvent	Corrales et al. (2008); Pingret et al. (2012)
Pulsed electric field	Less temperature increase; preserve structure of product	Grimi et al. (2010)
Super critical fluid	Reduction in extraction time and solvent consumption; less detrimental ecological effects; less degradation reactions	Adil et al. (2007); Massias et al. (2015)
Enzyme	High selectivity	Pinelo et al. (2008); Grigoras et al. (2013)
High voltage electric discharge	Less temperature increase; shorter treatment time; energy efficient	Boussetta et al. (2012); Gros et al. (2003)
Solvent diffusion	Easy operation	Bai et al. (2010); Guo et al., 2007
Ozone treatment	Stability of bioactive compounds; improve texture of food products	Alothman et al. (2009a)
Radiation processing	Preserve nutritional quality; less time required; energy efficient	Abdel Rahman Alothman and Muhammad Nazih (2009)

60, X-rays or high energy electron. Among the nonionizing radiation including electromagnetic radiation represented by visible light, ultraviolet, infrared and microwave, etc. it was reported by Alothman et al. (2009a) that IR causes modifications in color, taste, nutrients, flavor and many other attributes of fruits. Nevertheless modifications in quality attributes depend on food, irradiation dose and the type of irradiation used (Bhat and Sridhar, 2008). Inactivation of microbes by irradiation is because of damage to the DNA, which causes the destruction of the reproductive potential of the cells (De Ruiter and Dwyer, 2002). Irradiation of the fruits with ultraviolet has a positive impact by increasing the enzyme's activity for flavonoid biosynthesis. Moreover Kataoka et al. (1996) stated that application of IR to fruit and their products along with microbial safety using modern techniques causes an increase in health-promoting compounds. Advantages of modern techniques during bioactive compounds extraction from apples is described in Table 16.2.

16.6   DEGRADATION MECHANISM

16.6.1   THERMAL PROCESSING

In fruits, the degradation of the bioactive compounds is a complex process, and it is dependent upon the type and conditions during processing. To overcome oxidative degradation of bioactive compounds of fruits better understanding as well as finding the optimal conditions is needed to avoid loss of fruits and their products during processing and storage. Vitamins are highly sensitive compounds in fruits and their products which are affected by pasteurization and sterilization. Ascorbic acid degradation mechanism is particular for specific systems, and it is dependent on various factors including the type of fruit and the kind of processing following the aerobic and anaerobic methods. Xanthophylls and carotenes are in abundance in fruits. These bioactive compounds are also extremely sensitive to several factors such as heat, oxygen and light. Their stability is also influenced by pasteurization. There are large number of polyphenolic compounds that exist in fruits. These are easily affected by heat. The degradation of anthocyanin is directly related to the time of exposure of thermal processing (Sadilova *et al.*, 2007). For the purpose of maximum retention time of bioactive compounds, understanding of degradation mechanism in fruits and their products is needed.

16.6.2 NONTHERMAL PROCESSING

In most nonthermal processing technologies the loss of bioactive compounds is negligible in comparison with thermal techniques. Due to its strong oxidizing potential, ozone treatment may cause a loss of valuable phytochemicals up to a certain extent, but in the case of ultrasound technology promising results have been observed. However, the basic chemical reaction mechanism in sonication degradation is pyrolysis which is a major cause of polar bioactive compound degradation (Sivasankar *et al.*, 2007). According to Portenlänger and Heusinger (1992), during sonication the degradation of ascorbic acid is due to free radical formation. In addition to this (Loomis-Husselbee *et al.*, 1991), in PEF treatment ascorbic acid degradation may be due to electrochemical reactions which may lead to changes in pH, chemical composition and temperature.

16.7 MAJOR STRATEGIES TO IMPROVE RETENTION OF BIOACTIVE COMPOUNDS DURING PROCESSING

Fruits and their products experience commercial processing techniques. Among these techniques thermal and nonthermal techniques are both applied varying according to requirements. Common bioactive compounds such as anthocyanins polyphenols are degraded by enzymes polyphenol oxidase. By application of mild heat treatment this enzyme can be inactivated. Considering this, many authors have suggested the inclusion of heat treatment (blanching approx. 50°C) can be beneficial to stop anthocyanin retention. Moreover Rossi *et al.* (2003) suggested that the blanching step is vital juice processing, as blanching can inactivate polyphenol oxidases; a parallel approach can be utilized for fruit and their products. In addition, more stability of bioactive compounds can be obtained by process optimization. It was reported by many authors that thermal burden affects bioactive compounds more during storage. In order to overcome this, intelligent selection of the best extrinsic conditions for storage is necessary (Patras *et al.*, 2010). In terms of higher retention of bioactive compounds demands of innovative, research to streamline processes by practically applying the best combination of technologies is needed. The process optimization with a combination of thermal techniques and nonthermal such as pulsed electric field, ultrasound and high pressure processing have considerable potential for higher bioactive retention.

16.8 FOOD APPLICATION OF APPLE BIOACTIVE POLYPHENOL

16.8.1 FUNCTIONAL FOOD

Many bioactive polyphenols derived from apple have potential to cure various diseases such as hypertension, tumors, allergies, senility, mutation and dental caries and has the potential for improvement of enzyme activity (catalase, superoxidase, glutathione peroxidase, dismutase, etc.); they can play a role as functional food in the human body (Padmini *et al.*, 2008). By extracting polyphenol from apples with an application of thermal and nonthermal processing technologies many value-added and functional foods have been prepared (Wang and Bohn, 2012; Landete, 2012; Lea, 2010; Terao, 2009).

16.8.2 Meat Products

Apple polyphenols are favorable against oxidation and deterioration of the meat and meat products. It was reported by Beriain *et al.* (2018), Velasco and William (2011) and Yu *et al.* (2015) that shelf life, as well as freshness, of meat products can be enhanced by utilizing polyphenol derived from apples (Lakshmanan, 2000). Moreover, by inhibiting volatile trimethylamine production, the fishy smell in meat products can be reduced. According to Luciano *et al.* (2009) and Rojas and Brewer (2008), the stability of the red color in meat and its products can be ensured by using apple polyphenol. In addition, bioactive polyphenol is beneficial against fat oxidation in meat.

16.8.3 Beverage and Wine

One of the modern applications of apple polyphenol is as clarifying agents in wine and beverages (Chung *et al.*, 2016). The connection of aminde group in protein and phenolic OH group with H-bond may precipitate a component forming with tannin and gelatin and suspended solids can be removed (Marchal and Jeandet, 2009; Siebert, 2009). By applying modern processing techniques and extracting polyphenol from apples, the problem of vitamin C degradation can be solved and the fading of pigments such as beta carotene can be prevented (Huvaere and Skibsted, 2015). In addition, it was stated by Chung *et al.* (2016) and Sun-Waterhouse (2011) that 1% apple polyphenol can maintain the color of beverages (Górnaś *et al.*, 2015). Phlorizin and xylose phlorizin have been utilized on a large scale as characteristics indexes of the apple and apple products; at international trade this is used to identify product quality.

16.8.4 Fruits and Vegetable

It has been reported that polyphenols extracted from apples by applying modern processing techniques have antibacterial effects on E. coli, bacillus, pseudomonas and several other tested bacteria. Extracted polyphenol from apples have excellent thermal stability in bacteriostatic activity (Cai-wen *et al.*, 2008; Vivek *et al.*, 2012). Spraying a solution of apple polyphenols have beneficial effects for fruits preservation by preventing bacterial growth. In addition, this also has the potential to maintain fruit color (Zhang *et al.*, 2015; Marshall *et al.*, 2000).

16.8.5 Bubble Gum

According to Porciani and Grandini (2012) and Tao *et al.* (2013), apple polyphenol has the capacity to lessen the amount of methyl mercaptan produced and remove bacterial from bubble gum. In bubble gum, the apple polyphenol can be added due to its excellent inhibition of halitosis. In many food products such as pastries, cakes and breads, apple polyphenol has been utilized as an additive (Yilmaz, 2006; Sudha *et al.*, 2007).

16.8.6 Cosmetics Application

Apple polyphenol is a product with outstanding properties and chemical activities which has adhesion as well as convergence potential (Zillich *et al.*, 2015). Apple polyphenol has the ability to maintain collagen synthesis, inhibit elastase, improve skin elasticity, reduce wrinkle generation and keep delicate skin appearance. Moreover apple polyphenol is often utilized as an active ingredient in several cosmetic products and plays a vital role in anti-aging, anti-ultraviolet, antioxidation sunscreen and moisturizing (Baldisserotto *et al.*, 2012; Gaudout *et al.*, 2006). From apples extracted tannins can be anti-allergic and effectively stop histamine dissociation and hyaluronidase activity (Magrone and Jirillo, 2012). It was reported by Shoji *et al.* (2005) that by applying one of the modern processing application other bioactive compounds from apples can be derived and utilized to synthesize vitamin E and vitamin C complexes (whitening agents) as phenolic compounds of apples can restrain tyrosinase and catalase activity and are helpful in the removal of reactive oxygen species.

16.8.7 OTHER APPLICATIONS

According to Alberto *et al.* (2006) and Du *et al.* (2011), apple polyphenol have been utilized to formulate antibacterial edible films. Furthermore, on a polymer chain the grafting of apple polyphenol reduces environmental damage due to phenolic waste and has been utilized for wastewater treatment (Celik and Demirbaş, 2005; Chand *et al.*, 2015). For these reasons, bioactive phenolic compounds of apples have gained importance both in food industries as well as nonfood domains (Ćetković *et al.*, 2008). That is why the application of innovation processing techniques and intensification of phenolic compounds and its by-products such as pomace, seeds, peels and flesh have gained researchers' attention.

16.9 THERAPEUTIC PROPERTIES (APPLE PHYTOCHEMICALS AND THEIR HEALTH BENEFITS)

16.9.1 ANTI-CANCER

Previous studies found that consumption of apples is associated with lowering the risk of several diseases such as lung cancer. A study was conducted to investigate the reduced risk of cancer by apple consumption. For this purpose, Nurses' Health Study and Health Professionals' Follow-up Study engaged about 47000 men and 77000 women to the research and estimated that the risk of lung cancer was decreased by about 21% in women, but that no such link was observed in men (Feskanich *et al.*, 2000). Among all the fruits and vegetables, such a significant association was observed in case of apples only. It was evaluated that intake of at least one servings of apple and pear each day can lower the risk of lung cancer in women (Feskanich *et al.*, 2000). A previous study stated that the participants consumed greater amounts of apples, grapefruits and onions showed a 40% to 50% reduction of risk of lung cancer than the participants who consumed lower amounts of these significant fruits and vegetables. It was also found that women, men and all other groups also showed reduced risk of lung cancer. Hollman (2000) stated that both apples and onions are rich in flavonoids, mainly quercetin and its conjugates. Another study found that quercetin and lung cancer exhibited an inverse relationship between them but the trend did not show many significant results (Le Marchand *et al.*, 2000).

16.9.2 INHIBIT CARDIOVASCULAR DISEASES

The intake of apple on a regular basis is linked with lowering the risk of cardiovascular diseases. The Women's Health Study did follow ups of about 6.9 years (of 40000 women) and evaluated the relationship between cardiovascular diseases and flavonoids (Sesso *et al.*, 2003). A 35% decreased risk of cardiovascular diseases was observed in the women who consumed a greater amount of flavonoids. Flavonoid and quercetin consumption was not estimated to be linked with cardiovascular diseases, stroke and myocardial infarction. Consumption of both apples and broccoli is linked with a decrease in the risk of cardiovascular diseases. A research study conducted in Finland observed association between flavonoid consumption and coronary failure. The study found that consumption of flavonoid content is inversely linked with coronary diseases and death (in women but not in men) (Knekt *et al.*, 1996). Women were found to have lowered the risk of coronary diseases due to regular consumption of apples and onions. This study also showed that apple consumption and quercetin presence is linked with cerebrovascular disease (Knekt *et al.*, 2000). Risk of a stroke of thromobiotic was evaluated to reduce by the intake of apples at greater amount (Knekt *et al.*, 2000).

16.9.3 EFFECTIVENESS OF ASTHMA AND PULMONARY FUNCTION

Studies showed that the intake of apples at a regular basis is inversely related to asthma and pulmonary disease. In Australia, a study was conducted engaging 1600 adult, and it was observed that pear and apple consumption was linked with reduced risk of bronchial hypersensitivity and asthma disorders. A point to ponder is that the consumption of total fruits and vegetables is not linked with the risk of asthma (Woods et al., 2003). In the United Kingdom a study was conducted, and it was estimated that consumption of apples along with selenium proved to lower asthma problem in adults (Shaheen et al., 2001). Consumption of apples and pears was found to be effective for pulmonary functioning and inversely linked with pulmonary and chronic disorder.

16.9.4 PREVENTION OF DIABETES AND OBESITY

The intake of apples is not only associated with lowering the risk of asthma, heart diseases and cancer but it is also linked with reducing the risk of diabetes. A recent study in Finland involved 10000 people to research and evaluated if apple consumption was linked with lowering the risk of type 2 diabetes (Knekt et al., 2002). Greater consumption of quercetin was found to decrease the risk of type 2 diabetes. Other fruits and vegetables such as white cabbage, onion, grapefruit and orange were not investigated to reduce the risk of disease. In Brazil, a research study found that consumption of apples and pears reduced the weight of overweight middle aged women (de Oliveira et al., 2003).

16.9.5 ANTIOXIDANT ACTIVITY

Apples but its peels have been investigated to exhibit strong antioxidant potential and have showed a great ability to reduce the risk of colon and liver cancer (Wolfe et al., 2003; Eberhardt et al., 2000). A study found that apples, along with peels, showed antioxidant potential of 83 μmol vitamin C equivalent, which depicted that one serving of apple (about 100 grams of apple) showed antioxidant potential of approximately 1500 mg of vitamin C. Total contents of vitamin C in one serving of apple is approximately 5.7 mg (Eberhardt et al., 2000). Although vitamin C is a strong antioxidant, this study investigated that antioxidant ability of apples mainly associated with other compounds. Apples' vitamin C contents were found to be responsible for less than 0.4% antioxidant potential.

16.9.6 ANTI-PROLIFERATING ACTIVITY

Many studies investigated significant anti-proliferating ability in apples. It was found that apples contains a distinctive mixture of phytochemicals that are considered significant in lowering the proliferation of tumor cells (Eberhardt et al., 2000). By comparing 11 other fruits with apples, it was observed that apple as ranked as the third highest fruit in exhibiting anti-proliferating activity (Sun et al., 2002). Different apple varieties exhibited differently on the propagation of cancerous cells of the liver (Liu et al., 2001). Apple peels were investigated to lower the propagation of Hep G2 more efficiently than apples (without peels) (Wolfe et al., 2003).

16.9.7 INHIBITION OF LIPID OXIDATION

Apple consumption caused a reduction in DPHPC oxidation indicating strong antioxidant potential (in vivo) of apples (Mayer et al., 2001). After 3 hours of intake of apples, shielding the impact of LDL oxidation arrived at its peak (Mayer et al., 2001). A similar trend was measured by these researchers for lowering the albumin DPHPA oxidation after apple consumption. It is found that whole apples contains greater amount of phenolic compounds than apple juice, but apple juice is still considered an extensively used source of dietary antioxidants. A study was conducted to evaluate the effect of different varieties of apple juices including whole apple and peels and flesh alone on the LDL oxidation in human body. In this study it was found that prevention of LDL oxidation ranged from 9%–34% and among all whole apples as well as peels alone showed 34% LDL oxidation prevention but flesh alone exhibited only 21% inhibition (Pearson *et al.*, 1999).

16.9.8 CHOLESTEROL LOWERING EFFECTS

Apples showed strong cholesterol prevention abilities and thus proved effective against cardiovascular disorders. Research was performed by Aprikian *et al.* (2001) to investigate the effect of lyophilized apples (on rats) for lowering cholesterol levels. It was found that a reduction in liver cholesterol and plasma cholesterol was investigated, and HDL levels were found to increase in rats. In addition, excretion of cholesterol through feces was increased indicating lowered absorption of cholesterol. In another study, rats were fed peaches, apples and pears to investigate the cholesterol lowering ability of these fruits. It was observed that among all fruits, apple was evaluated to show a significant reduction in cholesterol levels and enhanced plasma antioxidant activity (Leontowicz *et al.*, 2002).

16.10 CONCLUSION

Apple is among fruit which is widely consumed throughout the word. Processing of apples by various conventional techniques may affect the availability of various bioactive compounds. In comparison to these innovative processing technologies such as dense phase carbon dioxide, pulsed electric field, ozone processing, ultrasound processing, high hydrostatic pressure processing and radiation processing, etc. are not only energy efficient, but these modern processing technologies show enhanced availability of bioactive compounds present in apples. In addition, these innovative technologies have the potential to maintain quality as well as shelf life of the produce. With suitable selection of processing technology, treatment time and conditions, the target of improved sustainable strategies for high quality produce with minimal processing losses can be achieved.

REFERENCES

Abdel Rahman Alothman, and Nazih, M. 2009. *The Changes in the Antioxidant Capacity of Selected Tropical Fruits Upon Treatment with Gaseous Ozone and Ultraviolet C Radiation [T1–995]* (Doctoral dissertation, Universiti Sains Malaysia).
Adil, I.H., Cetin, H.I., Yener, M.E. and Bayındırlı, A. 2007. Subcritical (carbon dioxide+ ethanol) extraction of polyphenols from apple and peach pomaces, and determination of the antioxidant activities of the extracts. *The Journal of Supercritical Fluids* 43:(1): 55–63.
Ajila, C.M., Brar, S.K., Verma, M., Tyagi, R.D. and Valéro, J.R. 2011. Solid-state fermentation of apple pomace using Phanerocheate chrysosporium – Liberation and extraction of phenolic antioxidants. *Food Chemistry* 126:(3): 1071–1080.
Akbas, M.Y. and Ozdemir, M. 2008. Application of gaseous ozone to control populations of Escherichia coli, Bacillus cereus and Bacillus cereus spores in dried figs. *Food Microbiology* 25:(2): 386–391.
Alberto, M.R., Rinsdahl Canavosio, M.A. and Manca de Nadra, M.C. 2006. Antimicrobial effect of polyphenols from apple skins on human bacterial pathogens. *Electronic Journal of Biotechnology* 9:(3).
Alibes, X., Munoz, F. and Rodriguez, J. 1984. Feeding value of apple pomace silage for sheep. *Animal Feed Science and Technology* 11:(3): 189–197.
Alothman, M., Bhat, R. and Karim, A.A. 2009a. Effects of radiation processing on phytochemicals and antioxidants in plant produce. *Trends in Food Science & Technology* 20:(5): 201–212.
Alothman, M., Bhat, R. and Karim, A.A. 2009b. UV radiation-induced changes of antioxidant capacity of fresh-cut tropical fruits. *Innovative Food Science & Emerging Technologies* 10:(4): 512–516.
Aprikian, O., Levrat-Verny, M.A., Besson, C., Busserolles, J., Rémésy, C. and Demigné, C. 2001. Apple favourably affects parameters of cholesterol metabolism and of anti-oxidative protection in cholesterol-fed rats. *Food Chemistry* 75:(4): 445–452.

Aramwit, P., Bang, N. and Srichana, T. 2010. The properties and stability of anthocyanins in mulberry fruits. *Food Research International* 43:(4): 1093–1097.

Attri, D. and Joshi, V.K. 2005. Optimization of apple pomace based medium and fermentation conditions for pigment production by Micrococcus species. *Journal of Scientific and Industrial Research* 64: 598–601.

Bai, X.L., Yue, T.L., Yuan, Y.H. and Zhang, H.W. 2010. Optimization of microwave-assisted extraction of polyphenols from apple pomace using response surface methodology and HPLC analysis. *Journal of Separation Science* 33:(23–24): 3751–3758.

Baldisserotto, A., Malisardi, G., Scalambra, E., Andreotti, E., Romagnoli, C., Vicentini, C.B., Manfredini, S. and Vertuani, S. 2012. Synthesis, antioxidant and antimicrobial activity of a new phloridzin derivative for dermo-cosmetic applications. *Molecules* 17:(11): 13275–13289.

Barboni, T., Cannac, M. and Chiaramonti, N. 2010. Effect of cold storage and ozone treatment on physico-chemical parameters, soluble sugars and organic acids in Actinidia deliciosa. *Food Chemistry* 121:(4): 946–951.

Beriain, M.J., Gómez, I., Ibáñez, F.C., Sarriés, M.V. and Ordóñez, A.I. 2018. Improvement of the functional and healthy properties of meat products. In *Food quality: Balancing health and disease* (1–74). Academic Press.

Berovič, M. and Ostroveršnik, H. 1997. Production of Aspergillus niger pectolytic enzymes by solid state bioprocessing of apple pomace. *Journal of Biotechnology* 53:(1): 47–53.

Beyers, M. and Thomas, A.C. 1979. Gamma irradiation of subtropical fruits. 4. Changes in certain nutrients present in mangoes, papayas, and litchis during canning, freezing, and gamma irradiation. *Journal of Agricultural and Food Chemistry* 27:(1): 48–51.

Bhat, R. and Sridhar, K.R. 2008. Nutritional quality evaluation of electron beam-irradiated lotus (Nelumbo nucifera) seeds. *Food Chemistry* 107:(1): 174–184.

Bhushan, S. and Joshi, V.K. 2006. Baker's yeast production under fed batch culture from apple pomace. *Journal of Scientific and Industrial Research* 65: 72–76.

Boussetta, N., Vorobiev, E., Le, L.H., Cordin-Falcimaigne, A. and Lanoisellé, J.L. 2012. Application of electrical treatments in alcoholic solvent for polyphenols extraction from grape seeds. *LWT-Food Science and Technology* 46:(1): 127–134.

Braga, C.M., Zielinski, A.A.F., da Silva, K.M., de Souza, F.K.F., Pietrowski, G.D.A.M., Couto, M., Granato, D., Wosiacki, G. and Nogueira, A. 2013. Classification of juices and fermented beverages made from unripe, ripe and senescent apples based on the aromatic profile using chemometrics. *Food Chemistry* 141:(2): 967–974.

Budak, H.N. and Guzel-Seydim, Z.B. 2010. Antioxidant activity and phenolic content of wine vinegars produced by two different techniques. *Journal of the Science of Food and Agriculture* 90:(12): 2021–2026.

Burda, S., Oleszek, W. and Lee, C.Y. 1990. Phenolic compounds and their changes in apples during maturation and cold storage. *Journal of Agricultural and Food Chemistry* 38:(4): 945–948.

Cai-wen, D.O.N.G., Shao-hua, L., Feng-yu, T. and Bian-na, X. 2008. Study on the extraction of total flavonoids from apple pomace and its antibacterial activity [J]. *Journal of Anhui Agricultural Sciences* 27.

Canteri-Schemin, M.H., Fertonani, H.C.R., Waszczynskyj, N. and Wosiacki, G. 2005. Extraction of pectin from apple pomace. *Brazilian Archives of Biology and Technology* 48: 259–266.

Celik, A. and Demirbaş, A. 2005. Removal of heavy metal ions from aqueous solutions via adsorption onto modified lignin from pulping wastes. *Energy Sources* 27:(12): 1167–1177.

Ćetković, G., Čanadanović-Brunet, J., Djilas, S., Savatović, S., Mandić, A. and Tumbas, V. 2008. Assessment of polyphenolic content and in vitro antiradical characteristics of apple pomace. *Food Chemistry* 109:(2): 340–347.

Chand, P., Bafana, A. and Pakade, Y.B. 2015. Xanthate modified apple pomace as an adsorbent for removal of Cd (II), Ni (II) and Pb (II), and its application to real industrial wastewater. *International Biodeterioration & Biodegradation* 97: 60–66.

Chin, S.T., Nazimah, S.A.H., Quek, S.Y., Man, Y.B.C., Rahman, R.A. and Hashim, D.M. 2010. Effect of thermal processing and storage condition on the flavour stability of spray-dried durian powder. *LWT-Food Science and Technology* 43:(6): 856–861.

Chung, C., Rojanasasithara, T., Mutilangi, W. and McClements, D.J. 2016. Stabilization of natural colors and nutraceuticals: Inhibition of anthocyanin degradation in model beverages using polyphenols. *Food Chemistry* 212: 596–603.

Coldea, T.E., Socaciu, C., Mudura, E., Socaci, S.A., Ranga, F., Pop, C.R., Vriesekoop, F. and Pasqualone, A. 2020. Volatile and phenolic profiles of traditional Romanian apple brandy after rapid ageing with different wood chips. *Food Chemistry* 320: 126643.

Corrales, M., Toepfl, S., Butz, P., Knorr, D. and Tauscher, B. 2008. Extraction of anthocyanins from grape by-products assisted by ultrasonics, high hydrostatic pressure or pulsed electric fields: A comparison. *Innovative Food Science & Emerging Technologies* 9:(1): 85–91.

Costa, M.G.M., Fonteles, T.V., de Jesus, A.L.T. and Rodrigues, S. 2013. Sonicated pineapple juice as substrate for L. casei cultivation for probiotic beverage development: Process optimisation and product stability. *Food Chemistry* 139:(1–4): 261–266.

Cullen, P.J., Tiwari, B.K., O'Donnell, C.P. and Muthukumarappan, K. 2009. Modelling approaches to ozone processing of liquid foods. *Trends in Food Science & Technology* 20:(3–4): 125–136.

Daeschel, M.A. 1989. Antimicrobial substances from lactic acid bacteria for use as food preservatives. *Food Technology (Chicago)* 43:(1): 164–167.

Dahiya, S. and Dhawan, S.S. 2003. Effect of drying methods on nutritional composition of dehydrated aonla fruit (Emblica officinalis Garten) during storage. *Plant Foods for Human Nutrition* 58:(3): 1–9.

Daigle, P., Gélinas, P., Leblanc, D. and Morin, A. 1999. Production of aroma compounds by Geotrichum candidum on waste bread crumb. *Food Microbiology* 16:(5): 517–522.

Del Pozo-Insfran, D., Balaban, M.O. and Talcott, S.T. 2006. Microbial stability, phytochemical retention, and organoleptic attributes of dense phase CO_2 processed muscadine grape juice. *Journal of Agricultural and Food Chemistry* 54:(15): 5468–5473.

De Oliveira, M.C., Sichieri, R. and Moura, A.S. 2003. Weight loss associated with a daily intake of three apples or three pears among overweight women. *Nutrition* 19:(3): 253–256.

DeRuiter, F.E. and Dwyer, J. 2002. Consumer acceptance of irradiated foods: Dawn of a new era? *Food Service Technology* 2:(2): 47–58.

Dhillon, G.S., Kaur, S. and Brar, S.K. 2013. Perspective of apple processing wastes as low-cost substrates for bioproduction of high value products: A review. *Renewable and Sustainable Energy Reviews* 27: 789–805.

Drogoudi, P.D., Michailidis, Z. and Pantelidis, G. 2008. Peel and flesh antioxidant content and harvest quality characteristics of seven apple cultivars. *Scientia Horticulturae* 115:(2): 149–153.

Du, W.X., Olsen, C.W., Avena-Bustillos, R.J., Friedman, M. and McHugh, T.H. 2011. Physical and antibacterial properties of edible films formulated with apple skin polyphenols. *Journal of Food Science* 76:(2): M149–M155.

Eberhardt, M.V., Lee, C.Y. and Liu, R.H. 2000. Antioxidant activity of fresh apples. *Nature* 405:(6789): 903–904.

El Belghiti, K. and Vorobiev, E. 2004. Mass transfer of sugar from beets enhanced by pulsed electric field. *Food and Bioproducts Processing* 82:(3): 226–230.

Elez-Martínez, P., Aguiló-Aguayo, I. and Martín-Belloso, O. 2006. Inactivation of orange juice peroxidase by high-intensity pulsed electric fields as influenced by process parameters. *Journal of the Science of Food and Agriculture* 86:(1): 71–81.

Escarpa, A. and Gonzalez, M.C. 1998. High-performance liquid chromatography with diode-array detection for the determination of phenolic compounds in peel and pulp from different apple varieties. *Journal of Chromatography A* 823:(1–2): 331–337.

Feskanich, D., Ziegler, R.G., Michaud, D.S., Giovannucci, E.L., Speizer, F.E., Willett, W.C. and Colditz, G.A. 2000. Prospective study of fruit and vegetable consumption and risk of lung cancer among men and women. *Journal of the National Cancer Institute* 92:(22): 1812–1823.

Figuerola, F., Hurtado, M.L., Estévez, A.M., Chiffelle, I. and Asenjo, F. 2005. Fibre concentrates from apple pomace and citrus peel as potential fibre sources for food enrichment. *Food Chemistry* 91:(3): 395–401.

Filannino, P., Azzi, L., Cavoski, I., Vincentini, O., Rizzello, C.G., Gobbetti, M. and Di Cagno, R. 2013. Exploitation of the health-promoting and sensory properties of organic pomegranate (Punica granatum L.) juice through lactic acid fermentation. *International Journal of Food Microbiology* 163:(2–3): 184–192.

Fraser, D. 1951. Bursting bacteria by release of gas pressure. *Nature* 167:(4236): 33–34.

Fu, L., Xu, B.T., Xu, X.R., Gan, R.Y., Zhang, Y., Xia, E.Q. and Li, H.B. 2011. Antioxidant capacities and total phenolic contents of 62 fruits. *Food Chemistry* 129:(2): 345–350.

García, Y.D., Valles, B.S. and Lobo, A.P. 2009. Phenolic and antioxidant composition of by-products from the cider industry: Apple pomace. *Food Chemistry* 117:(4): 731–738.

Garcia-Gonzalez, L., Geeraerd, A.H., Spilimbergo, S., Elst, K., Van Ginneken, L., Debevere, J., Van Impe, J.F. and Devlieghere, F. 2007. High pressure carbon dioxide inactivation of microorganisms in foods: The past, the present and the future. *International Journal of Food Microbiology* 117:(1): 1–28.

Gassara, F., Brar, S.K., Pelletier, F., Verma, M., Godbout, S. and Tyagi, R.D. 2011. Pomace waste management scenarios in Québec – Impact on greenhouse gas emissions. *Journal of Hazardous Materials* 192:(3): 1178–1185.

Gaudout, D., Megard, D., Inisan, C., Esteve, C. and Lejard, F., Diana Ingredients SA. 2006. *Phloridzin-rich phenolic fraction and use thereof as a cosmetic, dietary or nutraceutical agent.* U.S. Patent 7,041,322.

Gomis, D.B., Tamayo, D.M., Valles, B.S. and Mangas Alonso, J.J. 2004. Detection of apple juice concentrate in the manufacture of natural and sparkling cider by means of HPLC chemometric sugar analyses. *Journal of Agricultural and Food Chemistry* 52:(2): 201–203.

Gorinstein, S., Zachwieja, Z., Folta, M., Barton, H., Piotrowicz, J., Zemser, M., Weisz, M., Trakhtenberg, S. and Màrtín-Belloso, O. 2001. Comparative contents of dietary fiber, total phenolics, and minerals in persimmons and apples. *Journal of Agricultural and Food Chemistry* 49:(2): 952–957.

Górnaś, P., Mišina, I., Olšteine, A., Krasnova, I., Pugajeva, I., Lācis, G., Siger, A., Michalak, M., Soliven, A. and Segliņa, D. 2015. Phenolic compounds in different fruit parts of crab apple: Dihydrochalcones as promising quality markers of industrial apple pomace by-products. *Industrial Crops and Products* 74: 607–612.

Graham, M.B. and Tyman, H.P. 2002. Ozonization of phenols from Anacardium occidentale (cashew). *Journal of the American Oil Chemists' Society* 79:(7): 725–732.

Grigoras, C.G., Destandau, E., Fougère, L. and Elfakir, C. 2013. Evaluation of apple pomace extracts as a source of bioactive compounds. *Industrial Crops and Products* 49: 794–804.

Grimi, N., Mamouni, F., Lebovka, N., Vorobiev, E. and Vaxelaire, J. 2010. Acoustic impulse response in apple tissues treated by pulsed electric field. *Biosystems Engineering* 105:(2): 266–272.

Grimi, N., Mamouni, F., Lebovka, N., Vorobiev, E. and Vaxelaire, J. 2011. Impact of apple processing modes on extracted juice quality: Pressing assisted by pulsed electric fields. *Journal of Food Engineering* 103:(1): 52–61.

Gros, C., Lanoisellé, J.L. and Vorobiev, E. 2003. Towards an alternative extraction process for linseed oil. *Chemical Engineering Research and Design* 81:(9): 1059–1065.

Gullón, B., Falqué, E., Alonso, J.L. and Parajó, J.C. 2007. Evaluation of apple pomace as a raw material for alternative applications in food industries. *Food Technology and Biotechnology* 45:(4): 426–433.

Gunes, G., Blum, L.K. and Hotchkiss, J.H. 2005. Inactivation of yeasts in grape juice using a continuous dense phase carbon dioxide processing system. *Journal of the Science of Food and Agriculture* 85:(14): 2362–2368.

Gunes, G., Blum, L.K. and Hotchkiss, J.H. 2006. Inactivation of Escherichia coli (ATCC 4157) in diluted apple cider by dense-phase carbon dioxide. *Journal of Food Protection* 69:(1): 12–16.

Guo, Y.X., Zhang, D.J., Wang, H., Xiu, Z.L., Wang, L.X. and Xiao, H.B. 2007. Hydrolytic kinetics of litho-spermic acid B extracted from roots of Salvia miltiorrhiza. *Journal of Pharmaceutical and Biomedical Analysis* 43:(2): 435–439.

Haas, G.J., Prescott Jr, H.E., Dudley, E., Dik, R., Hintlian, C. and Keane, L. 1989. Inactivation of microorganisms by carbon dioxide under pressure. *Journal of Food Safety* 9:(4): 253–265.

Hang, Y.D. 1987. Production of fuels and chemicals from apple pomace. *Food Technol (United States)* 41:(3).

Hang, Y.D. and Woodams, E.E. 1984. Apple pomace: A potential substrate for citric acid production by Aspergillus niger. *Biotechnology Letters* 6:(11): 763–764.

Hang, Y.D. and Woodams, E.E. 2010. Influence of apple cultivar and juice pasteurization on hard cider and eau-de-vie methanol content. *Bioresource Technology* 101:(4): 1396–1398.

Heimler, D., Romani, A. and Ieri, F. 2017. Plant polyphenol content, soil fertilization and agricultural management: A review. *European Food Research and Technology* 243:(7): 1107–1115.

Hemwimon, S., Pavasant, P. and Shotipruk, A. 2007. Microwave-assisted extraction of antioxidative anthraquinones from roots of Morinda citrifolia. *Separation and Purification Technology* 54:(1): 44–50.

Henríquez, C., Speisky, H., Chiffelle, I., Valenzuela, T., Araya, M., Simpson, R. and Almonacid, S. 2010. Development of an ingredient containing apple peel, as a source of polyphenols and dietary fiber. *Journal of Food Science* 75:(6): H172–H181.

Hoffmann-Ribani, R., Huber, L.S. and Rodriguez-Amaya, D.B. 2009. Flavonols in fresh and processed Brazilian fruits. *Journal of Food Composition and Analysis* 22:(4): 263–268.

Hollman, P. 2000. Bioavailability of flavonoids. In *International Congress and Symposium Series-Royal Society of Medicine*. Vol. 226, pp. 45–52. Royal Society of Medicine Press Ltd.

Huvaere, K. and Skibsted, L.H. 2015. Flavonoids protecting food and beverages against light. *Journal of the Science of Food and Agriculture* 95:(1): 20–35.

Iacopini, P., Camangi, F., Stefani, A. and Sebastiani, L. 2010. Antiradical potential of ancient Italian apple varieties of Malus× domestica Borkh. in a peroxynitrite-induced oxidative process. *Journal of Food Composition and Analysis* 23:(6): 518–524.

Januszek, M., Satora, P. and Tarko, T. 2020. Oenological characteristics of fermented apple musts and volatile profile of brandies obtained from different apple cultivars. *Biomolecules* 10:(6): 853.

Jay, J.M., Loessner, M.J. and Golden, D.A. 2005. Staphylococcal gastroenteritis. *Modern Food Microbiology*: 545–566.

Jin, H., Kim, H.S., Kim, S.K., Shin, M.K., Kim, J.H. and Lee, J.W. 2002. Production of heteropolysaccharide-7 by Beijerinckia indica from agro-industrial byproducts. *Enzyme and Microbial Technology* 30:(6): 822–827.

Joshi, V.K., Sharma, S. and Devi, M.P. 2009. Influence of different yeast strains on fermentation behaviour, physico-chemical and sensory qualities of plum wine.

Ju, Z., Yuan, Y., Liu, C., Zhan, S. and Wang, M. 1996. Relationships among simple phenol, flavonoid and anthocyanin in apple fruit peel at harvest and scald susceptibility. *Postharvest Biology and Technology* 8:(2): 83–93.

Kalinowska, M., Bielawska, A., Lewandowska-Siwkiewicz, H., Priebe, W. and Lewandowski, W. 2014. Apples: Content of phenolic compounds vs. variety, part of apple and cultivation model, extraction of phenolic compounds, biological properties. *Plant Physiology and Biochemistry* 84: 169–188.

Kataoka, I., Beppu, K., Sugiyama, A. and Taira, S. 1996. Enhancement of coloration of "Satohnishiki" sweet cherry fruit by postharvest irradiation with ultraviolet rays. *Environment Control in Biology* 34:(4): 313–319.

Kennedy, M., List, D., Lu, Y., Foo, L.Y., Newman, R.H., Sims, I.M., Bain, P.J.S., Hamilton, B. and Fenton, G. 1999. Apple pomace and products derived from apple pomace: Uses, composition and analysis. In *Analysis of plant waste materials* (75–119). Springer Berlin, Heidelberg.

Khadre, M.A., Yousef, A.E. and Kim, J.G. 2001. Microbiological aspects of ozone applications in food: A review. *Journal of Food Science* 66:(9): 1242–1252.

Khanizadeh, S., Tsao, R., Rekika, D., Yang, R., Charles, M.T. and Rupasinghe, H.V. 2008. Polyphenol composition and total antioxidant capacity of selected apple genotypes for processing. *Journal of Food Composition and Analysis* 21:(5): 396–401.

Kim, S., Iglesias-Sucasas, M. and Viollier, V. 2013. The FAO Geopolitical Ontology: A reference for country-based information. *Journal of Agricultural & Food Information* 14:(1): 50–65.

Knekt, P., Isotupa, S., Rissanen, H., Heliövaara, M., Järvinen, R., Häkkinen, S., Aromaa, A. and Reunanen, A. 2000. Quercetin intake and the incidence of cerebrovascular disease. *European Journal of Clinical Nutrition* 54:(5): 415–417.

Knekt, P., Jarvinen, R., Reunanen, A. and Maatela, J. 1996. Flavonoid intake and coronary mortality in Finland: A cohort study. *BMJ* 312:(7029): 478–481.

Knekt, P., Kumpulainen, J., Järvinen, R., Rissanen, H., Heliövaara, M., Reunanen, A., Hakulinen, T. and Aromaa, A. 2002. Flavonoid intake and risk of chronic diseases. *The American Journal of Clinical Nutrition* 76:(3): 560–568.

Knorr, D., Zenker, M., Heinz, V. and Lee, D.U. 2004. Applications and potential of ultrasonics in food processing. *Trends in Food Science & Technology* 15:(5): 261–266.

Kolodziejczyk, K., Markowski, J., Kosmala, M., Król, B. and Plocharski, W. 2007. Apple pomace as a potential source of nutraceutical products. *Polish Journal of Food and Nutrition Sciences* 57:(4[B]).

Kong, K.W., Ismail, A., Tan, C.P. and Rajab, N.F. 2010. Optimization of oven drying conditions for lycopene content and lipophilic antioxidant capacity in a by-product of the pink guava puree industry using response surface methodology. *LWT-Food Science and Technology* 43:(5): 729–735.

Lakshmanan, P.T. 2000. *Fish Spoilage and Quality Assessment*. Society of Fisheries Technologists, India, pp. 28–45.

Landete, J.M. 2012. Updated knowledge about polyphenols: Functions, bioavailability, metabolism, and health. *Critical Reviews in Food Science and Nutrition* 52:(10): 936–948.

Łata, B. and Tomala, K. 2007. Apple peel as a contributor to whole fruit quantity of potentially healthful bio-active compounds. Cultivar and year implication. *Journal of Agricultural and Food Chemistry* 55:(26): 10795–10802.

Lea, M.A. 2010. 26 Fruit phenolics in colon cancer prevention and treatment. *Bioactive Foods and Extracts* 415.

Le Marchand, L., Murphy, S.P., Hankin, J.H., Wilkens, L.R. and Kolonel, L.N. 2000. Intake of flavonoids and lung cancer. *Journal of the National Cancer Institute* 92:(2): 154–160.Leontowicz, H., Gorinstein, S., Lojek, A., Leontowicz, M., Číž, M., Soliva-Fortuny, R., Park, Y.S., Jung, S.T., Trakhtenberg, S. and Martin-Belloso, O. 2002. Comparative content of some bioactive compounds in apples, peaches and pears and their influence on lipids and antioxidant capacity in rats. *The Journal of Nutritional Biochemistry* 13:(10): 603–610.

Leontowicz, M., Gorinstein, S., Leontowicz, H., Krzeminski, R., Lojek, A., Katrich, E., Číž, M., Martin-Belloso, O., Soliva-Fortuny, R., Haruenkit, R. and Trakhtenberg, S. 2003. Apple and pear peel and pulp and their influence on plasma lipids and antioxidant potentials in rats fed cholesterol-containing diets. *Journal of Agricultural and Food Chemistry* 51:(19): 5780–5785.

Li, S., Nie, Y., Ding, Y., Zhao, J. and Tang, X. 2015. Effects of Pure and Mixed Koji Cultures with S accharomy-ces cerevisiae on Apple Homogenate Cider Fermentation. *Journal of Food Processing and Preservation* 39:(6): 2421–2430.

Liu, R.H., Eberhardt, M.V. and Lee, C.Y. 2001. Antioxidant and antiproliferative activities of selected New York apple cultivars. *New York Fruit Quarterly* 9:(2): 15–17.

Lohani, U.C. and Muthukumarappan, K. 2016. Application of the pulsed electric field to release bound phenolics in sorghum flour and apple pomace. *Innovative Food Science & Emerging Technologies* 35: 29–35.

Longo, M.A. and Sanromán, M.A. 2006. Production of food aroma compounds: Microbial and enzymatic methodologies. *Food Technology and Biotechnology* 44:(3): 335–353.

Loomis-Husselbee, J.W., Cullen, P.J., Irvine, R.F. and Dawson, A.P. 1991. Electroporation can cause artefacts due to solubilization of cations from the electrode plates. Aluminum ions enhance conversion of inositol 1, 3, 4, 5-tetrakisphosphate into inositol 1, 4, 5-trisphosphate in electroporated L1210 cells. *Biochemical Journal* 277:(3): 883–885.

López, N., Puértolas, E., Condón, S., Raso, J. and Alvarez, I. 2009. Enhancement of the extraction of betanine from red beetroot by pulsed electric fields. *Journal of Food Engineering* 90:(1): 60–66.

Luciano, G., Monahan, F.J., Vasta, V., Biondi, L., Lanza, M. and Priolo, A. 2009. Dietary tannins improve lamb meat colour stability. *Meat Science* 81:(1): 120–125.

Madrera, R.R., Bedriñana, R.P. and Valles, B.S. 2015. Production and characterization of aroma compounds from apple pomace by solid-state fermentation with selected yeasts. *LWT-Food Science and Technology* 64:(2): 1342–1353.

Madrera, R.R., Lobo, A.P. and Alonso, J.J.M. 2010. Effect of cider maturation on the chemical and sensory characteristics of fresh cider spirits. *Food Research International* 43:(1): 70–78.

Magrone, T. and Jirillo, E. 2012. Influence of polyphenols on allergic immune reactions: Mechanisms of action. *Proceedings of the Nutrition Society* 71:(2): 316–321.

Marchal, R. and Jeandet, P. 2009. Use of enological additives for colloid and tartrate salt stabilization in white wines and for improvement of sparkling wine foaming properties. In *Wine chemistry and biochemistry* (127–158). Springer.

Marshall, Maurice R., Kim, Jeongmok and Wei, C. 2000. Enzymatic browning in fruits, vegetables and sea-foods. *FAO, Rome* 49.

Masoodi, F.A., Sharma, B. and Chauhan, G.S. 2002. Use of apple pomace as a source of dietary fiber in cakes. *Plant Foods for Human Nutrition* 57:(2): 121–128.

Massias, A., Boisard, S., Baccaunaud, M., Calderon, F.L. and Subra-Paternault, P. 2015. Recovery of pheno-lics from apple peels using CO2+ ethanol extraction: Kinetics and antioxidant activity of extracts. *The Journal of Supercritical Fluids* 98: 172–182.

May, C.D. 1990. Industrial pectins: Sources, production and applications. *Carbohydrate Polymers* 12:(1): 79–99.

Mayer, B., Schumacher, M., Brandstätter, H., Wagner, F.S. and Hermetter, A. 2001. High-throughput fluorescence screening of antioxidative capacity in human serum. *Analytical Biochemistry* 297:(2): 144–153.

McRae, K.B., Lidster, P.D., Demarco, A.C. and Dick, A.J. 1990. Comparison of the polyphenol profiles of apple fruit cultivars by correspondence analysis. *Journal of the Science of Food and Agriculture* 50:(3): 329–342.

Medeiros, A.B., Pandey, A., Freitas, R.J., Christen, P. and Soccol, C.R. 2000. Optimization of the production of aroma compounds by Kluyveromyces marxianus in solid-state fermentation using factorial design and response surface methodology. *Biochemical Engineering Journal* 6:(1): 33–39.

Meyvaci, K.B., Sen, F. and Uthaibutra, A. 2010. Optimization of magnesium phosphide treatment used to control major dried fig storage pests. *Horticulture Environment and Biotechnology* 51:(1): 33–38.

Miceli-Garcia, L.G. 2014. Pectin from apple pomace: Extraction, characterization, and utilization in encapsulating alpha-tocopherol acetate.

Minnocci, A., Iacopini, P., Martinelli, F. and Sebastiani, L. 2010. Micromorphological, biochemical, and genetic characterization of two ancient, late-bearing apple varieties. *European Journal of Horticultural Science* 75: 1–7.

Moscetti, R., Carletti, L., Monarca, D., Cecchini, M., Stella, E. and Massantini, R. 2013. Effect of alternative postharvest control treatments on the storability of 'Golden Delicious' apples. *Journal of the Science of Food and Agriculture* 93:(11): 2691–2697.

Mudura, E., Coldea, T.E., Socaciu, C., Ranga, F., Pop, C.R., Rotar, A.M. and Pasqualone, A. 2018. Brown beer vinegar: A potentially functional product based on its phenolic profile and antioxidant activity. *Journal of the Serbian Chemical Society* 83:(1): 19–30.

Murthy, Z.V.P. and Joshi, D. 2007. Fluidized bed drying of aonla (Emblica officinalis). *Drying Technology* 25:(5): 883–889.

Nicolini, G., Román, T., Carlin, S., Malacarne, M., Nardin, T., Bertoldi, D. and Larcher, R. 2018. Characterisation of single-variety still ciders produced with dessert apples in the Italian Alps. *Journal of the Institute of Brewing* 124:(4): 457–466.

Oey, I., Van der Plancken, I., Van Loey, A. and Hendrickx, M. 2008. Does high pressure processing influence nutritional aspects of plant based food systems? *Trends in Food Science & Technology* 19:(6): 300–308.

Oszmiański, J., Lachowicz, S., Głąwdel, E., Cebulak, T. and Ochmian, I. 2018. Determination of phytochemical composition and antioxidant capacity of 22 old apple cultivars grown in Poland. *European Food Research and Technology* 244:(4): 647–662.

Padmini, E., Prema, K., Vijaya Geetha, B. and Usha Rani, M. 2008. Comparative study on composition and antioxidant properties of mint and black tea extract. *International Journal of Food Science & Technology* 43:(10): 1887–1895.

Patras, A., Brunton, N.P., Da Pieve, S., Butler, F. and Downey, G. 2009. Effect of thermal and high pressure processing on antioxidant activity and instrumental colour of tomato and carrot purées. *Innovative Food Science & Emerging Technologies* 10:(1): 16–22.

Patras, A., Brunton, N.P., O'Donnell, C. and Tiwari, B.K. 2010. Effect of thermal processing on anthocyanin stability in foods; mechanisms and kinetics of degradation. *Trends in Food Science & Technology* 21:(1): 3–11.

Pearson, D.A., Tan, C.H., German, J.B., Davis, P.A. and Gershwin, M.E. 1999. Apple juice inhibits human low density lipoprotein oxidation. *Life Sciences* 64:(21): 1913–1920.

Peng, W., Meng, D., Yue, T., Wang, Z. and Gao, Z. 2021. Effect of the apple cultivar on cloudy apple juice fermented by a mixture of Lactobacillus acidophilus, Lactobacillus plantarum, and Lactobacillus fermentum. *Food Chemistry* 340: 127922.

Pereira, A.L.F., Maciel, T.C. and Rodrigues, S. 2011. Probiotic beverage from cashew apple juice fermented with Lactobacillus casei. *Food Research International* 44:(5): 1276–1283.

Pereira, R.N. and Vicente, A.A. 2010. Environmental impact of novel thermal and non-thermal technologies in food processing. *Food Research International* 43:(7): 1936–1943.

Pérez-Ilzarbe, J., Hernández, T. and Estrella, I. 1991. Phenolic compounds in apples: Varietal differences. *Zeitschrift für Lebensmittel-Untersuchung und Forschung* 192:(6): 551–554.

Perussello, C.A., Zhang, Z., Marzocchella, A. and Tiwari, B.K. 2017. Valorization of apple pomace by extraction of valuable compounds. *Comprehensive Reviews in Food Science and Food Safety* 16:(5): 776–796.

Picinelli Lobo, A., Fernández Tascón, N., Rodríguez Madrera, R. and Suárez Valles, B. 2005. Sensory and foaming properties of sparkling cider. *Journal of Agricultural and Food Chemistry* 53:(26): 10051–10056.

Pinelo, M., Zornoza, B. and Meyer, A.S. 2008. Selective release of phenols from apple skin: Mass transfer kinetics during solvent and enzyme-assisted extraction. *Separation and Purification Technology* 63:(3): 620–627.

Pingret, D., Fabiano-Tixier, A.S., Le Bourvellec, C., Renard, C.M. and Chemat, F. 2012. Lab and pilot-scale ultrasound-assisted water extraction of polyphenols from apple pomace. *Journal of Food Engineering* 111:(1): 73–81.

Porciani, P.F. and Grandini, S. 2012. The effect of zinc acetate and magnolia bark extract added to chewing gum on volatile sulfur-containing compounds in the oral cavity. *The Journal of Clinical Dentistry* 23:(3): 76–79.

Portenlänger, G. and Heusinger, H. 1992. Chemical reactions induced by ultrasound and γ-rays in aqueous solutions of L-ascorbic acid. *Carbohydrate Research* 232:(2): 291–301.

Prakash, S., Jha, S.K. and Datta, N. 2004. Performance evaluation of blanched carrots dried by three different driers. *Journal of Food Engineering* 62:(3): 305–313.

Rabetafika, H.N., Bchir, B., Blecker, C. and Richel, A. 2014. Fractionation of apple by-products as source of new ingredients: Current situation and perspectives. *Trends in Food Science & Technology* 40:(1): 99–114.

Rastogi, N.K., Raghavarao, K.S.M.S., Balasubramaniam, V.M., Niranjan, K. and Knorr, D. 2007. Opportunities and challenges in high pressure processing of foods. *Critical Reviews in Food Science and Nutrition* 47:(1): 69–112.

Rezaei, S., Rezaei, K., Haghighi, M. and Labbafi, M. 2013. Solvent and solvent to sample ratio as main parameters in the microwave-assisted extraction of polyphenolic compounds from apple pomace. *Food Science and Biotechnology* 22:(5): 1–6.

Ricci, A., Cirlini, M., Guido, A., Liberatore, C.M., Ganino, T., Lazzi, C. and Chiancone, B. 2019. From byproduct to resource: Fermented apple pomace as beer flavoring. *Foods* 8:(8): 309.

Roberts, J.S., Gentry, T.S. and Bates, A.W. 2004. Utilization of dried apple pomace as a press aid to improve the quality of strawberry, raspberry, and blueberry juices. *Journal of Food Science* 69:(4): SNQ181–SNQ190.

Rojas, M.C. and Brewer, M.S. 2008. Effect of natural antioxidants on oxidative stability of frozen, vacuum-packaged beef and pork. *Journal of Food Quality* 31:(2): 173–188.

Rosend, J., Kaleda, A., Kuldjärv, R., Arju, G. and Nisamedtinov, I. 2020. The effect of apple juice concentration on cider fermentation and properties of the final product. *Foods* 9:(10): 1401.

Rossi, M., Giussani, E., Morelli, R., Scalzo, R.L., Nani, R.C. and Torreggiani, D. 2003. Effect of fruit blanching on phenolics and radical scavenging activity of highbush blueberry juice. *Food Research International* 36:(9–10): 999–1005.

Rupasinghe, H.V., Wang, L., Huber, G.M. and Pitts, N.L. 2008. Effect of baking on dietary fibre and phenolics of muffins incorporated with apple skin powder. *Food Chemistry* 107:(3): 1217–1224.

Sadilova, E., Carle, R. and Stintzing, F.C. 2007. Thermal degradation of anthocyanins and its impact on color and in vitro antioxidant capacity. *Molecular Nutrition & Food Research* 51:(12): 1461–1471.

Salleh-Mack, S.Z. and Roberts, J.S. 2007. Ultrasound pasteurization: The effects of temperature, soluble solids, organic acids and pH on the inactivation of Escherichia coli ATCC 25922. *Ultrasonics Sonochemistry* 14:(3): 323–329.

Saxena, S., Gautam, S. and Sharma, A. 2010. Microbial decontamination of honey of Indian origin using gamma radiation and its biochemical and organoleptic properties. *Journal of Food Science* 75:(1): M19–M27.

Scotch Whisky Association v. Michael Klotz Regulation (EC) No 110/2008, Art. 16 (a) – (c), Annex III. 2018. "Scotch whisky association" decision of the European court of justice (fifth chamber), 7 June 2018 – Case No. C-44/17.

Sesso, H.D., Gaziano, J.M., Liu, S. and Buring, J.E. 2003. Flavonoid intake and the risk of cardiovascular disease in women. *The American Journal of Clinical Nutrition* 77:(6): 1400–1408.

Shaheen, S.O., Sterne, J.A., Thompson, R.L., Songhurst, C.E., Margetts, B.M. and Burney, P.G. 2001. Dietary antioxidants and asthma in adults: Population-based case – control study. *American Journal of Respiratory and Critical Care Medicine* 164:(10): 1823–1828.

Shoji, T., Masumoto, S., Moriichi, N., Kobori, M., Kanda, T., Shinmoto, H. and Tsushida, T. 2005. Procyanidin trimers to pentamers fractionated from apple inhibit melanogenesis in B16 mouse melanoma cells. *Journal of Agricultural and Food Chemistry* 53:(15): 6105–6111.

Sian, N.K. and Ishak, S. 1991. Carotenoid and anthocyanin contents of papaya and pineapple: Influence of blanching and predrying treatments. *Food Chemistry* 39:(2): 175–185.

Siebert, K.J. 2009. Haze in beverages. *Advances in Food and Nutrition Research* 57: 53–86.

Silva, M.L., Macedo, A.C. and Malcata, F.X. 2000. Steam distilled spirits from fermented grape pomace Revision: Bebidas destiladas obtenidas de la fermentación del orujo de uva. *Food Science and Technology International* 6:(4): 285–300.

Singh, B. and Narang, M.P. 1992. Studies on the rumen degradation kinetics and utilization of apple pomace. *Bioresource Technology* 39:(3): 233–240.

Sivasankar, T., Paunikar, A.W. and Moholkar, V.S. 2007. Mechanistic approach to enhancement of the yield of a sonochemical reaction. *AIChE Journal* 53:(5): 1132–1143.

Small, R.W., Couturier, M. and Godfrey, M. 2011. Beverage basics: Understanding and appreciating wine, beer, and spirits. *Beverage Basics: Understanding and Appreciating Wine, Beer, and Spirits*, 1st ed. Hoboken, NJ, Wiley.

Soliva-Fortuny, R., Balasa, A., Knorr, D. and Martin-Belloso, O. 2009. Effects of pulsed electric fields on bioactive compounds in foods: A review. *Trends in Food Science & Technology* 20:(11–12): 544–556.

Sudha, M.L., Baskaran, V. and Leelavathi, K. 2007. Apple pomace as a source of dietary fiber and polyphenols and its effect on the rheological characteristics and cake making. *Food Chemistry* 104:(2): 686–692.

Sun, J., Chu, Y.F., Wu, X. and Liu, R.H. 2002. Antioxidant and antiproliferative activities of common fruits. *Journal of Agricultural and Food Chemistry* 50:(25): 7449–7454.

Sun-Waterhouse, D. 2011. The development of fruit-based functional foods targeting the health and wellness market: A review. *International Journal of Food Science & Technology* 46:(5): 899–920.

Suthanthangjai, W., Kajda, P. and Zabetakis, I. 2005. The effect of high hydrostatic pressure on the anthocyanins of raspberry (Rubus idaeus). *Food Chemistry* 90:(1–2): 193–197.

Tao, D.Y., Shu, C.B., Lo, E.C.M., Lu, H.X. and Feng, X.P. 2013. A randomized trial on the inhibitory effect of chewing gum containing tea polyphenol on caries. *Journal of Clinical Pediatric Dentistry* 38:(1): 67–70.

Tarazona-Díaz, M.P. and Aguayo, E. 2013. Assessment of by-products from fresh-cut products for reuse as bioactive compounds. *Food Science and Technology International* 19:(5): 439–446.

Tarko, T., Duda-Chodak, A., Sroka, P. and Januszek, M. 2020. Effect of musts oxygenation at various stages of cider production on oenological parameters, antioxidant activity, and profile of volatile cider compounds. *Biomolecules* 10:(6): 890.

Terao, J. 2009. Dietary flavonoids as antioxidants. *Food Factors for Health Promotion* 61: 87–94.

Tian, H.L., Zhan, P. and Li, K.X. 2010. Analysis of components and study on antioxidant and antimicrobial activities of oil in apple seeds. *International Journal of Food Sciences and Nutrition* 61:(4): 395–403.

Turk, M.F., Vorobiev, E. and Baron, A. 2012. Improving apple juice expression and quality by pulsed electric field on an industrial scale. *LWT-Food Science and Technology* 49:(2): 245–250.

Valles, B.S., Bedriñana, R.P., Queipo, A.L. and Alonso, J.J.M. 2008. Screening of cider yeasts for sparkling cider production (Champenoise method). *Food Microbiology* 25:(5): 690–697.

Van der Sluis, A.A., Dekker, M. and van Boekel, M.A. 2005. Activity and concentration of polyphenolic antioxidants in apple juice. 3. Stability during storage. *Journal of Agricultural and Food Chemistry* 53:(4): 1073–1080.

Velasco, V. and Williams, P. 2011. Improving meat quality through natural antioxidants. *Chilean Journal of Agricultural Research* 71:(2): 313.

Verón, H.E., Cano, P.G., Fabersani, E., Sanz, Y., Isla, M.I., Espinar, M.T.F., Ponce, J.V.G. and Torres, S. 2019. Cactus pear (Opuntia ficus-indica) juice fermented with autochthonous Lactobacillus plantarum S-811. *Food & Function* 10:(2): 1085–1097.

Versini, G.I.U.S.E.P.P.E., Franco, M.A., Moser, S., Barchetti, P. and Manca, G. 2009. Characterisation of apple distillates from native varieties of Sardinia island and comparison with other Italian products. *Food Chemistry* 113:(4): 1176–1183.

Vieira, M.C., Teixeira, A.A. and Silva, C.L.M. 2000. Mathematical modeling of the thermal degradation kinetics of vitamin C in cupuaçu (Theobroma grandiflorum) nectar. *Journal of Food Engineering* 43:(1): 1–7.

Villas-Bôas, S.G., Esposito, E. and de Mendonça, M.M. 2002. Novel lignocellulolytic ability of Candida utilis during solid-substrate cultivation on apple pomace. *World Journal of Microbiology and Biotechnology* 18:(6): 541–545.

Villas-Bôas, S.G., Esposito, E. and de Mendonca, M.M. 2003. Bioconversion of apple pomace into a nutritionally enriched substrate by Candida utilis and Pleurotus ostreatus. *World Journal of Microbiology and Biotechnology* 19:(5): 461–467.

Virot, M., Tomao, V., Le Bourvellec, C., Renard, C.M. and Chemat, F. 2010. Towards the industrial production of antioxidants from food processing by-products with ultrasound-assisted extraction. *Ultrasonics Sonochemistry* 17:(6): 1066–1074.

Vivek, M.N., Sunil, S.V., Pramod, N.J., Prashith, K.T.R., Mukunda, S. and Mallikarjun, N. 2012. Anticariogenic activity of Lagerstroemia speciosa (L.). *Science, Technology and Arts Research Journal* 1:(1): 53–56.

Vorobiev, E. and Lebovka, N. 2010. Enhanced extraction from solid foods and biosuspensions by pulsed electrical energy. *Food Engineering Reviews* 2:(2): 95–108.

Wang, L. and Bohn, T. 2012. Health-promoting food ingredients and functional food processing. *Nutrition, Well-being and Health*: 201–224.

Wang, L., Boussetta, N., Lebovka, N. and Vorobiev, E. 2018. Selectivity of ultrasound-assisted aqueous extraction of valuable compounds from flesh and peel of apple tissues. *LWT* 93: 511–516.

Wang, X., Kristo, E. and LaPointe, G. 2019. The effect of apple pomace on the texture, rheology and microstructure of set type yogurt. *Food Hydrocolloids* 91: 83–91.

Wegrzyn, T.F., Farr, J.M., Hunter, D.C., Au, J., Wohlers, M.W., Skinner, M.A., Stanley, R.A. and Sun-Waterhouse, D. 2008. Stability of antioxidants in an apple polyphenol – milk model system. *Food Chemistry* 109:(2): 310–318.

Whangchai, K., Saengnil, K. and Uthaibutra, J. 2006. Effect of ozone in combination with some organic acids on the control of postharvest decay and pericarp browning of longan fruit. *Crop Protection* 25:(8): 821–825.

Wijngaard, H. and Brunton, N. 2009. The optimization of extraction of antioxidants from apple pomace by pressurized liquids. *Journal of Agricultural and Food Chemistry* 57:(22): 10625–10631.

Wiktor, A., Gondek, E., Jakubczyk, E., Nowacka, M., Dadan, M., Fijalkowska, A. and Witrowa-Rajchert, D. 2016. Acoustic emission as a tool to assess the changes induced by pulsed electric field in apple tissue. *Innovative Food Science & Emerging Technologies* 37: 375–383.

Wolfe, K.L. and Liu, R.H. 2003. Apple peels as a value-added food ingredient. *Journal of Agricultural and Food Chemistry* 51:(6): 1676–1683.

Wolfe, K.L., Wu, X. and Liu, R.H. 2003. Antioxidant activity of apple peels. *Journal of Agricultural and Food Chemistry* 51:(3): 609–614.

Woods, R.K., Walters, E.H., Raven, J.M., Wolfe, R., Ireland, P.D., Thien, F.C. and Abramson, M.J. 2003. Food and nutrient intakes and asthma risk in young adults. *The American Journal of Clinical Nutrition* 78:(3): 414–421.

Worrall, J.J. and Yang, C.S. 1992. Shiitake and oyster mushroom production on apple pomace and sawdust. *HortScience* 27:(10): 1131–1133.

Wu, C., Li, T., Qi, J., Jiang, T., Xu, H. and Lei, H. 2020. Effects of lactic acid fermentation-based biotransformation on phenolic profiles, antioxidant capacity and flavor volatiles of apple juice. *LWT* 122: 109064.

Yagiz, Y., Lim, S.L. and Balaban, M.O. 2005. *Continuous high-pressure CO2 processing of mandarin juice: IFT annual meeting book of abstracts*. Institute of Food Technologists New Orleans.

Ye, M., Yue, T. and Yuan, Y. 2014. Changes in the profile of volatile compounds and amino acids during cider fermentation using dessert variety of apples. *European Food Research and Technology* 239:(1): 67–77.

Yen, G.C. and Lin, H.T. 1996. Comparison of high pressure treatment and thermal pasteurization effects on the quality and shelf life of guava puree. *International Journal of Food Science & Technology* 31:(2): 205–213.

Yilmaz, Y. 2006. Novel uses of catechins in foods. *Trends in Food Science & Technology* 17:(2): 64–71.

Yu, H., Qin, C., Zhang, P., Ge, Q., Wu, M., Wu, J., Wang, M. and Wang, Z. 2015. Antioxidant effect of apple phenolic on lipid peroxidation in Chinese-style sausage. *Journal of Food Science and Technology* 52:(2): 1032–1039.

Yue, T., Shao, D., Yuan, Y., Wang, Z. and Qiang, C. 2012. Ultrasound-assisted extraction, HPLC analysis, and antioxidant activity of polyphenols from unripe apple. *Journal of Separation Science* 35:(16): 2138–2145.

Zabetakis, I., Leclerc, D. and Kajda, P. 2000. The effect of high hydrostatic pressure on the strawberry anthocyanins. *Journal of Agricultural and Food Chemistry* 48:(7): 2749–2754.

Zarein, M., Samadi, S.H. and Ghobadian, B. 2015. Investigation of microwave dryer effect on energy efficiency during drying of apple slices. *Journal of the Saudi Society of Agricultural Sciences* 14:(1): 41–47.

Zenker, M., Heinz, V. and Knorr, D. 2003. Application of ultrasound-assisted thermal processing for preservation and quality retention of liquid foods. *Journal of food Protection* 66:(9): 1642–1649.

Zepka, L.Q. and Mercadante, A.Z. 2009. Degradation compounds of carotenoids formed during heating of a simulated cashew apple juice. *Food Chemistry* 117:(1): 28–34.

Zhang, J., Davis, T.A., Matthews, M.A., Drews, M.J., LaBerge, M. and An, Y.H. 2006. Sterilization using high-pressure carbon dioxide. *The Journal of Supercritical Fluids* 38:(3): 354–372.

Zhang, Z., Huber, D.J., Qu, H., Yun, Z., Wang, H., Huang, Z., Huang, H. and Jiang, Y. 2015. Enzymatic browning and antioxidant activities in harvested litchi fruit as influenced by apple polyphenols. *Food Chemistry* 171: 191–199.

Zillich, O.V., Schweiggert-Weisz, U., Eisner, P. and Kerscher, M. 2015. Polyphenols as active ingredients for cosmetic products. *International Journal of Cosmetic Science* 37:(5): 455–464.

Zulueta, A., Esteve, M.J. and Frígola, A. 2010. Ascorbic acid in orange juice – milk beverage treated by high intensity pulsed electric fields and its stability during storage. *Innovative Food Science & Emerging Technologies* 11:(1): 84–90.

17 Effect of Preharvest and Postharvest Technologies on Bioactive Compounds

Jessica Pandohee, Sajad Ahmad Wani,
Shafiya Rafiq and Basharat N. Dar

CONTENTS

17.1 INTRODUCTION

The apple (*Malus domestica Borkh*), which is part of the *Rosaceae* family, is a nutrient-dense fruit containing hundreds of compounds, many of which have important roles in human health (Spengler, 2019). The contemporary cultivated apple is thought to be the result of centuries of hybridization. Therefore, the scientific name *Malus domestica Borkh.* for apples is preferred over from the old name *Malus pumila* (Korban and Skirvin, 1984). The main parent of the domesticated apple is thought

DOI: 10.1201/9781003239925-17

to be a wild apple species (*Malus sieversii Lebed.*), which is native to Central Asia (Hancock et al., 2008; Musacchi and Serra, 2018). It is no surprise that apples are grown in temperate, subtropical, and tropical climates throughout the world. Usually, commercial production is restricted to latitudes of 25° to 52° (Palmer et al., 2003).

Most of the apple fruit is edible. The stem, core, and seeds of the apple are usually not consumed but are used to make other products. Other common products from apples include juices, ciders, wine, jams, jellies, tea, and dried apples. Consumer acceptability is influenced by quality attributes that can widely be categorized as internal and external indicators. Examples of external indicators are the apples' height, breadth, shape, and weight whereas internal quality indicators are fruit firmness, soluble solid concentration, starch degradation index, maturity index, percentage acidity, and the ratio of the soluble solid concentration to the total acid concentration (Musacchi and Serra, 2018). In addition to these indicators, the quality of apple is also determined by their total phenolic content, as they contribute to taste characteristics including flavor, astringency, and bitterness (Bengoechea et al., 1997).

Phytochemicals are a diverse group of chemical compounds and are also known as bioactive metabolites or specialized metabolites. Based on the chemical structure and physiological activity, phytochemicals are categorized into carotenoids, phenolic compounds, glucosinolate, saponin, sulfide, and phytosterol groups (Rafiq et al., 2018). Distinct apple types and sections of the apple have varied polyphenolic component profiles, which are often considerably different (Le Deun et al., 2015). The interest in studying the phytochemical content in apples is due to the increasing evidence of the health benefits of phytochemical compounds. Moreover, in recent years there has been increased interest in the quantity and diversity of phytochemicals in our diet and their effect on our health (Drogoudi and Pantelidis, 2011). The proverb "An apple a day keeps the doctor away" in this case is not just a saying to encourage children to eat more fruits; apples are essential in human nutrition because they boost immunity, improve stress tolerance, and contain several bioactive compounds that are helpful to humans.

17.2   PREHARVEST AND POSTHARVEST TECHNOLOGIES

Prior to being sold to the general public, apples from orchards undergo various treatments in order to keep them free of pests and diseases or to extend their shelf life. Preharvest technologies are tools and procedures that are applied to agricultural raw materials prior to harvest of apples in order to improve apple quality and shelf life. Some examples of preharvest technologies include application of chemicals such as growth regulators, ethylene inhibitors, or gibberellic acid. Postharvest technologies, on the other hand, are treatments and measures performed after apples have been picked from the trees and include processing into food products, storage, and transportation.

The diversity of bioactive compounds in apples, like in other fruits, is affected by cultivar, maturity stage, planting season, environmental conditions, geographic location, production processes, and storage conditions (Tsao et al., 2005; Duda-Chodak et al., 2010; Francini and Sebastiani, 2013). In this chapter we present a literature review and discussion on the up-to-date research in order to provide a better understanding of the effect of preharvest and postharvest technologies on the bioactive compounds in apples.

17.3   PHENOLIC COMPOUNDS

17.3.1   PREHARVEST

A conventional orcharding practice involves high yield crop cultivars, agrochemicals, watering, and mechanization. Although old practices provide dependable high yield crops, there is concern regarding the negative environmental and biological implications and long-term sustainability linked with these practices. Donno et al. (2012) reported old varieties contain more polyphenols than commercial varieties, and in few studies the amounts in old and new varieties were similar (Wojdyło et al., 2008). In the study of Feliciano et al. (2010) both traditional and exotic apple varieties from Portugal showed high amounts of polyphenols. In contrast to this, few recent studies have shown apples grown organically have higher polyphenolic content and different polyphenolic profiles than conventional ones (Veronica et al., 2020; Barański et al., 2014; Średnicka-Tober et al., 2020).

Santarelli et al. (2020) reported organic or conventional apples have different polyphenolic profile but the content was similar. Organic and conventional orcharding practices have different protocols including the application of fertilizers and crop rotation, among others. Such contrasting practices in organic and conventional systems may have an impact on the metabolism of plants, which leads to differences in the composition of secondary metabolites (Cooper et al., 2011; Tétard-Jones et al., 2013). Rempelos et al. (2013) also reports replacement of synthetic with natural fertilizers affects the profile of phenolics in plants by changing the protein expressions. In addition to this, Lattanzio et al. (2018) has attributed phenolics accumulation in organic plant tissues as a defense mechanism triggered by biotic and abiotic stresses associated with the absence of synthetic pesticides.

Stracke et al. (2009) also found that organically produced apples had a greater phenolic content than conventionally grown apples. In their study, however, the crops were planted on two separate farms, which might have generated a variance unrelated to management method, even though the farms were geographically nearby. Veberic et al. (2005) also reported greater phenolic content in organically cultivated apple pulp when compared to conventionally farmed apple pulp. According to Jakopic et al. (2013) and Jakobek and Barron (2016), old apple cultivars have the same polyphenol groups as commercial varieties. A study published by Slatnar et al. (2012) found that the antimicrobial defense mechanisms for apple scab disease of plants results into the formation of phenolic substances. Infection with *Venturia inaequalis* also led to an increase in total phenolic content (Mikulic-Petkovsek et al., 2008). As a result of the increased polyphenol content, resistant cultivars showed greater natural resistance than susceptible ones.

In contrast to the findings of the previous study, Iacopini et al. (2010) found that ancient apple germplasm can be used for producing fruit with higher polyphenolic content and antioxidant scavenging capabilities. Furthermore, the antioxidant activity of apple extract was favorably associated with the overall amount of polyphenolics as well as the concentration of the three main phenolic compounds found in apple extract: catechin, epicatechin, and chlorogenic acid. When two late-bearing apple cultivars, 'Diacciata' and 'Limoncella', were compared to two current commercial cultivars ('Gala' and 'Golden Delicious'), similar results were observed by Minnocci et al. (2010). 'Limoncella' had a particularly high phenolic concentration in this study (around twice that of 'Diacciata' and nearly three times that of 'Gala'), confirming the importance of preserving and screening ancient germplasm as a potential source of genes for apple breeding programs that alter polyphenols in apple fruit.

17.3.2 Environmental Factors

Environmental factors and production techniques have been shown to influence apple fruit phenolic content, and their effects can sometimes outweigh genotype effects (Jakopic et al., 2009; Stopar et al., 2002; van der Sluis et al., 2001). The climate may have a significant impact on the physiology of trees and the quality of their fruits (Musacchi and Serra, 2018). It is possible that not all the places where apples are grown provide the best conditions for the fruit's growth. The least productive orchards are those that develop in an area where the days and nights are warm. Proper irrigation may make up for a shortage of water, but temperature, day length, and light exposure are all tied to a specific area.

When it comes to the cuticula of the apple, factors like as the amount of tissue present, how thick it is, its composition, how flexible it is, and how well it can heal itself are all affected by the climate in the orchard. To keep germs, viruses, and insects out, the apple's cuticle acts as a barrier. It is also essential for the fruit's long-term storage. Because the temperature in the orchard during fruit growth affects photosynthesis, metabolism, water relations, membrane integrity, plant hormone levels, and levels of primary and secondary metabolites, this alone must have an impact on fruit storage (Magan et al., 2011). According to Zoratti et al. (2015) temperature is the most important environmental element influencing polyphenol content and composition. Fruits cultivated at a maximum daily temperature of 25°C had the highest content of polyphenols, which drastically decreased at a maximum daily temperature of 35°C.

17.3.3 ALTITUDE

The concentration of some physiologically active chemicals in fruit is influenced by altitude, which has a significant impact on the environment and climate-related traits. Anthocyanins and flavonols production are more favored at higher altitudes than at lower ones (Xing et al., 2015). The shape and content of fruits were shown to be influenced by altitude as well (Silvanini et al., 2014). Research by Hamadziripi et al. (2014) revealed that apples growing in locations well exposed to light and at higher temperatures had greater levels of phenolic compounds and antioxidant capacity when exploring relationships between bioactive compound content and apple position on the tree. Apple peel is more resistant to photo-damage diseases owing to these chemicals, which perform a photo-protective role. It was found that apples grown in cooler climates and at higher elevations had a higher antioxidant capacity and polyphenolic compound concentrations in their peels and flesh, according to the study by Yuri et al. (2009) of four cultivars from four different locations. These locations differed in altitude, longitude, vegetation period length, and temperature. In the peels of apples cultivated in colder climates, levels of some phenolics such quercetin glycosides were found to be greater. Apple polyphenol content varied by growing site according to Awad et al. (2000). Apples grown in New Zealand have previously shown region-dependent variations in polyphenolic compound levels (McGhie et al., 2005). Yuri et al. (2009) found the lowest polyphenol concentration in apples cultivated in orchards in the southernmost section at the greatest elevation.

17.3.4 POSITION IN CANOPY

The position of the apple fruit on the canopy also affects the amount of polyphenols and other antioxidants. Although cultivar-dependent modifications occur, good light conditions are critical for growing apples with health-promoting phytochemicals in skin and flesh tissue. Fruit near the top of the canopy receives more light with more UV and red light and less fared light than fruit at other elevations (Jakopic et al., 2009; Looney, 1968). Awad et al. (2001a, 2001b) and Jakopic et al. (2009) reported higher levels of quercetin glycosides and cyanidin glycosides in the skin of sun-exposed apples imply that enhanced light conditions stimulate polyphenol production. Different crop load levels also have a significant effect on fruit quality (Serra et al., 2016). In fact, a large canopy provides a wider range of quality characteristic variations than a small tree, resulting in higher variability (Rudell et al., 2017).

17.3.5 HARVEST TIME

Harvest timing is another critical factor in order to ensure a good final fruit quality for consumers. Both early and late harvest can result in a variety of negative effects that reduce fruit quality (Vanoli and Buccheri, 2012). Kondo et al. (2002), Takos et al. (2006), and Renard et al. (2007) reported polyphenol accumulation has an inverse relationship with growth and maturity of fruit suggesting that delay in harvesting can cause a dilution of an initially accumulated polyphenol store. Similarly, Drogoudi and Pantelidis (2011) reported a decreasing trend in polyphenols during the last 28 days before harvest, which might be due to dilution effect associated with growth and maturity of fruit.

17.3.6 Cultivation Methods

Phenolic compound variations can also arise among different types of cultivation methods (Jakopic et al., 2013). Pruning, thinning, fertilization with phosphorus and potassium, and deficit irrigation can be used to increase the content of phenolic compounds (Reščič et al., 2015; Pena et al., 2013; Topalovic et al., 2011). The use of physical or chemical inducers is another way to increasing phenolic content including chitosan, methyl jasmonate, arachidonic acid, jasmonic acid, methyl salicylate, and salicylic acid (Valverde et al., 2015; Portu et al., 2016; Preciado-Rangel et al., 2019). These inducers maintain higher levels of polyphenolic and other bioactive compounds by increasing the activity of some key enzymes (chalcone synthase, alcone isomerase, phenylalanine ammonia-lyase, among others) of the flavonoid biosynthesis pathway (Dokhanieh et al., 2013; Gimenez et al., 2014; Valverde et al., 2015).

17.3.7 Postharvest Storage Temperature

Postharvest treatments of fruits and vegetables have been evaluated for their efficiency in preserving quality characteristics such as phytochemicals including phenolics and flavonoids. To lengthen the postharvest life of a fruit or vegetable, most postharvest treatments entail changing the fruits or vegetables standard conditions. Storage is critical for preserving high antioxidant levels in mature fruits since proper storage may greatly increase the life of fresh apples (Radenkovs and Juhnevica-Radenkova, 2017). Apples are kept at low temperatures and typically under controlled air conditions to lengthen postharvest storage. Low-temperature storage is the most common strategy for extending the storage period and reducing apple quality loss. Low temperatures slow down the metabolism of fruit cells, respiration, and ethylene formation and as a result maintain the integrity of cells (Bessemans et al., 2016). Disruption of cell's integrity may begin to accelerate the conversions and degradations of phenolic substances with the help of enzymes such as decarboxylases, glycosidases, and esterases (Cheynier, 2005). According to Carbone et al. (2011) cold storage (1°C for a period of 3 months) also affects total phenol content dramatically in both flesh (50%) and peels (20%) of 'Hillwell' apples while as other cultivars including the 'Fuji clone Kiku8' and the 'Golden clone B' were unaffected. Begić-Akagić et al. (2011) reported that phenolic content in apples stored for 60 days at 1°C were slightly decreased in three commonly found apple cultivars, namely 'Topaz', 'Pink Lady', 'Pinova', 'Ljepocvjetka', 'Ruzmarinka', and 'Paradija'.

Apple cultivars ('Golden Delicious', 'Pinova', 'Mairac', and 'Honeycrisp' cultivars) stored at ambient temperature (20°C) for 14 days showed no change in their total phenolic contents, but in contrast to this, the apple cultivar ('Topazin') showed a 50% reduction in total phenolic content, and the cultivar ('Wellant') showed significant positive effects (Matthes and Schmitz-Eiberger, 2009). Napolitano et al. (2004) also reported an increase in polyphenols including catechin and phloridzin during cold storage in a common Italian apple, 'Annurca'. Łysiak et al. (2020) collected the 'Jonagold' cultivar from four orchards with different microclimates, latitudes, and longitudes and stored them in a controlled atmosphere storage for 180 days. They reported polyphenols were more strongly affected by storage than by location of orchard. Furthermore, changes in polyphenols differ individually, and the degradation rate was found highest for flavonols in apples grown in the southern region.

17.3.8 Other Postharvest Treatments

Pre-treatments, which are commonly used to protect the quality of frozen products by enzyme inactivation, can damage cell structure resulting in quality deterioration such as color changes, softening, drip loss, and nutritional and bioactive component loss (Rickman et al., 2007; Chassagne-Berces et al., 2009). Veronica et al. (2020) reported that fresh organic and conventional fruits had equal total phenolic content and antioxidant activity, but the polyphenol profiles were different.

After pre-treatments (dipping and vacuum impregnation) organic and conventional apples revealed distinct physiological responses to physical stresses, which were reflected in their phenolic profiles. Both pre-treatments increased the amount of free epicatechin, coumarin, and rutin in the samples but had no effect on conjugated polyphenols. Furthermore, the amount of free epicatechin and rutin, as well as all bound polyphenols except catechin, was negatively affected by freezing and frozen storage. The differences in free and bound polyphenol content found on organic and conventionally processed samples after freezing might be attributed to sugar residue hydrolysis, changes in polyphenol distribution, and polyphenol oxidase activity inside the cells (Sacchetti et al., 2008). The combined impact of cultivation method and processing procedures, such as dipping, vacuum impregnation, and freezing on polyphenols suggests that organic apples' tissue responds differently to thermal and mechanical stresses generated by processing than conventional apples. Other studies (Mullen et al., 2002; Sacchetti et al., 2008; Leong and Oey, 2012) have shown that freezing can improve the quality and functional characteristics of plant foods by releasing bioactive components such bound phenolic acids and proanthocyanidins, leading to an increase in antioxidant activity.

17.4 ANTHOCYANINS

Anthocyanin belongs to the one of the largest group of pigments. They are among the water-soluble pigments (He and Giusti, 2010; Andersen and Jordheim, 2006). They are characterized by the blue, red, and purple colors. They are mostly found in the skin of fruits with certain exceptions of red fruits, where they are found in the flesh part of fruit such as strawberries and cherries (Manach and Donovan, 2004). Anthocyanins are most reactive, which makes them sensitive to the degradation process. The main anthocyanins are displayed in Figure 17.1 and Figure 17.2 (Kahkonen and Heinonen, 2003; Giusti et al., 1999). Chemically, they are glycosylated polymethoxy or polyhydroxy derivatives of two phenylbenzopyrilium (He and Giusti, 2010) with two benzyl rings. Intensity and predominate color of anthocyanins depends on the number of –OH groups–CH_3O groups (He and Giusti, 2010); when the –OH groups are in abundance then the perceived color would be bluish, and it would be reddish for the –CH_3O groups. In fruits, anthocyanins occur in glycosidic forms. In fruits they are derived from aglycone bases.

The red color of apple fruit has been associated with the accumulation of the pigment known as anthocyanin, although complete color development of the whole apple is the result of the interaction between the pigments such as anthocyanin, chlorophyll, and carotenoid (Lancaster et al., 1994). Cyanidin, in particular, cyanidin-3-galactoside, is the major derivative of apple fruits (Dussi et al., 1995; Macheix et al., 1990). The antioxidant properties of apples are greatly due to anthocyanins. Apple peels are rich in anthocyanins, along with the flesh and whole fruit. In apples, anthocyanins have been found to be responsible for red pigmentation and constitutes about 1%–3% of the total

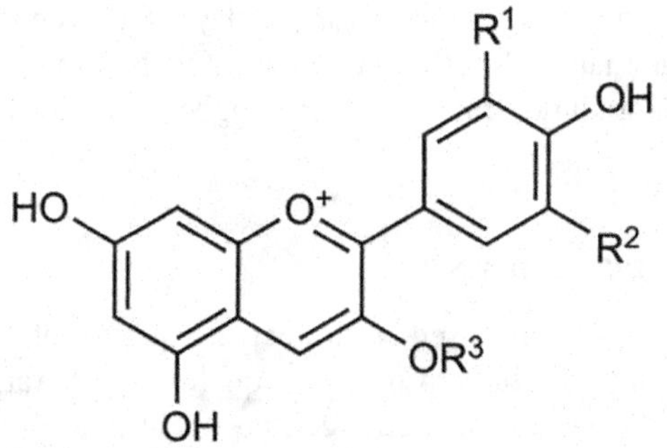

FIGURE 17.1   Skeletal structure of the main anthocyanins-3-O-glucoside present in fruits.

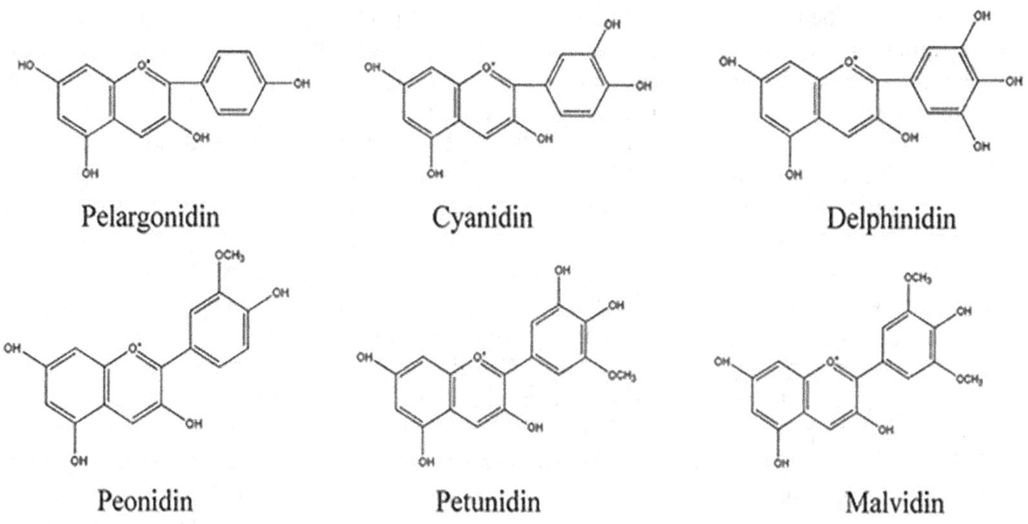

FIGURE 17.2 Structure of the main anthocyanins.

phenolic content (Vrhovsek et al., 2004). Generally, anthocyanins exist as glycosides monoside (Golding et al., 2001). Therefore, intake of apples along with the peels on a daily basis is highly recommended due to its free radical scavenging properties that protects us from reactive oxygen species.

Anthocyanin product in apple fruit occurs in two steps. The first one occurs during cell division in young fruit; it is generally considered as insignificant. The second one happens during ripening and maturation, and it is the period during which the concentration of anthocyanin raises five times more as compared to first one (Knee, 1972; Lancaster, 1992; Reay et al., 1998). There occurs a change in the anthocyanin concentration in the apples because of preharvest and postharvest treatments. The effect of following preharvest and postharvest treatment on the anthocyanin has been studied in this case.

17.4.1 EFFECT OF PREHARVEST TREATMENT ON ANTHOCYANINS

In the peel of fruit, the anthocyanin synthesis has been found to affected by various preharvesting factors such as temperature, ethylene, light, soil, fruit maturity, and tree factors (Ubi, 2004). The various preharvesting factors involved in the synthesis of anthocyanins have been reported (Lancaster, 1992; Saure, 1990). Among the preharvest factors, temperature effect on anthocyanin has been found to be of particular interest. As high temperature conditions have been associated with low pigmentation and vice-versa in apples (Lin-Wang et al., 2011; Palmer et al., 2012). This can lead to a reduction in market value since consumers are attracted toward the red color of apples, which is also directly related to the high market price.

17.4.2 TEMPERATURE

Apple reddening due to anthocyanin has been found to be affected greatly by temperature. Low temperature has associated with anthocyanin synthesis, whereas high temperature has been found to be responsible for the anthocyanin inhibition in apples (Blankenship, 1987; Marais et al., 2001; Reay and Lancaster, 2001). Irrespective of light, low temperature has been observed to promote anthocyanin synthesis. Loss of sugars in the peel by respiration due to the effect of reduced temperature has been suggested to be the main reason that causes increased photosynthate that would flow into the anthocyanin biosynthesis (Creasy, 1966; Lancaster, 1992). Low temperature combinations increase the anthocyanin amount and chalcone synthase transcripts in corollas in petunia grown at 17°C and 12°C (day/night) or 32°C and 27°C (day/night) (Shvarts et al., 1997). The minute eliminated corollas were kept in the dark on sucrose solutions; the chalcone synthase gene expression was quickly induced at 10°C or 15°C but not at 25°C. The changing effect of temperature on the apple color varies during the developmental stage, solar radiation, and variety used (Marais et al., 2001; Reay and Lancaster, 2001; Saure, 1990).

A study was conducted to observe the effects of temperature on 'Gala' and 'Royal Gala'. It was found that only the immature 'Royal Gala' and 'Gala' apples as compared to mature fruits responds to the lesser temperatures. Upon exposure to UV-B-visible light at 10°C, the exposed portion of fruit responded less compared to the unexposed side. This showed that fruits from outer portion of tree, which are well-exposed and colored, do not respond to reduced temperature just as the green apple and the green skin apple (Reay and Lancaster, 2001).

High anthocyanin concentrations in immature apples on or off the tree were obtained when the condition of day and night temperature was set as lowest. At the time of ripening, inhibition in the anthocyanin by a lower temperature had been observed in attached and detached apples. The riper the apple, the more optimal temperature range for development of color following enlightenment. Therefore, at every phase of ripening there is an optimal temperature regime, which does not depend upon cultivar but upon the fruit ripening phase of the cultivar (Diener and Naumann, 1981; Faragher, 1983). Temperature rise limits not only the anthocyanin biosynthetic genes expression levels (Ubi et al., 2006) and MYB transcription factor genes (Ban et al., 2007; Lin-Wang et al., 2011; Palmer et al., 2012), but also the activity of a biosynthetic enzyme, phenylalanine ammonia-lyase, (Iglesias et al., 1999), thus reducing the anthocyanin biosynthesis in the skin of apple.

On a practical note, it is necessary to know the dependency of anthocyanin accumulation in the apples in order to obtain the effective color management of fruits, for instance through evaporation cooling (Iglesias et al., 2005). So various researchers have reported the optimal temperatures for pigmentation of fruits for different detached apples (Arakawa, 1991; Curry, 1997; Faragher, 1983; Marais et al., 2001). Though the comparison between the results was difficult because the experimental conditions were also different.

17.4.3 Ethylene

Ethylene has been reported to be the main plant hormone involved in apple ripening (Saure, 1990). In 'Pink Lady' fruit ethylene was reported to be the main factor for the anthocyanin synthesis, as ethylene production was significantly related with the biosynthesis of anthocyanin and not with other pigments (Whale and Singh, 2007). Application of ethylene externally has been reported to promote anthocyanin production during ripening in the fruit skin (Larrigaudiere et al., 1996; Li et al., 2002). Poor red pigmentation of 'Royal Gala' apples with the antisense suppression of MdACO1, which is a gene for ethylene biosynthetic, that reduced the concentration of ethylene fruit (Johnston et al., 2009). This indicated that ethylene production is necessary for the synthesis of anthocyanin in the apple peel during ripening.

Some studies have reported that low temperature treatment after bloom for the period of 6 weeks suppresses the production of ethylene in 'Elstar' fruit until a later stage (Tromp, 1997). Genetic and molecular reports showed that ethylene receptors are negative regulators of ethylene-dependent responses (Hua and Meyerowitz, 1998). Blankenship and Dole (2003) reported that the compound 1-MCP impedes the sensitivity of ethylene by competing for ethylene receptors in fruit and that when it is applied to any fruit it decreases the ethylene-dependent reactions such as the evolution of ethylene (Tatsuki et al., 2011). A number of findings have examined the inhibitory effect of ethylene in apples by the preharvest use of 1-MCP on shelf life, the red color of the peel, and ethylene production.

17.4.4 LIGHT

Light is the most significant preharvesting factor affecting apple color development. Light stimulates genes related to the photosynthesis of plants. Generally, the red color development in apples does not occur in dark or low light conditions. Anthocyanin synthesis is dependent on light in case of mature fruits (Lancaster, 1992). The biosynthesis of anthocyanin due to light is generally ascribed to a prerequisite for a photomorphogenic signal, arbitrated by photoreceptors or due to a shortfall of carbohydrates occurred by delimited photosynthesis in dark or low light (Kawabata et al., 1999). Carbohydrate has been found to play two important roles in anthocyanin biosynthesis: first as a substrate for biosynthesis of flavonoid through shikimic acid and phenylpropanoid pathways and second as a gene inducer for anthocyanin biosynthesis (Murray et al., 1994). For the initiation of anthocyanin synthesis, it is necessary to use light with enough energy and spectral composition, and number of studies have reported on the photoregulation of apple color.

Light of sufficient energy and spectral composition is necessary to initiate anthocyanin synthesis, and many studies on the photoregulation of apple fruit color have been reported in the literature (Dong et al., 1995; Lancaster, 1992; Reay and Lancaster, 2001; Saure, 1990). An experiment was conducted with apples detached from the tree and skin of apples with energy reflux. Result showed a rise in anthocyanin synthesis (Arakawa et al., 1986). A linear relationship was observed between light intensity and anthocyanin accumulation by various studies (Lancaster, 1992; Saure, 1990). The most active portion of the spectrum for the synthesis of anthocyanin in apple peels was also observed from the research (Saure, 1990). Use of blue-violet and ultraviolet radiation has been found to be most effective as compared to far-red. The coloration of apples was more intensive at higher altitudes, and during the rainy weather condition the color of apples have been ascribed to more UV radiation (Saure, 1990).

17.4.5 FRUIT MATURITY

Mature fruits, as compared to immature fruits, have been found to accumulate a higher percentage of anthocyanin at the peak of harvest season (Reay and Lancaster, 2001). Fruit maturity along with temperature has positive effect on the postharvest color stimulation of apples by radiation (Reay and Lancaster, 2001; Marais et al., 2001; Reay, 1999). The increase in ethylene content in ripened apples has shown a positive effect on anthocyanin synthesis as compared to effects shown by temperature alone (Faragher, 1983). Anthocyanin accumulation was greater in mature fruits as reported by Marais et al. (2001) and Chalmers et al. (1973).

17.4.6 SOIL CONDITION

Soil contains nutrients required by plants for their growth and development. However, certain nutrients play important role in the color development of fruits. Nutrients in soil such as nitrogen and potassium and water have been related to the synthesis of anthocyanin. Nitrogen has been found to be the most important nutrient for the anthocyanin development. However, excess of nitrogen has been related to the decrease in the percentage of fruit color at the time of harvest even if the overall yield of colored fruits would be higher (Saure, 1990). Higher amounts of potassium have been reported to develop color formation in the case of apples but only in connection with other factors. Potassium has been found to antagonize the reduced effect of nitrogen on the color development, and a higher supply of potassium can complement the increased effect of a lower nitrogen supply on anthocyanin synthesis. Thus, potassium has been linked indirectly with the anthocyanin synthesis by the promotion of fruit development (Saure, 1990). The anthocyanidin synthase gene has been reported to be induced by calcium under white light (Gallop et al., 2001). Moisture in the soil has been reported to endorse anthocyanin formation during the dry weather season or in dry areas providing satisfactory conditions for the development of fruit. When the environmental conditions are under water stress, which has an effect on fruit development, it may cause negative effect on color formation (Saure, 1990).

17.4.7 Tree Condition

The tree on which the fruits are grown plays an important role in the translocation of water, minerals, sugars, and other important constituents to the fruits. Delay in ripening was noticed in high crop apples as observed by a rise in ethylene production as well as rise in anthocyanin content as compared to low crop apples at the same place. There is a lack of recent studies on the influence of tree on the production of anthocyanin. Several studies from past literature have showed that apples which are highly exposed to light after harvest have more coloration than apples which are still on the tree even if they are highly exposed to light. In the case of the younger apples, attached fruits had good coloration as compared to detached fruits, transitory suppressing of color development by red bags. In comparison to vigorous rootstock, the rootstocks such as semi-dwarfing, dwarf, and growth reducing interstocks have been reported to promote anthocyanin development. The effect on anthocyanin formation of the dwarfing rootstock may not be only because of the indirect effect of solar illumination due to lesser coverings of the tree but also ascribed of the rootstock. The higher percentage of red color on apples on dwarfing rootstocks was observed as compared to apples on vigorous rootstocks when place under identical light conditions. Pruning of root have been found to be associated with a higher percentage of color development. Leaf number has also been correlated with the apple color development; color formation was not found to be sufficient in various cultivars with as few as 10 to 20 leaves for each fruit although the exposure of light was perfect (Saure, 1990).

Harvested 'Jonathan' apples from severely dense branches have been not reported to have noticeable color formation even if exposed for 10 days to full light. 'Jonagold' trees, with high quantities of colored fruits, were found to differ significantly from the trees with lesser quantities of colored fruits in the same farm but only differ in slightly thicker stems, lower fruits in each tree, bigger and more abundant leaves, and thus a greater leaf area for each fruit and tree (Heinecke, 1966).

17.4.8 Application of Chemicals

Chemicals have been reported to be used for the development of color. Various types of thiocyanates were found to be used to raise the color of fruits even though the fruit is inside the tree canopy (Saure, 1990). On the other hand, thiocyanates and other related chemicals that have been tested have given unpredictable or not sufficient effects and negative side effects. Few growth regulators such as paclobutrazol, daminocide, ethephon, and seniphos were used commercially to stimulate apple color development (Gomez-Cordoves et al., 1996). The effect of an ethephon growth regulator on color development may be due to their advancing maturity. Ethephon is an ethylene releasing component that has been found to increase the process of ripening, however, it is not capable of developing red color without adequate light (Saure, 1990). Conversely, studies have shown that growth regulators have no influence on fruits maturity/ripening, but anthocyanin development and color was evidently stimulated by growth regulator ethephon and the accumulation of anthocyanin was deferred about 14 days by (S)-*trans*-2-amino-4-(2-aminoethoxy)-3-butenoic acid hydrochloride (ABG-3168). Anthocyanin accumulation was significantly reduced by GA_3. So it can be concluded that anthocyanin development in apple peels is not only related to ripening, but at the same time regulated equally by the non-ethylene components as well as ethylene components (Awad and Jager, 2002).

The application of preharvest spray of various chemicals such as L-phenylalanine, D-phenylalanine, p-coumaric acid, naringenin, and cinnamic acid with a concentration of 0 mg L^{-1}, 50 mg L^{-1}, 100 mg L^{-1}, or 200 mg L^{-1} on red color development and anthocyanins formation on the peel of 'Cripps Pink' apple were investigated. Harvested apples when sprayed with L-phenylalanine or cinnamic acid showed higher levels of total anthocyanins and cyanidin 3-galactoside as compared to control. L* a* b*, chroma, and hue was studied on the exterior of fruits at maturity, and red blush development was noticed in response to the phenylpropanoids spray. Higher values of a* and lowest hue were reported in response to L-phenylalanine or cinnamic acid showing a significant trend of red blush intensification on the surface of fruit (Shafiq and Singh, 2018).

17.4.9 EFFECT OF POSTHARVEST TREATMENT ON ANTHOCYANINS

The anthocyanin synthesis has been found to be affected by various postharvesting factors. Decline in the quality of fruits has been found to be associated with color change, texture, and nutrient losses (Pramanik et al., 2004; Lee and Kader, 2000). Although, a lot of research has been conducted on the effects of postharvesting treatment on nutrients, but little is known about the effects of postharvesting treatment on anthocyanin.

17.4.10 STORAGE

To minimize the quality loss, fruits are stored either in modified or controlled atmospheric conditions or in cold conditions. Loss of phenolic compounds takes place during the postharvest conditions by enzymatic browning which are caused by oxidation of phenolics in the presence of enzymes such as peroxidases and polyphenol oxidase. However, anthocyanins, which are not phenolic compounds, increase in postharvest conditions, and this is not a negative quality characteristics. Therefore, an increase in anthocyanin content not only increases physical characteristics such as color of fruit but also antioxidant property. Anthocyanins have been found to increase with ripening in several fruits, especially apples, and using proper storage conditions after harvest can continue to increase the anthocyanins (Connor et al., 2002). One of the storage conditions, water loss, has been found to decrease the anthocyanin formation (Nunes et al., 2005) and may be due to the loss of cellular material which is linked with the increase in activity of polyphenol oxidase. Therefore, it is better to maintain high humidity during storage.

When the storing atmosphere had high carbon dioxide concentration, a significant reduction in the anthocyanin content was reported (Tomas-Barberan and Espin, 2001; Gil et al., 1997), especially in internal tissues. Likewise, modified atmosphere packaging also lowers the rise in anthocyanins when stored at 5°C and 10 kPa carbon dioxide (Holcroft and Kader, 1999). Five apple varieties cultivated in Tekirdağ, Turkey, were examined for anthocyanins storing at 0°C −1°C and a relative humidity of 90°C −95% for a period of 4 months with an interval of 30 days. Variation in the anthocyanin content was reported in this study. A decrease in anthocyanin contents was observed (Bal, 2017).

17.4.11 PROCESSING

A decline in phenolic compounds has been observed during processing (Tomas-Barberan and Espin, 2001), but an increase in anthocyanin content was reported during minimal processing (Ferreres et al., 1997). However, more severe processing treatment can lead to more decline. A decline of 68% of anthocyanin was reported during juicing because of activation of hydrolysing enzymes and about 20% of anthocyanin was discarded as skins and other wastes (Kalt, 2005). During the process of cooking in water, a decrease in anthocyanin content was observed as they are soluble in water, which causes them to leach out (Moreno et al., 2003). But at the same time frying did not have any effect on anthocyanins (Philpott et al., 2003).

When 'Cripps Pink' apples were subjected to 72 hours of postharvest treatment with radiation using high pressure sodium with a hue angle of 56.5°, an improved red blush was reported as compared with UV-B plus incandescent lamps with hue angle of 70.7°. The rise in red color in 'Braeburn' apples using high pressure sodium with a hue angle decrease of 14.9° kept at −0.5°C for a period of 2 months prior to treatment was lower as compared to the fruit kept for 1 month, in this case with the hue angle decrease of 23°. Irradiation for a period of 168 hours with high pressure sodium at 20/20°C and 20/6°C and an absence of leaves lowers the reduction in hue angle; however, this was due to yellowing, but not due to red blush development. Therefore, it can be concluded that the development of anthocyanin in apples was stimulated due to postharvest irradiation with high pressure sodium lights (Marais et al., 2001).

17.5   PHYTOESTROGENS

Phytoestrogens are the set of nonsteroidal plant metabolites having similar functionality and structure with 17b-estradiol (Kuhnle et al., 2009; Branham et al., 2002). They can induce biotic responses that also mimic the effects of endogenous oestrogens; they can even bind to the receptors of estrogens (Committee on Toxicity of Chemicals in Food, 2003; Shutt and Cox, 1972). Studies have also shown that phytoestrogens show antioxidant activity (Wei et al., 1995) and act as enzyme inhibitor like DNA topoisomerase (Markovits et al., 1989) and tyrosine kinase (Akiyama et al., 1987). Phytoestrogens show beneficial effects on human, for instance with osteoporosis (Dang and Lowik, 2005), heart disease (Anthony, 2002), cancer (Duffy et al., 2007), climacteric problems (Krebs et al., 2004), male fertility problems (Phillips and Tanphaichitr, 2008), obesity, and sugar problems (Bhathena and Velasquez, 2002).

The leading food sources of phytoestrogens are from the plant kingdom such as fruits and vegetables. They mainly occur as glycosides in plants, guard against light, and especially serve as defensive mediators against predators (Mazur and Adlercreutz, 1998). The main important classes of phytoestrogens are lignans, isoflavones, and coumestans. Lignans are found in fruits and vegetables (Committee on Toxicity of Chemicals in Food, 2003). Isoflavones have been found in a wide range of fruits and vegetables. While genistein has been reported mostly in soy-based foods, it has been also noticed in a few fruits and vegetables in low amounts (Boker et al., 2002).

The presence of phytoestrogens in apples has been reported to be approximately 12 µg/100 g (Kuhnle et al., 2009). Though apples have been reported to contain genistein at very low concentrations (Boker et al., 2002), the other phytoestrogens have also been reported (Lee et al., 2003). However, the effects of preharvest and postharvest technologies on phytoestrogens of apples have not yet been studied.

17.6   PHYTOSTEROLS

Phytosterols are the compounds found in plants. These are similar to that of cholesterol. More than 200 phytosterols have been reported. Nuts, legumes, and vegetable oils are the richest sources of phytosterols (Katcher et al., 2009). Phytosterols belong to a vast family of 3β-hydroxy triterpenoids on the basis of substituted perhydro-,1,2-cyclopentanophenanthrene ring system (Hartmann, 1998). Among the various sterols reported in plants campesterol; stigmasterol; sitosterol; and 4-desmethyl-sterol and 14-desmethylsterol are the dominant ones. These occur not only as free sterols but also as conjugates (Moreau et al., 2002). Phytosterols are the important constituent of plant plasma membrane with lesser chance of presence in other membranes (Hartmann and Benveniste, 1987). They play an important role in plasma membrane such as the modification of membrane permeability and fluidity (Schuler et al., 1990), modulation of membrane protein (Wojciechowski, 1991), and regulation of the proliferation of cells (Hartmann, 1998).

Apples have been reported to be the source of components that have beneficial physiological effects (Anderson et al., 2000; Woyengo et al., 2009; Awad and Fink, 2000). Supercritical fluid extraction of apple pomace revealed substantial amounts of β-sitosterol, oleanolic acid, and ursolic acid (Wozniak et al., 2018). There are two main bioactive compounds of low polarity groups reported in apples as triterpenic acids as well as phytosterols. Among them phytosterols are the essential structural constituents of plant membranes. Phytosterols have been reported to lower the cholesterol level. A reduction in serum cholesterol level has been reported clinically with the daily intake of 1 g–3 g of phytosterols and their derivatives such as phytostanols. Reduction in cholesterol results in a reduction of the risk of heart diseases (Wasserman, 2011).

Environmental factors influence the phytosterols. Temperature variation can incite the relative amounts of phytosterols in apple fruit (Whitaker et al., 1997). Change in phytosterol metabolism has been related to stressors such as chemical and cold temperature stress (Moreau et al., 2002). Chilling in apples has been reported to incite the necrotic peel disorder known as superficial scald. Variations in the tissue of apple peel levels of conjugates campesterol and β-sitosterol such as free sterols, steryl esters, steryl glycosides, and acylated steryl glycosides were determined during the course of superficial scald development. Prestorage treatments such as ethylene inhibitor, diphenylamine (which is also an antioxidant), quick elevation of temperature, and cold acclimation using irregular warming behaviors were assessed. The level of acylated steryl glycosides increased, whereas steryl esters decreased in the fruits which were untreated during storage. With the removal of fruits from cold conditions to ambient temperature conditions, a rapid swing in acylated steryl glycosides and steryl esters fatty acyl moieties from unsaturated to saturated was noticed. The level of free sterols and steryl glycosides remains relatively steady during storage period, but steryl glycosides rise following a temperature rise after storage. Levels of acylated steryl glycosides, steryl esters, and steryl glycosides did not rise for a period of 6 months of storage in fruits placed in intermittent warming. It can be concluded that apple peel phytosteryl were affected during storage, temperature variation, ethylene, and oxidative stress (Rudell et al., 2011).

17.7 CONCLUSION

This chapter presents how preharvest practices and postharvest technologies affect the bioactive compounds in apples. Preharvest and postharvest customs are widely used around the world to ensure that high quality apples with a long shelf life postharvest are obtained. The differences in phytochemical compounds in apples grown at different altitudes and in different climates and that were exposed to various chemicals show that meticulous considerations are needed when choosing an apple farming site. Moreover, careful choice of postharvest technologies has been shown to extend the content of phytochemicals in apples from harvest to storage and sales to customers. The importance of knowing the fate of phytochemicals from the field to the fork is to provide awareness to the food chain processes that there is no nutrient loss as well as to provide certainty to customers that the apples are as nutritious as apples that have been freshly picked.

REFERENCES

Akiyama, T., Ishida, J., Nakagawa, S., Ogawara, H., Watanabe, S., Itoh, N., Shibuya, M., Fukami, Y. 1987. Genistein, a specific inhibitor of tyrosine-specific protein kinases. *Journal of Biological Chemistry* 262(12): 5592–5595.

Andersen, O. M., Jordheim, M. 2006. The anthocyanins. In *Flavonoids: Chemistry, Biochemistry and Applications*. Andersen, O. M., Markham, K. R. (Eds.), Boca Raton, FL: Taylor and Francis, pp. 471–551.

Anderson, J. W., Allgood, L. D., Lawrence, A. 2000. Cholesterol-lowering effects of psyllium intake adjunctive to diet therapy in men and women with hypercholesterolemia: Meta-analysis of 8 controlled trials. *The American Journal of Clinical Nutrition* 71(2): 472–479.

Anthony, M. S. 2002. Phytoestrogens and cardiovascular disease: Where's the meat? *Arteriosclerosis, Thrombosis, and Vascular Biology* 22(8): 1245–1247.

Arakawa, O. 1991. Effect of temperature on anthocyanin accumulation in apple fruit as affected by cultivar, stage of fruit ripening and bagging. *The Journal of Horticultural Sciences* 66: 763–768.

Arakawa, O., Hori, Y., Ogata, R. 1986. Characteristics of color development and relationship between anthocyanin synthesis and phenylalanine ammonia-lyase activity in 'Starking Delicious', 'Fuji' and 'Mutsu' apple fruits. *The Japanese Society for Horticultural Science* 54: 424–430.

Awad, M. A., de Jager, A. 2002. Formation of flavonoids, especially anthocyanins and chlorogenic acid in 'Jonagold' apple skin: Influences of growth regulators and fruit maturity. *Scientia Horticulturae* 93: 257–266.

Awad, M. A., de Jager, A., van der Plas, L. H. W., van der Krol, A. R. 2001a. Flavonoid and chlorogenic acid changes in skin of 'Elstar' and 'Jonagold' apples during development and ripening. *Scientia Horticulturae* 90: 69–83.

Awad, M. A., de Jager, A., van Westing, L. M. 2000. Flavonoid and chlorogenic acid levels in apple fruit: Characterisation of variation. *Scientia Horticulturae* 83: 249–263.

Awad, M. A., Fink, C. 2000. Phytosterols as anticancer dietary components: Evidence and mechanism of action. *Journal of Nutrition* 130(9): 2127–2130.

Awad, M. A., Wagenmakers, P. S., de Jager, A. 2001b. Effects of light on flavonoid and chlorogenic acid levels in the skin of 'Jonagold' apples. *Scientia Horticulturae* 88: 289–298.

Bal, E. 2017. Changes in phenolic compounds, anthocyanin and antioxidant capacity of some apple culti-vars during cold storage. *2nd International Balkan Agriculture Congress*. Electronic Book. ISBN: 978-605-4265-49-7

Ban, Y., Honda, C., Hatsuyama, Y., Igarashi, M., Bessho, H., Moriguchi, T. 2007. Isolation and functional analysis of a MYB transcription factor gene that is a key regulator for the development of red coloration in apple skin. *Plant Cell Physiology* 48: 958–970.

Barański, M., Średnicka-Tober, D., Volakakis, N., Seal, C., Sanderson, R., Stewart, G. B., Benbrook, C., Biavati, B., Markellou, E., Giotis, C., Gromadzka-Ostrowska, J., Rembiałkowska, E., Skwarło-Sońta, K., Tahvonen, R., Janovská, D., Niggli, U., Nicot, P., Leifert. 2014. Higher antioxidant and lower cad-mium concentrations and lower incidence of pesticide residues in organically grown crops: A systematic literature review and meta-analyses. *British Journal of Nutrition* 112(5): 794–811.

Begić-Akagić, A., Spaho, N., Oručević, S., Drkenda, P., Kurtović, M., Gaši, F., Kopjar, M., Piližota, V. 2011. Influence of cultivar, storage time, and processing on the phenol content of cloudy apple juice. *Croatian Journal of Food Science and Technology* 3: 1–8.

Bengoechea, M. L., Sancho, A. I., Bartolome, B. 1997. Phenolic composition of industrially manufactured purees and concentrates from peach and apple fruits. *Journal of the Agriculture and Food Chemistry* 45: 4071–4075.

Bessemans, N., Verboven, P., Verlinden, B. E., Nicolai, B. M. 2016. A novel type of dynamic controlled atmo-sphere storage based on the respiratory quotient (RQ-DCA). *Postharvest Biology and Technology* 115: 91–102.

Bhathena, S. J., Velasquez, M. T. 2002. Beneficial role of dietary phytoestrogens in obesity and diabetes. *American Journal of Clinical Nutrition* 76(6): 1191–1201.

Blankenship, S. M. 1987. Night temperature effects on rate of apple fruit maturation and fruit quality. *Scientia Horticulturae* 33: 205–212.

Blankenship, S. M., Dole, J. M. 2003. 1-Methylcyclopropene: A review. *Postharvest Biology and Technology* 28(1): 1–25.

Boker, L. K., Van der Schouw, Y. T., De Kleijn, M. J., et al. 2002. Intake of dietary phytoestrogens by Dutch women. *Journal of Nutrition* 132: 1319–1328.

Branham, W. S., Dial, S. L., Moland, C. L., et al. 2002. Phytoestrogens and mycoestrogens bind to the rat uterine estrogen receptor. *The Journal of Nutrition* 132(4): 658–664.

Carbone, K., Giannini, B., Picchi, V., Lo Scalzo, R., Cecchini, F. 2011. Phenolic composition and free radical scavenging activity of different apple varieties in relation to the cultivar, tissue type and storage. *Food Chemistry* 127(2): 493–500.

Chalmers, D. J., Faragher, J. D., Raff, J. W. 1973. Changes in anthocyanins synthesis as a index of maturity in red apple varieties. *Journal of Horticultural Science* 48: 387–392.

Chassagne-Berces, S., Poirier, C., Devaux, M. F., et al. 2009. Changes in texture, cellular structure and cell wall composition in apple tissue as a result of freezing. *Food Research International* 42(7): 788–797.

Cheynier, V. 2005. Polyphenols in foods are more complex than often thought. *American Journal of Clinical Nutrition* 81: 223–229.

Committee on Toxicity of Chemicals in Food, Consumer Products and the Environment. 2003. *Phytoestrogens and Health*. London: Food Standards Agency.

Connor, A. M., Luby, J. J., Hancock, J. F., Berkheimer, S., Hanson, E. J. 2002. Changes in fruit antioxidant activity among blueberry cultivars during cold temperature storage. *Journal of Agricultural and Food Chemistry* 50(4): 893–898.

Cooper, J., Sanderson, R., Cakmak, I., Ozturk, L., Shotton, P., Carmichael, A., Sadrabadi Haghighi, A., Tetard-Jones, C., Volakakis, N., Eyre, M., Leifert, C. 2011. Effect of organic and conventional crop rotation, fertilization, and crop protection practices on metal contents in wheat (*Triticum aestivum*). *Journal of Agricultural and Food Chemistry* 59: 4715–4724.

Creasy, L. L. 1966. The effect of temperature on anthocyanin synthesis in 'McIntosh' apple apple skin. *Proc. New York State Horticulture Society* 111: 93–96.

Curry, E. A. 1997. Temperature for optimum anthocyanin accumulation in apple tissue. *Journal of Horticulture Science* 72: 723–729.

Dang, Z. C., Lowik, C. 2005. Dose-dependent effects of phytoestrogens on bone. *Trends in Endocrinology and Metabolism* 16(5): 207–213.

Diener, H. A., Naumann, W. D. 1981. Influence of day and night temperatures on anthocyanins synthesis in apple skin. *Gartenbau* 46: 125–132.

Dokhanieh, A. Y., Aghdam, M. S., Fard, J. R., Hassanpour, H. 2013. Postharvest salicylic acid treatment enhances antioxidant potential of cornelian cherry fruit. *Scientia Horticulturae* 154: 31–36.

Dong, Y. H., Mitra, D., Kootstra, A., Lister, C., Lancaster, J. 1995. Postharvest stimulation of skin color in Royal Gala apple. *Journal of American Society of Horticulture Science* 120: 95–100.

Donno, D., Beccaro, G. L., Mellano, M. G., Torello Marinoni, D., Cerutti, A. K., Canterino, S., Bounous, G. 2012. Application of sensory, nutraceutical and genetic techniques to create a quality profile of ancient apple cultivars. *Journal of Food Quality* 35: 169–181.

Drogoudi, P. D., Pantelidis, G. 2011. Effects of position on canopy and harvest time on fruit physico-chemical and antioxidant properties in different apple cultivars. *Scientia Horticulturae* 129: 752–760.

Duda-Chodak, A., Tarko, T., Satora, P., Sroka, P., Tuszyński, T. 2010. The profile of polyphenols and antioxidant properties of selected apple cultivars grown in Poland. *Journal of Fruit and Ornamental Plant Research* 18: 39–50.

Duffy, C., Perez, K., Partridge, A. 2007. Implications of phytoestrogen intake for breast cancer. *CA: A Cancer Journal for Clinicians* 57(5): 260–277.

Dussi, M. C., Sugar, D., Wrolstad, R. E. 1995. Characterizing and quantifying anthocyanins in red pears and the effect of light quality on fruit color. *Journal of American Society of Horticulture Science* 120: 785–789.

Faragher, J. D. 1983. Temperature regulation of anthocyanin accumulation in apple skin. *Journal of Experimental Botany* 34: 1291–1298.

Feliciano, R. P., Antunes, C., Ramos, A., Serra, A. T., Figueira, M. E., Duarte, C. M. M., de Carvalho, A., Bronze, M R. 2010. Characterization of traditional and exotic apple varieties from Portugal. Part 1 – Nutritional, phytochemical and sensory evaluation. *Journal of Functional Foods* 2: 35–45.

Ferreres, F., Gil, M. I., Castaner, M., Tomas-Barberan, F. A. 1997. Phenolic metabolites in red pigmented lettuce (*Lactuca saliva*). Changes with minimal processing and cold storage. *Journal of Agricultural and Food Chemistry* 45: 4249–4254.

Francini, A., Sebastiani, L. 2013. Phenolic compounds in apple (Malus x domestica Borkh.): compounds characterization and stability during postharvest and after processing. *Antioxidants* 2(3): 181–193.

Gallop, R., Farhi, S., Perl, A. 2001. Regulation of the *leucoanthocyanidin dioxygenase* gene expression in *Vitis vinifera*. *Plant Science* 161: 579–588.

Gil, M. I., Holcroft, D. M., Kader, A. A. 1997. Changes in strawberry anthocyanins and other polyphenols in response to carbon dioxide treatments. *Journal of Agricultural and Food Chemistry* 45: 1662–1667.

Gimenez, M. J., Valverde, J. M., Valero, D., Guillén, F., Martínez-Romero, D., Serrano, M., Castillo, S. 2014. Quality and antioxidant properties on sweet cherries as affected by preharvest salicylic and acetylsalicylic acids treatments. *Food Chemistry* 160: 226–232.

Giusti, M. M., Rodríguez-Saona, L. E., Griffin, D., Wrolstad, R. E. 1999. Electrospray and tandem mass spectroscopy as tools for anthocyanin characterization. *Journal of Agricultural and Food Chemistry* 47(11): 4657–4664.

Golding, J. F., Mueller, A. G., Gresty, M. A. 2001. A motion sickness maximum around the 0.2 Hz frequency range of horizontal translational oscillation. *Aviation, Space, and Environmental Medicine* 72(3): 188–192.

Gomez-Cordoves, C., Varela, F., Larrigaudiere, C., Vendrell, M. 1996. Effect of ethephon and seniphos treatments on anthocyanins composition of Starking apples. *Journal of Agricultural and Food Chemistry* 44: 3449–3452.

Hamadziripi, E. T., Theron, K. I., Muller, M., Steyn, W. J. 2014. Apple compositional and peel color differences resulting from canopy microclimate affect consumer preference for eating quality and appearance. *HortScience* 49: 384–392.

Hancock, J. F., Luby, J. J., Brown, S. K., Lobos, G. A. 2008. Apples. In *Temperate Fruit Crop Breeding*. Hancock, J. F. (Ed.). Dordrecht: Springer, pp. 1–37.

Hartmann, M.-A. 1998. Plant sterols and the membrane environment. *Trends in Plant Science* 3: 170–175.

Hartmann, M.-A., Benveniste, P. 1987. Plant membrane sterols: Isolation, identification and biosynthesis. *Methods Enzymology* 148: 632–650.

He, J., Giusti, M. 2010. Anthocyanins: Natural colorants with health-promoting properties. *Annual Review of Food Science and Technology* 1(1): 163–187.

Heinecke, D. R. 1966. Characteristics of McIntosh and Red Delicious apples as influenced by exposure to sunlight during the growing season. *Proceedings of the American Society for Horticultural Science* 89: 10–13.

Holcroft, D. M., Kader, A. A. 1999. Carbon dioxide-induced changes in colour and anthocyanin synthesis of stored strawberry fruit. *HortScience* 34: 1244–1248.

Hua, L., Meyerowitz, E. M. 1998. Ethylene responses are negatively regulated by a receptor gene family in Arabidopsis thaliana. *Cell* 94: 261–271.

Iacopini, P., Camangi, F., Stefani, A., Sebastiani, L. 2010. Antiradical potential of ancient Italian apple varieties of Malus x domestica Borkh. In a peroxynitrite-induced oxidative process. *Journal Food Composition and Analysis* 23: 518–524.

Iglesias, I., Graell, J., Echeverria, G., Vendrell, M. 1999. Differences in fruit color development, anthocyanin content, yield and quality of seven' Delicious' apple strains. *Fruit Varieties Journal*.

Iglesias, I., Salvia, J., Torguet, L., Montserrat, R. 2005. The evaporative cooling effects of overtree microsprinkler irrigation on 'Mondial Gala' apples. *Scientia Horticulturae* 103: 267–287.

Jakobek, L., Barron, A. R. 2016. Ancient apple varieties from Croatia as a source of bioactive polyphenolic compounds. *Journal of Food Composition and Analysis* 45: 9–15. https://doi.org/https://doi.org/10.1016/j.jfca.2015.09.007. https://www.sciencedirect.com/science/article/pii/S0889157515001970

Jakopic, J., Slatnar, A., Mikulic-Petkovsek, M., Veberic, R, Stampar, F., Bavec, F., Bavec, M. 2013. Effect of different production systems on chemical profiles of dwarf French bean (*Phaseolus vulgaris* L. Cv. Top Crop) pods. *Journal of the Agriculture and Food Chemistry* 61: 2392–2399.

Jakopic, J., Stampar, F., Veberic, R. 2009. The influence of exposure to light on the phenolic content of 'Fuji' apple. *Scientia Horticulturae* 123: 234–239.

Johnston, J. W., Gunaseelan, K., Pidakala, P, Wang, M., Schaffer, R. J. 2009. Co-ordination of early and late ripening events in apples is regulated though differential sensitivities to ethylene. *Jouranl of Experimental Botany* 60: 2689–2699.

Kahkonen, M. P., Heinonen, M. 2003. Antioxidant activity of anthocyanins and their aglycons. *Jouranl of Agriculture and Food Chemistry* 51(3): 628–633.

Kalt, W. 2005. Effects of production and processing factors on major fruit and vegetable antioxidants. *Journal of Food Science* 70(1): R11–R19.

Katcher, H. I., Hill, A. M., Lanford, J. L., Yoo, J. S., Kris-Etherton, P. M. 2009. Lifestyle approaches and dietary strategies to lower LDL-cholesterol and triglycerides and raise HDL-cholesterol. *Endocrinology and Metabolism Clinics of North America* 38(1): 45–78.

Kawabata, S., Kusuhara, Y., Li, Y., Sakiyama, R. 1999. The regulation of anthocyanin biosynthesis in *Eustoma grandiflorum* under low light conditions. *The Japanese Society for Horticultural Science* 68(3): 519–526.

Knee, M. 1972. Anthocyanin, carotenoid, and chlorophyll changes in the peel of 'Cox's Orange Pippin' apples during ripening on and of the tree. *Journal of Experimental Botany* 23: 184–196.

Kondo, S., Tsuda, K., Muto, N., Ueda, J. 2002. Antioxidative activity of apple skin or flesh extracts associated with fruit development on selected apple cultivars. *Scientia Horticulturae* 96: 177–185.

Korban, S. S., Skirvin, R. M. 1984. Nomenclature of the cultivated apple. *HortScience* 19(2): 177–180.

Krebs, E. E., Ensrud, K. E., MacDonald, R., Wilt, T. J. 2004. Phytoestrogens for treatment of menopausal symptoms: A systematic review. *Obstetrics and Gynecology* 104(4): 824–836.

Kuhnle, G. G. C., Aquila, C. D., Aspinall, S. M., Runswick, S. A., Joosen, A. M. C. P., Mulligan, A. A. M., Bingham, S. A. 2009. Phytoestrogen content of fruits and vegetables commonly consumed in the UK based on LC – MS and ^{13}C-labelled standards. *Food Chemistry* 116: 542–554.

Lancaster, J. E. 1992. Regulation of skin colour in apples: A review. *Critical Review in Plant Science* 10: 487–502.

Lancaster, J. E., Grant, J. E., Lister, C. E., Taylor, M. C. 1994. Skin color in apples – Influence of copigmentation and plastid pigments on shades and darkness of red color in five genotypes. *Journal of the American Society of Horticulutal Science* 119: 63–69.

Larrigaudiere, C., Pinto, E., Vendrell, M. 1996. Differential effects of ethephon and seniphos on color development of 'Starking Delicious' apple. *Journal of the American Society of Horticulutal Science* 121: 746–750.

Lattanzio, V., Caretto, S., Linsalata, V., Colella, G., Mita, G. 2018. Signal transduction in artichoke [Cynara cardunculus L. subsp. scolymus (L.) Hayek] callus and cell suspension cultures under nutritional stress. *Plant Physiology and Biochemistry* 127: 97–103. https://doi.org/https://doi.org/10.1016/j.plaphy.2018.03.017. https://www.sciencedirect.com/science/article/pii/S0981942818301402

Le Deun, E., van der Werf, R., Le Bail, G., Le Quere, J. M., Guyot, S. 2015. HPLC-DAD-MS profiling of polyphenols responsible for the yellow-orange color in apple juices of different French cider apple varieties. *Journal of the Agriculture and Food Chemistry* 63: 7675–7684.

Lee, K. W., Kim, Y. J., Kim, D. O., Lee, H. J., Lee, C. Y. 2003. Major phenolics in apple and their contribution to the total antioxidant capacity. *Jouranl of the Agriculture and Food Chemistry* 51: 6516–6520.

Lee, S. K., Kader, A. A. 2000. Preharvest and postharvest factors influencing vitamin C content of horticultural crops. *Postharvest Biology and Technology* 20: 207–220.

Leong, S. Y., Oey, I. 2012. Effects of processing on anthocyanins, carotenoids and vitamin C in summer fruits and vegetables. *Food Chemistry* 133(4): 1577–1587.

Li, Z., Gemma, H., Iwahori, S. 2002. Simulation of 'Fuji' apple skin color by ethephon and phosphorus-calcium mixed compounds in relation to flavonoid synthesis. *Scientia Horticultrae* 94: 193–199.

Lin-Wang, K., Micheletti, D., Palmer, J., Volz, R., Lozano, L., Espley, R., Hellens, R. P., Chagnè, D., Rowan, D. D., Troggio, M., Iglesias, I., Allan, A. C. 2011. High temperature reduces apple fruit colour via modulation of the anthocyanin regulatory complex. *Plant Cell Environment* 34: 1176–1190.

Looney, N. E. 1968. Light regimes within standard size apple trees as determined spectrophotometrically. *Proceedings of the American Society for Horticultural Science* 93: 1–6.

Łysiak, G. P., Michalska, A., Wojdyło, A. 2020. Postharvest changes in phenolic compounds and antioxidant capacity of apples cv. Jonagold growing in different locations in Europe. *Food Chemistry* 310: 125912.

Macheix, J., Fleuriet, A., Billot, J. 1990. *Fruit Phenolics*. Boca Raton: CRC Press/Taylor & Francis.

Magan, N., Medina, A., Aldred, D. 2011. Possible climate-change effects on mycotoxin contamination of food crops pre- and postharvest. *Plant Pathology* 60: 150–163.

Manach, C., Donovan, J. L. 2004. Pharmacokinetics and metabolism of dietary flavonoids in humans. *Free Radical Research* 38(8): 771–785.

Marais, E., Jacobs, G., Holcroft, D. M. 2001. Postharvest irradiation enhances anthocyanin synthesis in apples but not in pears. *Hortscience* 36(4): 738–740.

Markovits, J., Linassier, C., Fosse, P., Couprie, J., Pierre, J., Jacquemin-Sablon, A. S. 1989. Inhibitory effects of the tyrosine kinase inhibitor genistein on mammalian DNA topoisomerase II. *Cancer Research* 49(18): 5111–5117.

Matthes, A., Schmitz-Eiberger, M. 2009. Polyphenol content and antioxidant capacity of apple fruit: Effect of cultivar and storage conditions. *The Journal of Applied Botany and Food Quality* 82: 152–157.

Mazur, W., Adlercreutz, H. 1998. Naturally occurring oestrogens in food. *Pure and Applied Chemistry* 70(9): 1759–1776.

McGhie, T. K., Hunt, M., Barnett, L. E. 2005. Cultivar and growing region determine the antioxidant polyphenolic concentration and composition of apples grown in New Zealand. *Journal of Agricultural and Food Chemistry* 53: 3065–3070.

Mikulic-Petkovsek, M., Stampar, F., Veberic, R. 2008. Increased phenolic content in apple leaves infected with the apple scab pathogen. *The Journal of Plant Pathology* 90(1): 49–55.

Minnocci, A., Iacopini, P., Martinelli, F., Sebastiani, L. 2010. Micromorphological, biochemical, and genetic characterization of two ancient, late-bearing apple varieties. *European Journal of Horticulture Science* 75: 1–7.

Moreau, R. A., Whitaker, B. D., Hicks, K. B. 2002. Phytosterols, phytostanols, and their conjugates in foods: Structural diversity, quantitative analysis, and healthpromoting uses. *Progress in Lipid Research* 41: 457–500.

Moreno, Y. S., Bustos, F. M., Hernández, M. S., Paczka, R. O., Arellano Vázquez, J. L. 2003. Efecto de la nixtamalización sobre las antocianinas del grano de maíces pigmentados. *Agrociencia* 37(6): 617–628.

Mullen, W., Stewart, A. J., Lean, M. E. J., Gardner, P., Duthie, G. G., Crozier, A. 2002. Effect of freezing and storage on the phenolics, ellagitannins, flavonoids, and antioxidant capacity of red raspberries. *Journal of the Agricultural and Food Chemistry* 50(18): 5197–5201.

Murray, J. R., Smith, A. J., Hackett, W. P. 1994. Differential dihydroflavonol reductase transcription and antho-cyanin pigmentation in the juvenile and mature phases of ivy (*Hedera helix* L.). *Planta* 194: 102–109.

Musacchi, S., Serra, S. 2018. Apple fruit quality: Overview on pre-harvest factors. *Scientia Horticulturae* 234: 409–430.

Napolitano, A., Cascone, A., Graziani, G., Ferracane, R., Scalfi, L., Vaio, C. D., Ritieni, A., Fogliano, V. 2004. Influence of variety and storage on the polyphenol composition of apple flesh. *Journal of the Agricultural and Food Chemistry* 52: 6526–6531.

Nunes, M. C. N., Brecht, J. K., Morais, A. M., Sargent, S. A. 2005. Possible influences of water loss and polyphenol oxidase activity on anthocyanin content and discolouration in fresh ripe strawberry (cv Oso Grande) during storage at 1°C. *Journal of Food Science* 70(1): S79–S84.

Palmer, J. W., Lozano, L., Chagné, D., Volz, R., Lin-Wang, K., Bonany, J., Micheletti, D., Troggio, M., White, A., Kumar, S., Allan, A. C., Iglesias, I. 2012. *Physiological, Molecular and Genetic Control of Apple Skin Colouration Under Hot Temperature Environments.* In XXVIII International Horticultural Congress on Science and Horticulture for People (IHC2010): International Symposium on 929, 81–87.

Palmer, J. W., Prive, J. P., Tustin, D. S. 2003. Temperature. In *Apples: Botany, Production and Uses.* Ferree, D. C., Warrington, I. J. (Eds.). Cabi Publishing, Wallingford, pp. 217–236.

Pena, M. E., Artes-Hernandez, F., Aguayo, E. 2013. Effect of sustained deficit irrigation on physicochemi-cal properties, bioactive compounds and postharvest life of pomegranate fruit (cv.'Mollar de Elche'). *Postharvest Biology and Technology* 86: 171–180.

Phillips, K. P., Tanphaichitr, N. 2008. Human exposure to endocrine disrupters and semen quality. *Journal of Toxicology and Environmental Health Part B* 11(3): 188–220.

Philpott, M., Gould, K. S., Markham, K. R., Lewthwaite, S. L., Ferguson, L. R. 2003. Enhanced colouration reveals high antioxidant potential in new sweet potato cultivars. *Journal of the Science of Food and Agriculture* 83(10): 1076–1082.

Portu, J., Lopez, R., Baroja, E., Santamaría, P., Garde-Cerdan, T. 2016. Improvement of grape and wine phe-nolic content by foliar application to grapevine of three different elicitors: methyl jasmonate, chitosan, and yeast extract. *Food Chemistry* 201: 213–221.

Pramanik, B. K., Matsui, T., Suzuki, H., Kosugi, Y. 2004. Changes in activities of sucrose synthase and sucrose phosphate synthase and sugar content during postharvest senescence in two broccoli cultivars. *Asian Journal of Plant Science* 3: 398–402.

Preciado-Rangel, P., Reyes-Perez, J. J., Ramírez-Rodríguez, S. C., Salas-Pérez, L., Fortis-Hernández, M., Murillo-Amador, B., Troyo-Diéguez, E. 2019. Foliar aspersion of salicylic acid improves phenolic and flavonoid compounds, and also the fruit yield in cucumber (*Cucumis sativus* L.). *Plants* 8: 44.

Radenkovs, V., Juhnevica-Radenkova, K. 2017. Effect of storage technology on the chemical composition of apples of the cultivar 'Auksis'. *Zemdirbyste-Agriculture* 104(4): 359–368.

Rafiq, S., Kaul, R., Sofi, S. A., Bashir, N., Nazir, F., Nayik, G. A. 2018. Citrus peel as a source of functional ingredient: A review. *Journal of the Saudi Society of Agricultural Sciences* 17(4): 351–358.

Reay, P. F. 1999. The role of low temperatures in the development of the red blush on apple fruit ('Granny Smith'). *Scientia Horticulturae* 79: 113–119.

Reay, P. F., Fletcher, R. H., Thomas, V. J. 1998. Chlorophylls, carotenoids and anthocyanin concentrations in the skin of 'Gala' apples during maturation and the influence of foliar applications of nitrogen and magnesium. *Journal of the Science of Food and Agriculture* 76: 63–71.

Reay, P. F., Lancaster, J. E. 2001. Accumulation of anthocyanins and quercetin glycosides in 'Gala' and 'Royal Gala' apple fruit skin with UV-B – visible irradiation: Modifying effects of fruit maturity, fruit side, and temperature. *Scientia Horticulturae* 90: 57–68.

Rempelos, L., Cooper, J., Wilcockson, S., Eyre M., Shotton, P., Volakakis, N., Orr, C. H., Leifert, C., Gatehouse, A. M. R., Tétard-Jones, C. 2013. Quantitative proteomics to study the response of potato to contrasting fertilisation regimes. *Molecular Breeding* 31: 363–378.

Renard, C. M. G. C., Dupont, N., Guillermin, P. 2007. Concentrations and characteristics of procyanidins and other phenolics in apples during fruit growth. *Phytochemistry* 68: 1128–1138.

Reščič, J., Mikulič-Petkovsek, M., Stampar, F., Zupan, A., Rusjan, D. 2015. The impact of cluster thinning on fertility and berry and wine composition of'Blauer Portugieser'(*Vitis vinifera* L.) grapevine variety. *Oeno One* 49: 275–291.

Rickman, J. C., Barrett, D. M., Bruhn, C. M. 2007. Nutritional comparison of fresh, frozen and canned fruits and vegetables. Part 1. Vitamins C and B and phenolic compounds. *Journal of the Science of Food and Agriculture* 87(6): 930–944.

Rudell, D. R., Buchanan, D. A., Leisso, R. S., Whitaker, B. D., Mattheis, J. P, Zhu, Y., Varanasi, V. 2011. Ripening, storage temperature, ethylene action, and oxidative stress alter apple peel phytosterol metabolism. *Phytochemistry* 72: 1328–1340.

Rudell, D. R., Serra, S., Sullivan, N., Mattheis, J., Musacchi, S. 2017. Survey of 'Anjou' pear metabolic profile following harvest from different canopy positions and fruit tissues. *HortScience* 52(11): 1–10.

Sacchetti, G., Cocci, E., Pinnavaia, G., Mastrocola, D., Dalla Rosa, M. 2008. Influence of processing and storage on the antioxidant activity of apple derivatives. *International Journal of Food Science and Technology* 43(5): 797–804.

Santarelli, V., Nerri, L., Sacchetti, G., Carla, D., Mattia, D., Mastrocola, D., Pittia, P. 2020. Response of organic and conventional apples to freezing and freezing pre-treatments: Focus on polyphenols content and antioxidant activity. *Food Chemistry* 308: 125570.

Saure, M. C. 1990. External control of anthocyanin formation in apple: A review. *Scientia Horticulturae* 42: 181–218.

Schuler, I., Duportail, G., Glasser, N., Benveniste, P., Hartmann, M.-A. 1990. Soybean phosphatidylcholine vesicles containing plant sterols: A fluorescence anisotropy study. *Biochimica et Biophysica Acta* 1028: 82–88.

Serra, S., Leisso, R., Giordani, L., Kalcsits, L., Musacchi, S. 2016. Crop load influences fruit quality, nutritional balance, and return bloom in 'Honeycrisp' apple. *HortScience* 51(3): 236–244.

Shafiq, M., Singh, Z. 2018. Pre-harvest spray application of phenylpropanoids influences accumulation of anthocyanin and flavonoids in 'Cripps Pink' apple skin. *Scientia Horticulturae* 233: 141–148.

Shutt, D. A., Cox, R. I. 1972. Steroid and phyto-oestrogen binding to sheep uterine receptors in vitro. *Journal of Endocrinology* 52(2): 299–310.

Shvarts, M., Borochov, A., Weiss, D. 1997. Low temperature enhances petunia flower pigmentation and induces chalcone synthase gene expression. *Physiologia Plantarum* 99(1): 67–72. https://doi.org/ https://doi.org/10.1111/j.1399-3054.1997.tb03432.x. https://onlinelibrary.wiley.com/doi/abs/10.1111/ j.1399-3054.1997.tb03432.x

Silvanini, A., Dall'Asta, C., Morrone, L., Cirlini, M., Beghè, D., Fabbri, A., Ganino, T. 2014. Altitude effects on fruit morphology and flour composition of two chestnut cultivars. *Scientia Horticulturae* 176: 311–318.

Slatnar, A., Mikulic-Petkovsek, M., Halbwirth, H., Stampar, F., Stich, K., Veberic, R. 2012. Polyphenol metabolism of developing apple skin of a scab resistant and a susceptible apple cultivar. *Trees – Structure and Function* 26(1): 109–119.

Spengler, R. N. 2019. Origins of the apple: The role of megafaunal mutualism in the domestication of Malus and rosaceous trees. *Frontiers in Plant Science* 10: 617.

Średnicka-Tober, D., Barański, M., Kazimierczak, R., Ponder, A., Kopczyńska, K., Hallmann, E. 2020. Selected antioxidants in organic vs. conventionally grown apple fruits. *Applied Sciences* 10(9): 2997.

Stopar, M., Bolcina, U., Vanzo, A., Vrhovsek, U. 2002. Lower crop load for Cv. Jonagold apples (Malus × domestica Borkh.) increases polyphenol content and fruit quality. *Journal of Agricultural* and *Food Chemistry* 50: 1643–1646.

Stracke, B. A., Rufer, C. E., Weibel, F. P., Bub, A., Watzl, B. 2009. Three-year comparison of the polyphenol contents and antioxydant capacities in organically and conventionally produced apples (*Malus domestica* Bork. Cultivar 'Golden Delicious). *Journal of Agricultural and Food Chemistry* 57: 4598–4605.

Takos, A. M., Ubi, B. E., Robinson, S. P., Walker, A. R. 2006. Condensedtanninbiosynthesis genes are regulated separately from other flavonoid biosynthesis genes in apple fruit skin. *Plant Science* 170: 487–499.

Tatsuki, M., Hayama, H., Yoshioka, H., Nakamura, Y. 2011. Cold pre-treatment is effective for 1-MCP efficacy in 'Tsugaru' apple fruit. *Postharvest Biology and Technology* 62: 282–287.

Tétard-Jones, C., Edwards, G. M., Rempelos, L., Gatehouse, A. M. R., Eyre, M., Wilcockson, S. J., Leifert, C. 2013. Effects of previous crop management, fertilization regime and water supply on potato tuber proteome and yield. *Agronomy* 3: 59–85.

Tomas-Barberan, F. A., Espin, J. C. 2001. Phenolic compounds and related enzymes as determinants of quality in fruits and vegetables. *Journal of the Science of Food and Agriculture* 81: 853–876.

Topalovic, A., Slatnar, A., Štampar, F., Knezevic, M., Veberic, R. 2011. Influence of foliar fertilization with P and K on chemical constituents of grape cv.'Cardinal'. *Journal of Agriculture and Food Chemistry* 59: 10303–10310.

Tromp, J. 1997. Maturity of apple cv. Elstar as affected by temperature during a six week period following bloom. *Jouranl of Horticultural Science* 72: 811–819.

Tsao, R., Yang, R., Xie, S., Sockovie, E., Khanizadeh, S. 2005. Which polyphenolic compounds contribute to the total antioxidant activities of apple? *Journal of Agriculture and Food Chemistry* 53: 4989–4995.

Ubi, B. E. 2004. The genetics of anthocyanin reddening in apple fruit skin. *Jouranl of Food, Agriculture and Environment* 2: 163–165.

Ubi, B. E., Honda, C., Bessho, H., Kondo, S., Wada, M., Kobayashi, S., Moriguchi, T. 2006. Expression analysis of anthocyanin biosynthetic genes in apple skin: Effect of UV-B and temperature. *Plant Science* 170(3): 571–578.

Valverde, J. M., Giménez, M. J., Guilléna, F., Valeroa, D., Martínez-Romeroa, D., Serrano, M. 2015. Methyl salicylate treatments of sweet cherry trees increase antioxidant systems in fruit at harvest and during storage. *Postharvest Biology and Technology* 109: 106–113.

Van der Sluis, A. A., Dekker, M., de Jager, A., Jongen, W. M. 2001. Activity and concentration of polyphenolic antioxidants in apple: Effect of cultivar, harvest year, and storage conditions. *Journal of Agriculture and Food Chemistry* 49: 3606–3613.

Vanoli, M., Buccheri, M. 2012. Overview of the methods for assessing harvest maturity. *Stewart Postharvest Review's* 8(1): 1–11.

Veberic, R., Trobec, M., Herbinger, K., et al. 2005. Phenolic compounds in some apple (Malus domestica Borkh) cultivars of organic and integrated production. *Journal of the Science of Food and Agriculture* 85: 1687–1694.

Veronica, S., Lilia, N., Giampiero, S., Di Mattia, C. D., Mastrocola, D., Pittia, P. 2020. Response of organic and conventional apples to freezing and freezing pre-treatments: focus on polyphenols content and anti-oxidant activity. *Food Chemistry* 308(5): 125570.

Vrhovsek, U., Rigo, A., Tonon, D., Mattivi, F. 2004. Quantitation of polyphenols in different apple varieties. *Journal of Agricultural and Food Chemistry* 52(21): 6532–6538.

Wasserman, S. I. 2011. Approach to the person with allergic or immunologic disease. In *Cecil Medicine*. 24th ed. Philadelphia, PA: Saunders Elsevier, p. 257.

Wei, H., Bowen, R., Cai, Q., Barnes, S., Wang, Y. 1995. Antioxidant and antipromotional effects of the soybean isoflavone genistein. *Proceedings of the Society for Experimental Biology and Medicine* 208(1): 124–130.

Whale, S. K., Singh, Z. 2007. Endogenous ethylene and colour development in the skin of 'Pink Lady' apple. *Journal of the American Society Horticultural Science* 132: 20–28.

Whitaker, B. D., Klein, J. D., Conway, W. S., Sams, C. E. 1997. Influence of prestorage heat and calcium treatments on lipid metabolism in 'Golden Delicious' apples. *Phytochemistry* 45: 465–472.

Wojciechowski, Z. A. 1991. Biochemistry of phytosterol conjugates. In: Patterson, G. W., Nes, W. D. (Eds.), *Physiology and Biochemistry of Sterols. The Journal of the American Oil Chemists' Society Champagne*: 361–395.

Wojdyło, A., Oszmianski, J., Laskowski, P. 2008. Polyphenolic compounds and antioxidant activity of new and old apple varieties. *Journal of Agriculture and Food Chemistry* 56: 6520–6530.

Woyengo, T. A., Ramprasath, V. R., Jones, P. J. H. 2009. Anticancer effects of phytosterols. *European Journal of Clinical Nutrition* 63: 813–20.

Wozniak, L., Szakiel, A., Paczkowski, C., Marszałek, K., Skąpska, S., Kowalska, H., Jędrzejczak, R. 2018. Extraction of triterpenic acids and phytosterols from apple pomace with supercritical carbon dioxide: Impact of process parameters, modelling of kinetics, and scaling-up study. *Molecules* 23(11): 2790.

Xing, R.-R., He, F., Xiao, H.-L., Duan, C.-Q., Pan, Q.-H. 2015. Accumulation pattern of flavonoids in Cabernet Sauvignon grapes grown in a low-latitude and high-altitude region. *South African Journal of Enology and Viticulture* 36: 32–43.

Yuri, J. A., Neira, A., Quilodran, A., Motomura, Y., Palomo, I. 2009. Antioxidant activity and total phenolics concentration in apple peel and flesh is determined by cultivar and agroclimatic growing regions in Chile. *Journal of Food, Agriculture and Environment* 7: 513–517.

Zoratti, L., Jaakola, L., Häggman, H., Giongo, L. 2015. Anthocyanin profile in berries of wild and cultivated *Vaccinium* spp. along altitudinal gradients in the Alps. *Journal of Agricultural and Food Chemistry* 63: 8641–8650.

18 Apples and Apple By-Products
Properties and Health Benefits

Asif Ahmad and Chamman Liaqat

CONTENTS

18.1 INTRODUCTION

Fruits are an essential part of a daily diet and play a crucial role in maintaining good health. Apples are considered one of the juiciest, most nutritious, and highly enriched fruit among all fruits, and they consumed by people of every age group. Moreover, in the list of fruit consumption, it's ranked fourth (Zhang et al., 2021). Apples are mainly cultivated in tropical, subtropical, and temperate regions (Musacchi & Serra, 2018). According to the Food and Agriculture Organization (FAO), the worldwide mass production of apples was over 87 metric tons in 2019 while in the 1990s the global production of apples was just 47 metric tons (FAO, 2021). Apple (*Malus domestica*) belongs to the *Rosaceae* family. This family also includes other fruits like plums, cherries, pears, apricots, and peaches, but in the juice industry, among all fruits, apple is the most significant raw material (Bolarinwa et al., 2015). Apple is an excellent source of nutritional components, mainly fiber and polyphenols. Different classes of phenolic compounds have been evaluated in apple tissues. Procyanidin (catechin polymers and oligomers) and flavanol monomers normally consist of more than 80% of the total phenolic content. It is followed by the hydroxycinnamic acids, flavonols, and dihydrochalcones (Li et al., 2019). In commercial-scale production of apple juice, 75% of the total processed apples represent the raw juice while the remaining 25% is in the pomace. The pomace of apple is the main by-product of the juice industry (Vukušić et al., 2021). It is commonly considered a waste and thrown away which results in different health hazards and also causes environmental problems. In recent years, different research on apple pomace has shown that apple pomace is an excellent source of nutritional components such as carbohydrates, phenolic components, many minerals, and dietary fiber (Lyu et al., 2020). All these nutritional components can be recovered from apple pomace, and pomace can also be used in different kinds of products to enhance their economic values and health benefits

DOI: 10.1201/9781003239925-18

Apples are principally composed of 85% water; 14% carbohydrates; sugar, mainly fructose; and dietary fiber. They are also the excellent source of minerals, vitamins (chiefly vitamin C and vitamin E), and polyphenols. But, apples lack an anti-nutritional component (phytic acid) due to which the incorporation of minerals like calcium/zinc/iron get affected and result in the deficiency of minerals (Bondonno et al., 2017; Sudha et al., 2016). The main micronutrients found in apples are calcium, sodium, potassium, and magnesium. Trace elements like zinc, copper, iron, manganese, and selenium are also found (Wilczyński et al., 2020). Apples are not just consumed whole as a fruit, but different products also have been made from them. The principal food products that have been made from them are baby foods, apple juices, apple butter, cider, vinegar, jelly, and canned fresh apple cubes or slices (Sudha et al., 2016). Two types of apple juices are primarily prepared by the juice industries: clear and cloudy. Cloudy apple juice usually has low sugar, high polyphenol, and pectin as compared to clear juice, but in comparison to whole apples, it has much lower polyphenol and pectin content (Bondonno et al., 2017).

The distribution and nature of phytochemicals vary between the different apple parts. The apple flesh contains caffeic acid, catechins, phloridzin, phloretin glycosides, and procyanidins; however, the peel of the apple contains all these phytochemical compounds, and the peel also contains some flavonoids like cyaniding and quercetin glycosides (Shehzadi et al., 2020). The apple pomace is categorized by nutritional compounds like fiber and different phytochemicals, including many phenolic compounds like catechin, quercetin, chlorogenic, and gallic acid. All of these substances can decrease prolonged health ailment risks (Kruczek et al., 2021). Moreover, many researchers have found that apple pomace improved antioxidant activity, gastrointestinal activity, and lipid metabolism and also showed a positive effect on different metabolic ailments like insulin resistance and hyperglycemia, etc. (Skinner et al., 2018). Similarly, seeds are also considered valuable apple by-products. In some studies, it was observed that fresh apples constitute up to 0.7% of seeds in certain varieties. Also, the normal seed oil in apples is among 15% to almost 22%, but it can go as high as 29%. Recent research on apple seeds show that they contain strong antimicrobial, antiproliferative and antioxidant properties. Moreover, in cosmetic industries seed oil is used in several products like soaps, body lotion, perfumes, conditioners, skin cream, lip balm, butter, shampoos, and toothpaste (Montañés et al., 2018). Furthermore, the apple peels in the fruit industry is also considered wastage, but peels are also a great source of different phytochemical compounds, essentially phenolic compounds. The main polyphenols that are present in the peel of apples are quercetin, procyanidin, trimer epicatechin, glycosides, chlorogenic acid, trimer, etc. Due to the presence of phytochemicals, apple peel holds anti-inflammatory properties, anti-carcinogenic, antioxidant, and antiproliferative characteristics. Polyphenols also have radical scavenging properties. Apple peels are also used in controlling different diseases like inflammatory bowel disease, cancer, diabetes, arthritis, cardiovascular diseases, etc. (Shehzadi et al., 2020). Likewise, the juice of apples are considered a commonly treated food all over the world. The main components that involved the quality of apple juice are phenolic compounds, sugars, different pigments, acids, and some preserving compounds, but the level of these components is greatly influenced by the apple juice production process in the industry (Ramos-Aguilar et al., 2017). So it has been observed that apples and apple products are an excellent source of bioactives and phytochemicals. In Figure 18.1, different properties of apple parts have been explained.

18.1 PROPERTIES OF FRUIT PULP

Apple pulp is a great source of nutritional components. A study was carried out to check the composition of custard apples. The chemical composition of custard apples was as follows: [Moisture (74%), crude protein (2.8%), sugar (22.7%), carbohydrates (21.5%), ash (1.05%), fat (0.39%), acidity (0.63%) and crude fiber (3.3%), iron (0.47 mg/100 g), calcium (22 mg/100 g) and phosphorus (25 mg/100 g)] (Srivastava et al., 2017). A study was conducted for the characterization of fatty acids, phenolics, antioxidant potential, and amino acids in 'Himalayan' apple pulp and seeds. The concentration of

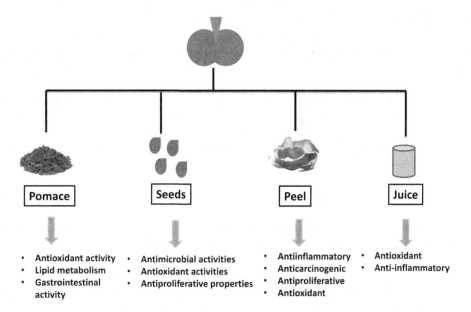

FIGURE 18.1 Properties and health benefits of different apple parts.

phloridzin (83.0 µg/mg) and phloretin (88.3 µg/mg) were observed to be greater compared to other phenolic components measured by using ultra-performance liquid chromatography (UPLC). Apple pulp showed great antioxidant properties as compared to the seed part. While high-performance liquid chromatography (HPLC) for amino acids showed the following amino acids: [Cysteine (76.8 µg/mg), serine (9.0 µg/mg), tyrosine (10.3 µg/mg) and alanine (8.0 µg/mg)]. The high-performance anion-exchange chromatography technique showed the sucrose and fructose in a fair quantity of 17.3 mg/g and 21 mg/g. It was concluded that apple fruits are an excellent source of nutritional components and phenolic components (Dadwal et al., 2018). Similarly, in another research study fresh, cell wall, and freeze-dried samples of apples were used for the determination of structural, rheological, and chemical properties by using attenuated total reflectance Fourier transform spectroscopy (ATR-FTIR). The basic objective was to prepare the puree samples according to the expected quality properties. For this purpose, 36 apple sets were used, while 72 purees were used to construct models including multivariate calibration, exploratory investigation, and supervised organization. The spectroscopy techniques on purees predicted the textural characteristics of particle average volume and size, while freeze drying improved the rheological properties and assessment of chemical properties using least squares regression (Lan et al., 2020). Moreover, a study was conducted to check the concentration of ascorbic acid (vitamin C) in peel, pulp, and juices obtained from 64 apple varieties. For the determination of ascorbic acid content, high-performance liquid chromatography (HPLC) was used. It was observed that old varieties showed the highest amount of ascorbic acid in peel and pulp, and the red-fleshed varieties maintained the most of ascorbic acid amount during processing. It was concluded that red-fleshed and old varieties could be used for nutritional juice and processed food components (Bassi et al., 2017).

In the present years, many pieces of research have been performed to study the potential health properties of fruit pulp. A study was conducted to check the biological properties of ethanol based extract from 'Mela Rosa Marchigiana' in the culture of apple pulp. In relation to antioxidant potential, the extract showed the ABTS assay and DPPH activity as 67% and 39%, respectively. The genoprotective effect of pulp culture was examined by using the DNA assay, which showed a prominent defense of up to 70%. Also, the anti-inflammation response was observed at 0.5 mg/ml through the discharge of nitric oxide using bacterial RAW/LPS cells. By observing the biological effect of callus, it was suggested that it could be used in food and cosmetic industries along with medicinal industries for the production of nutraceuticals (Potenza et al., 2020). Moreover, in another research extraction, characterization, purification, isolation of phenolic components, and antioxidant characteristics of 5-hydroxymaltol from apple pulps were carried out. The high-performance liquid chromatography study showed the existence of 16 components including the main components of fumaric acid, vanillic acid, 4-hydroxybenzoic acid, morine, and gentisic. Also, the antioxidant properties were examined with reducing power, metal chelating, total antioxidant potential, inhibition of lipid peroxidation, H_2O_2, and free radical scavenging properties. It was concluded that apple pulps extracted with ethyl acetate and 5-HM have strong antioxidant properties and are suitable to apply in vivo tests (Demirci et al., 2018).

18.2   PROPERTIES OF APPLE PEEL

An apple peel almost contains one-tenth of an apple mass and is usually thrown away as waste. It contains flavonoid compounds like quercetins and anthocyanins; in addition the main phytochemicals that are present in apple peels are polyphenols including anthocyanidins, flavan-3-ols, small quantities of phenolic acids, and flavonols dihydrochalcones. Chlorogenic acid, procyanidins, phloridzin, quercetin, and epicatechin are also present in apple peels (El-sayed & El-Shemy, 2021; Riaz et al., 2020). In addition, apple peels are also a rich source of minerals, fiber, and polyphenols, and apple peel powder is used in the manufacturing of different bakery items (Nakov et al., 2020).

In recent years different research has been conducted to study the different properties of apple peels. The research was conducted to develop the chitosan-based food film by the incorporation of apple peel phenolic compounds to study their antimicrobial properties. Scanning electron microscopy, thermogravimetric analyzer, and Fourier transform infrared spectroscopy technique were carried out to check the thermal stability, possible interaction, and structure of chitosan-based films. Results showed that the films with 0.50% apple peel polyphenols exhibited amazing antimicrobial and mechanical properties. It was concluded that apple peels could be used as food packaging material with strong antimicrobial properties in the food industry (Riaz et al., 2018). Likewise, in a similar research study, citric acid red peel was utilized in the manufacturing of multifunctional smart packaging of a food item. The characterization of the film was carried out by using Fourier transform infrared spectroscopy, mechanical and dynamic analyzer, scanning electron microscopy, and thermogravimetric analyzer. After the film preparation, it was applied to original samples like cheese and pasteurized milk. These films showed strong antimicrobial properties. It was concluded that citric acid red peel is very beneficial in the production of biomedical materials and food packaging along with many other applications in the food industry (Alpaslan et al., 2021). In addition, a study was carried out to check the significance of CUPRAC, FRAP, and DPPH assays for the determination of antioxidant properties in apple fruit extracts. For this purpose, 13 apple cultivars were used to analyze their total flavonoid compounds, ascorbic acid (vitamin C), total phenolic components, and anthocyanins in methanol extracts of both cortex and peel fractions. The total phenolic was higher in peels at 2.8 times and also had a higher flavonoid content of 2.68% as compared to the cortex. Principle analysis can efficiently explain 84.27% and 76.86% variability in antioxidant assays in the cortex and peel parts of apple varieties, correspondingly. Based on the different analysis and the concept of relative antioxidant capacity index (RACI), the ferric reducing ability of plasma (FRAP) antioxidant assessment is highly suggested for the determination of antioxidant properties in apples (Sethi et al., 2020).

Apple peel powder has also been utilized in the development of different products and their potential effects on the health of human beings have also been investigated. For example, a research study was conducted to check the effect of apple peel powder on the nutritional and physicochemical characteristics of wheat cookies. For this purpose, six different kinds of wheat cookies with continuously increasing peel powder percentages were developed. Results showed that wheat cookies had considerably high lipid, moisture, fiber, ash, antioxidant property, and total polyphenols as compared to the control wheat cookies. From the six percentages, 24% addition of peel powder produced the cookies with maximum good quality (Nakov et al., 2020). Moreover, in another study, microencapsulation of phenolic components extracted from the apple peel was performed, and then it was used in yogurt. Polyphenol components from apple peel were extracted by using 80% ethanol. For the encapsulation of polyphenolic components, physical methods were used. The coating agents like whey protein, gum arabic mixture, and maltodextrin were used and homogenized by ultrasonication and Ultra Turrax. Then encapsulated polyphenolic compounds were added to the yogurt, but they did not have any prominent effect on the texture and physicochemical properties of the yogurt (El-Messery et al., 2019).

18.3   PROPERTIES OF APPLE SEEDS

Apple seeds are mainly composed of amino acids, fat, and protein; they also contain cryogenic glycoside amygdalin. Apple seeds are considered waste after the production of apple juice and become part of apple pomace. Generally, apple seeds have a slightly bad effect due to the incidence of glycoside amygdalin. It is present in a relatively small amount, as 0.6 mg g^{-1} of dry seeds. It is unable to cause toxic effects unless a great quantity of seeds has been taken (Chatzimarkou et al., 2018; Khalil & Mustafa, 2020). A study on apple seed oils showed that it mainly contains unsaturated fatty acids (90%) belonging to omega series [oleic acid (33.4%), linolenic acid (1.2%), and linoleic acid (55%)]. Oils also have great levels of tocopherols (total 1280 mg/kg tocopherols) with beta-tocopherol (794% mg/kg) being the richest followed by alpha-tocopherol (439 mg/kg). These compounds are mainly related to DPPH scavenging activity. In defatted apple seeds, there is 20.3% fiber and 37.5% protein (Madrera et al., 2018). Moreover, it has been observed that beta-glucan is the most valuable component that can easily be extracted from natural foods. It plays an important role in controlling different body functions such as control of postprandial glucose level, lowering cholesterol level, and immune modulation effects. In the food production industry, it could be used for the development of different nutraceutical food products that can help improve physiological functions (Ahmad & Kaleem, 2018).

Different studies have been carried out in recent years to study the potential and beneficial uses of apple seeds. For example, apple seeds were used for the synthesis of carbon nanodots for the recognition of 4-nitrophenol (4-NP) in human urine and environmental water. It was observed that carbon nanodots could discharge intense blue light that can efficiently be quenched by 4-nitrophenol. For the detection of this polyphenol, the Forster resonance energy transfer (FRET) method was used. The basic advantages of this technique were the low-value cost of analysis and synthesis and particular sensing. It was concluded that carbon nanodots (CNDs) have the ability and potential for biosensing and bioimaging applications (Chatzimarkou et al., 2018). Moreover, in other research, coumarins, compounds from 'Delicious' red apple seeds, were extracted and their phytochemical and antimicrobial properties were studied. For the extraction purpose, four different kinds of solvents were used: water, n-hexane, methanol, and chloroform. Ultrasound, microwaves, and kinetic maceration techniques were used for the removal of coumarins from seeds. The antibacterial activities of extracts against the following strains were studied: *Klebsiella pneumonia*, *Escherichia coli*, *Pseudomonas aeruginosa,* and *Haemophilus influenza*. Results showed that extracted furanocoumarins have strong antimicrobial potential against the tested microbes. It was concluded that these coumarins can be useful for the development of different antimicrobial compounds (Mohammed & Mustafa, 2020). Furthermore, in another study, apple seeds were subjected to the super critical carbon dioxide (CO_2) extraction process and comprehensively categorized to study the effects of the supercritical extraction procedure on their nutritional value and physicochemical properties. The thermal breakdown of apple seeds was performed by thermogravimetric analyzer attached with a Fourier transform infrared (FTIR) spectrometer for the analysis of evolved gas to evaluate their energetic potential. The results of this study showed that lipid fraction was the main component of the extracted sample (Paini et al., 2021). Likewise, a study was conducted for the evaluation of apple seed flour for its nutraceutical and structural properties. The antimicrobial, antioxidant, anti-obesity and anti-diabetic properties of 'Golden'/'Red Delicious' apple seed varieties were primarily studied. The morphological study of flours showed starch granules of different dimensions and shapes with agglomerated protein forms. X-ray diffraction (XRD) showed high crystallinity, while Fourier transforms infrared spectroscopy (FTIR) showed high bands intensity at 1639 cm−1 and 1528 cm−1 for 'Red Delicious' flour. High phenolic content in 'Red Delicious' apple flour was observed. Also, both flours revealed strong antimicrobial properties against *Staphylococcus aureus* and *Bacillus subtilis* (Manzoor et al., 2021). In Figure 18.2, the polyphenolic substances and health benefits of apple seed flour have been explained.

Different research studies have been conducted to study the beneficial effect of apple seeds on health and body regulatory function. A study on rats was conducted to check the dietary and beneficial health effects of a normal diet with apple seed meal. For this purpose, rats were divided into three distinct groups. In amygdalin and control groups, casein and cellulose were the main sources of protein and dietary fiber, while in the seed meal group protein and fiber extracted from the endosperms of seed were used. After the 14 days of diet feeding, it was observed that rats fed with apple seed meal lost body weight. Also, it was observed that the apple seed meal diet increased the high lipoprotein cholesterol concentration and antioxidant capability of water-soluble compounds. It was concluded that apple seed meals have comparatively low nutritional value (Opyd et al., 2017). Moreover, defatted apple seed cakes were used for the fortification of wheat bread. The apple seed oil cake was prepared from the cold-pressed method and added up to 20% in an amount to wheat flour for the enrichment of bread. After that, the antioxidant, chemical, textural, and sensory properties of wheat bread were evaluated. It was observed that by adding 5% and 20% defatted seed cake to the total wheat flour of three varieties as 'Idared', 'Golden Delicious' and 'Sumatovka' predominantly increased the protein content and insoluble fibers tested wheat bread samples, mainly in the samples accompanied with 20% defatted seed cake of 'Sumatovka' (15.9% and 7.21% respectively). It was concluded that the percentages of apple seed cake and varieties greatly influenced the tested characteristics of enriched wheat bread. So these enriched breads can play an important role in

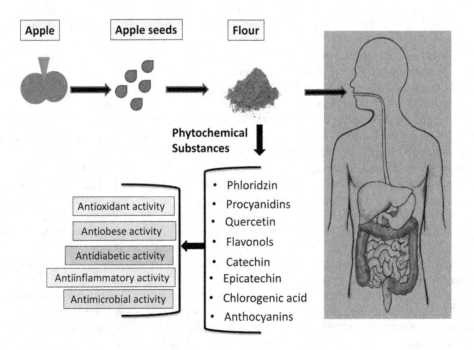

FIGURE 18.2 Polyphenolic substances and health benefits of apple seed flour.

fulfilling the nutritional requirements of the body (Purić et al., 2020). Furthermore, a study was conducted in which defatted seeds were incorporated in chewing bubble gum and dissolution kinetics of phloridzin was studied. Chewing gum is commonly consumed, and it has great potential for the delivery of bioactive substances. As well, its different formulations have also been studied for different plant materials and many drugs to improve their bioavailability and releasing power. According to the results, ABTS scavenging activity, total phenolic compounds (TPC), and DPPH radical scavenging assays of flours were determined between 291–391 µmol Trolox/g, 2861 and 5141 mg GAE/kg and 21–43 µmol, respectively. It was also studied that amount of phloridzin represented 52% to 67% and 75% to 83% of the total polyphenolic content that was calculated by Folin-Ciocalteu and high-performance liquid chromatography, respectively. It was determined that for the delivery of phloridzin, chewing gum can be suitable delivery material (Gunes et al., 2019).

18.4   PROPERTIES OF APPLE JUICE

Due to the amazing health benefits and organoleptic properties, the juice of apples is considered the most delicious and common among juices. For apple juice, cloud stability, flavor, and color are the most prominent sensory properties that determine the acceptability and enjoyment of the consumer. These properties are susceptible to different changes during the processing of juice and storage due to biochemical reactions. In recent years, customers have been demanding minimally processed juices with a fresh flavor and natural appearance. For this reason, novel technologies, pulsed electric field, and high pressure processing have gained much attention (Lee et al., 2017). In a study, organoleptic properties, proximate composition, and vitamin C of fresh apple juice were examined. The composition of the apple revealed that it contains 87.73% moisture content with 0.13% lipid, 1.21% fiber, and 0.30% moisture content. Apple juice was also an excellent source of vitamin C with 22.15/100 mg concentration. It was also observed that organoleptic properties like taste, color, and flavor of the apple juice with different preservatives like sodium benzoate, garlic, and ginger were acceptable at refrigerator temperature (4°C) for almost 8 days storage time (Okokon et al., 2019).

Different studies have been conducted to check the antioxidant, anti-inflammatory, and antimicrobial properties of apple juice. A study was conducted to check the effect of cryoconcentration on the antioxidant property and polyphenol compounds in apple juice. The results showed that the levels of polyphenol components significantly increased with every freeze concentration cycle. It was also found that the concentration of phenolic components increased in the first (1.9 times), second (2.9 times), and third (3.8 times) concentration steps. The antioxidant ability of apple juice was also increased with each freezing step. It was concluded that the antioxidant activity of apple juice can be preserved by using cryoconcentration (Zielinski et al., 2019). Moreover, in another study, the effect of four different lactic acid bacteria (LAB) was evaluated on the flavor of fermented apple juice. The overall aromatic components, concentrations of organic compounds, and bacterial populations were evaluated during fermentation and after different days at refrigeration storage period. All four lactic acid bacteria strains *L. Plantarum, Lactobacillus acidophilus, L. casei, and L. rhamnosus* showed the amazing capability for growth in apple juice. *L. rhamnosus* and *L. casei* produced higher concentrations of acids and also showed greater sustainability as compared to other strains. Fermented juices were further categorized by the presence of 53 volatile components for aromatic profiles (Chen et al., 2019). Furthermore, primary and secondary metabolite substances were used as a tool for the distinction of apple juice regarding geographical area and cultivar. Three cultivars were used: 'Idared', 'Topaz', and 'Golden Delicious' from five different states in Slovenia: Dinaric, Mediterranean, Alpine, Submediterranean, and Pannonian. The minor phenolic substances were confirmed as consistent markers of apple juice including flavonols compounds (quercetin-3-glucoside + quercetin-3-galactoside and quercetin-3-rhamnoside) and flavanols (epicatechin, procyanidin B2+B4, catechin, and procyanidin B1). The overall ability for the prediction was 60.9%, where the apple juice samples from the Alpine state were parted from the Mediterranean region (Bat et al., 2018). Likewise, in another study, the consequence of press construction on the quality and yield of apple juice was studied. In this research study, three apple varieties were used: 'Jonaprince', 'Rubin', and 'Mutsu'. The screw press and basket press were tested in this study. Usually a higher yield of apple juice can be obtained by screw press as compared to the basket press. It was observed that apple juices obtained by using a screw press had a great number of soluble solids, strong antioxidant properties, the maximum content of phenolic compounds, a low acidity level, and high viscosity. It was concluded that press construction plays a key role in enhancing the quality of juice (Wilczyński et al., 2019). Similarly, a study was carried out to check the anti-inflammatory and anti-platelet properties of cider bioactive substances and Irish apple juice. It was observed that Irish cider and Irish apple juice were found to have bioactive polar lipid substances with strong anti-inflammatory and anti-platelet potential, while the fermentation process is considered a major factor in their bioactivities, lipid content, and structures (Tsoupras, Moran, Pleskach, et al., 2021). In addition, in another study, anti-inflammatory and antioxidant neuroprotective effects of apple-based drinks and beverages were determined in the lipopolysaccharide inflammation of mice model. In this study, four different concentrated apple-based drinks (apple juice and three ciders) were given to mice for almost 7 days in replacement of water. It was observed that the mice showed less inflammation and oxidative stress after the treatment with beverages. It was concluded that apple-based drinks have strong anti-inflammatory and antioxidant neuroprotective properties (Alvariño et al., 2020).

Apple juice drinks have also been studied for their health properties. In recent years, numerous research has shown the health benefits of apples; these health benefits are attributed to phytochemicals and bioactive substances. For example, a study was performed to explore the anti-cancer effect of polyphenol present in apples called phloretin. A wide range of preclinical studies has been conducted on phloretin due to its anti-cancer, anti-inflammatory, and antioxidant properties. In this study, it was observed that phloretin can inhibit xenograft tumors in mice that were implanted with different human being cancer cells. The ability of phloretin to interfere with the signaling of cancer cells has increased its potential for the development of anti-cancer drugs (Cho et al., 2019). Moreover, the anti-cancer and photoprotective effects of apple extracts was also studied in dermal cells in other research. The basic purpose of this study was to examine and characterize the protective capability of extracts in contrast to the damage of DNA caused by ultraviolet (UV) radiations in human fibroblasts and also to prove the anti-carcinogenic effects of apple extracts melanoma cells and murine. The results showed that polyphenol extracts from apples have great potential and capability for the formulations of medicines and different cosmetic items for the shield of DNA against ultraviolet (UV) radiations and for the treatment/cure of melanoma (de Oliveira Raphaelli et al., 2021). Furthermore, in another study, different possible uses of polyphenol substances extracted from the Korean apple varieties in food additives and cosmetic products were investigated by determining their functional characteristics. Mainly, the polyphenolic content of 'Fuji' and 'Arisoo' apples was used to study their antioxidant and anti-wrinkle properties. Ethanol and water extracted polyphenolic substances yields were 2.14 mg/g and 1.8 mg/g of fresh apple weight, respectively. The overall results showed that extracts obtained from the 'Arisoo' apples have amazing anti-diabetic, antioxidant, and anti-wrinkle properties as compared to extracts obtained from 'Fuji' apples (Cho et al., 2019). Likewise, in another study, the effect of apple juice (cloudy) fermented with *Lactobacillus* in controlling obesity through intestinal health and gut microbiota was observed. In recent years it has been observed that hyperglycemia and obesity have become the two major serious diseases, and their patients are growing day by day. In this study cloudy apple juice and fermented cloudy juice (rich in phenolic components) were prepared. The effect of these juices on gut microbiota, weight, intestinal health, and lipid level were examined for obese mice provided by a high-fat diet. It was observed that fermented apple juice with high phenolic content reduced the weight of mice; properly regulate the lipid levels, protected the health of the intestine, reduced the accumulation of fat, and also improved the health of gut microbiota. Finally, it was concluded that the fermented cloudy juice of apples can be developed into a healthy food product for controlling hyperglycemia and obesity (Han et al., 2021). It has been observed that *Lactobacillus* helps to control the bacteria that produces harmful and toxic substances in the gut. It is considered one of the main probiotics that help to improve the growth of microbiota and also plays a great role in controlling the different body functions by producing different bioactive and regulatory compounds (Ahmad and Khalid, 2018). However, for the proper functioning of probiotics, prebiotics is required. Prebiotics provide food to probiotics presents in the human gut and improve their functioning that ultimately leads to multiple health benefits (Ahmed and Ahmad, 2017).

18.5   PROPERTIES OF APPLE POMACE AND WASTE

Apple pomace is the major by-product of any juice processing industry. It is left over after the extraction of juice from apples. It is mainly composed of four physiological parts: 0.5–1% stem, 60%–65% flesh, 3%–4.5% seed, and 30%–35% peel (Montañés et al., 2018). It is estimated that every year several million metric tons of pomace are produced worldwide (Lyu et al., 2020). Apple pomace is mainly comprised of such compounds that remain indigestible by human digestive enzymes. Such compounds that remain undigested in the small intestine and remain intact are called dietary fiber or roughage. The analysis on 11 apple varieties showed that there was 43.6% dietary fiber was present on average. Moreover, in addition to indigestible polymers, apple pomace also has almost 14% starch content. In addition, dry apple pomace contains sucrose (3.4%–24%), fructose (18%–31%), and glucose (2.5%–12.4%) per weight (Waldbauer et al., 2017). Fruit pomace is also a great source of different bioactive substances such as phenolic components, phytochemicals, dietary fiber, polysaccharides, carbohydrates, and different natural antioxidants. These bioactive substances help in improving metabolic and cell activities due to their anti-carcinogenic, antioxidant, anti-inflammatory, and anti-allergenic properties (Iqbal et al., 2021).

In recent years different studies have been carried out on apple pomace to check their properties and health benefits. A study was conducted to enhance the antioxidant activity and lipid solubility of phenolic extracts from apple pomace by using the encapsulation method. In that study optimization of nanocapsules of phenolic extracts from red and 'Golden' apple pomace in soy protein and chitosan was performed. The characteristics of nanocapsules were confirmed by spectrophotometric analysis, Fourier transform infrared spectroscopy spectra, fluorescence spectra, and transmission electron microscopy. FTIR spectra showed that the extracted phenolic compounds were successfully encapsulated (Riaz et al., 2020). Moreover, in another research study the effects of gamma rays on the microstructural, physiochemical, microbiological, and thermogravimetric attributes of apple pomace flour were studied during a 9-month long storage. Results showed that irradiated samples of pomace flour stayed stable during the storage period regarding the contents of total sugars, protein, dietary fiber, lipids, zinc, manganese, iron, and potassium (Ito et al., 2017). Furthermore, a study was carried out to optimize the extraction time, extraction temperatures, and sonication powers for the extraction of antioxidant substances from apple pomace by using the ultrasound technique. Results showed that the conditions that are best for the extraction of polyphenols were found at 50% ultrasound amplitude, 90°C, and 20 minutes (Egüés et al., 2021). In addition, another research study was performed to check the physicochemical properties of apple pomaces. The physicochemical analysis of apple pomaces were performed to determine the ash content, lignocellulose, elemental composition, and sugar concentration. Scanning electron microscopy technique was used to study the morphology of apple pomaces. The results from that research provided important information for different value-added uses of apple pomaces. The study concluded that the diversion of apple pomace from wastage material to economic stream can help in sustainable product development and also reduce their burden on the environment (Gowman et al., 2019). Likewise, research was conducted in which apple pomace was used for the extraction of pectin by using radiofrequency-assisted techniques. For the optimization of this technique, response surface methodology was used. The conditions for the optimization include holding time (19 minutes), pH (2.2), and temperature (88°C). The pectin yield by different extraction techniques was as follows: [conventional extraction (7.46%), radio frequency assisted extraction

(11.24%) and microwave-assisted extraction (10.58%)]. Also, X-ray diffraction (XRD), Nuclear magnetic and Fourier transform infrared spectroscopy analysis also showed that samples obtained from extraction methods contained the properties of functional groups present in pectin, but no prominent differences were noticed in the chemical structures. Overall, it was concluded that radio frequency assisted extraction is best regarding quality and pectin yield as compare to conventional extraction and microwave-assisted extraction (Zheng et al., 2021).

In present years, apple pomace also has been used in different products to enhance its antioxidant properties and health relevant properties. For example, the effects of decreasing fat levels and the use of apple pomace powder on oxidative stability, emulsion stability, and crude fiber content of low-fat meat emulsions were studied. Results showed that emulsion formulations containing pomace powder had considerably high fiber content as linked to control emulsion formulations. Therefore, it was summarized that the addition of pomace powder can be a healthier alternative for the production of meat products with low fat. Additionally, use of apple pomace powder is a very economical and environment-friendly approach (Rather et al., 2018). Moreover, in another research study, fermented apple pomace was included in lambs' diet to evaluate their effect on lipid oxidation and physicochemical properties of meat at 4°C storage temperature. Results showed that lipid oxidation in meat from the sheep that were previously fed fermented apple pomace was lower as linked to the control group. So, it was summarized that fermented apple pomace has amazing antioxidant properties (Alarcon-Rojo et al., 2019). The effects of apple pomace powder as an alternative for sucrose in different chocolate based products were determined in another study. The dried apple pomace concentrations used between 4 g/100 g and 20 g/100 g. The results showed that increasing dried apple pomace chocolate formulations increased the phenolic substances, particle size, and hue angle. However, some parameters like color, hardness, and overall acceptance of chocolate formulations decreased (Büker et al., 2021). Likewise, apple pomace was also used in the development of fiber-rich acidophilus yogurt. In this study, fiber-rich yogurt was prepared by adding 2.5%, 5%, 7.5%, and 10% pomace in milk. All the samples were inoculated with the culture of *Bifidobacterium longum* and *Lactobacillus acidophilus*. The fiber, pH, total solids, sensory properties, and acidity of the yogurts were examined. It was observed that the addition of fiber resulted in the decrease of fat content and acidity. Further, yogurt with 5% fiber was considered the best and also optimized for the development of fiber-rich acidophilus yogurt with required quality/ sensory properties (Issar et al., 2017).

18.6   HEALTH BENEFITS OF APPLES AND THEIR BY-PRODUCTS

Fruits are an excellent source of nutritional components, biologically active substances, and phytochemicals that possess strong antioxidant properties and play an great role in improving the growth of the body and quality of life. Multiple studies have been performed to study the beneficial health properties of fruits, and it has been observed that due to the presence of antioxidants they play an excellent role in reducing oxidative stress and also protect against several diseases like inflammatory diseases, cancer, coronary diseases, and diabetes type 2 that ultimately play an important role in preventing aging (Dhalaria et al., 2020). Apples play a significant role in the intake of dietary substances linked with coronary heart disease prevention. Different studies on apples have shown that they have a strong beneficial influence on lipids, hyperglycemia, vascular functioning, inflammation, and blood pressure. A study was carried out to evaluate the beneficial health properties of whole apple and isolated substances. It was observed that apples are a great source of polyphenolic compounds such as epicatechin, chlorogenic acid, procyanidins, quercetin glycosides, catechin, and phloridzin. The chemical structure of these phenolic compounds is shown in Figure 18.3. It was concluded that there is a significant mutual association between the flavonoids and fiber present in the whole apple, which plays a significant role in the growth of gut microbiota. It was also investigated how apples play an important role in controlling cardiovascular risk factors (Bondonno et al., 2017).

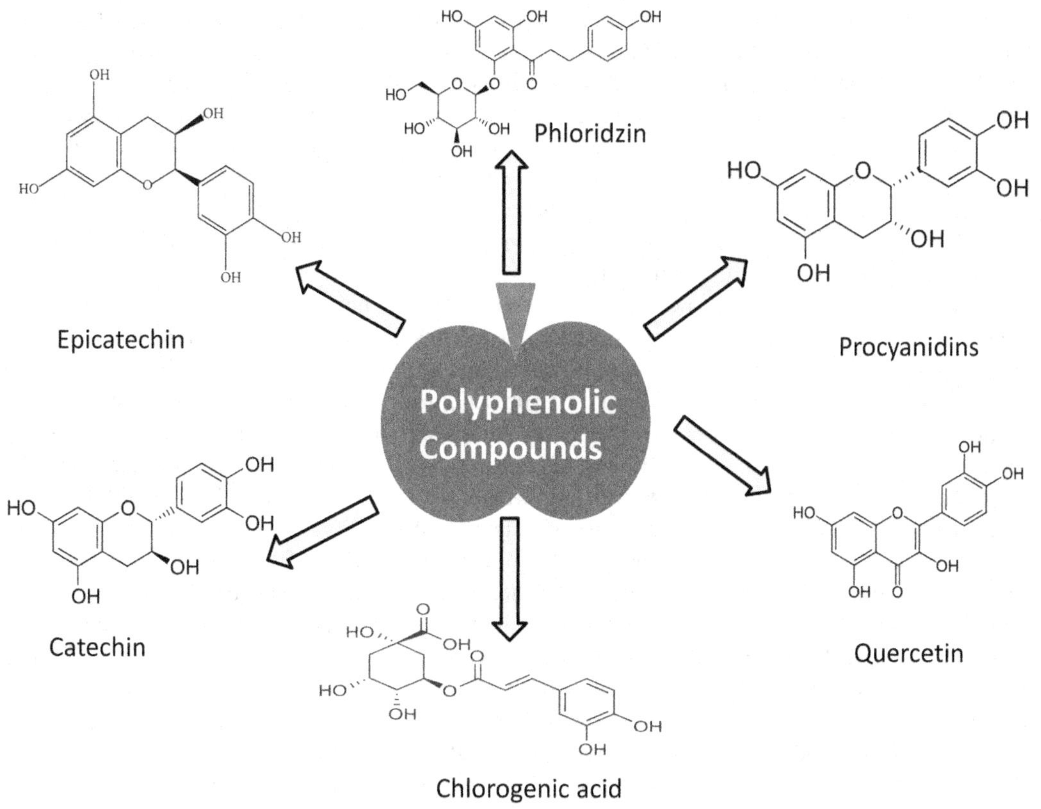

FIGURE 18.3 Chemical structures of polyphenolic substances.

Moreover, research was carried out to check the functional characteristics of 'Summer King' apples. In this study, the peels of apples were extracted by using ethanol and water, and the extracts showed comparatively high levels of polyphenolic compounds at 7.39 mg/g and 4.41 mg/g, respectively. The ethanol and water extracts also exhibited the antioxidant protection factor levels of 2.65 protection factor and 3.37 protection factors, respectively. The antioxidation effect of ethanol and water extracts toward thiobarbituric acid reactive substances (TBARS) was 30% and 38% at 100 $\mu g \cdot mL^{-1}$. It was concluded that the apple peel of 'Summer King' has multiple beneficial characteristics such as an anti-diabetic effect, anti-inflammation, anti-wrinkle, and antioxidation (Lee et al., 2020). Furthermore, in another research study, the effect of different apple varieties on the composition of human gut microbiota metabolic output by using a colonic model was investigated. In this study, three apple varieties ('Pink Lady', 'Renetta Canada', and 'Golden Delicious') were absorbed and further fermented by using a colonic model (37°C and 5.5–6.0 pH) inculcated with feces from three different healthy donors. From the results, it was concluded that the 'Renetta Canada' has great potential in inducing the changes in the composition of the microbiota and metabolic pathway, which can be directly associated with improving the health properties (Koutsos et al., 2017). Likewise, research was conducted for the characterization of biological activities and phytochemical analysis of Italian ancient apples. For this purpose, nine apple samples were used and their anti-inflammatory and antioxidant properties were studied. High-performance liquid chromatography attached with mass spectrometry and photodiode detector was used for the analysis of 20 different phytochemicals in extracts. It was observed that extracts obtained from lyophilized material were rich in phenolic substances as compare to dried ones. The strong antioxidant properties apples have been revealed by ABTS, DPPH, Folin-Ciocalteau, and FRAP assays. The given results greatly contributed to the exploration of the old apple variety as a source of nutraceutical components (Wandjou et al., 2020). In addition, antiproliferative, phytochemical, and antioxidant properties of four red-fleshed varieties were investigated in China. It was observed that as compared to 'Fuji' apples, the red-fleshed apples were rich in flavonoid and phenolic substances, ranging from 1.4–2.4 fold and 1.5–2.6 fold, respectively. In all antioxidant assays (ABTS scavenging activity, DPPH scavenging activity, cell antioxidant ability, and ferric reducing power), 'A-38' showed the high antioxidant value, while 'Fuji' showed the low antioxidant value. It was concluded that 'Meihong' and 'A-38' showed higher antiproliferative and antioxidant properties because of their high levels of phenolic compounds and strong potential for utilization and development (Li et al., 2020). Moreover, research was conducted to study the role of apple phloridzin, phytochemical compounds, and phloretin in regulating the processes related to the inflammation of the intestine. In this study, their inhibitory effects on the formation of glycation products and their consequence on the synthesis of pro-inflammatory molecules in 1 L-1β treated with myofibroblasts of colon cell line were investigated. Overall, it was concluded that phloridzin and phloretin compounds are significantly found in compounds, and also associated with beneficial and healthy effects of apple consumption (Zielinska et al., 2019). Furthermore, the phytochemical and antioxidant properties of 'Kei' apple fruit were investigated. The main polyphenolic compounds identified in fruit were pyrogallol and chlorogenic acid. Among all acids, malic acid was the most plentiful. The whole fruit showed a high antioxidant property as compared to flesh extract. It was detected that the red and white blood cells, hemoglobin, and mean hemoglobin values didn't greatly change when female and male rats ingested 1000 mg/kg whole fruit extract (Taher et al., 2018).

Apples have been used in the production of different food products for their proper utilization beside eating as fruit. A study was performed to observe the technological characteristics of apples for the production of jam. From the results of tests performed, it was observed that the fruits of the 'Zhigulevskoe' cultivar showed the best set of indicators: [total pectin substances (6.4%), water-soluble pectin (1.1%), vitamin C (13.9 mg/100 g), organic acids (1%), R-active substances (110 mg/100 g), dry matter content (15.4%) and sugar-acid (12.3 units)] (Ryadinskaya et al., 2021). In another research study, color and rheological properties of apple-based jellies improved with inulin with different degrees of polymerization were studied. It was observed that both the addition of oligofructose and inulin predominantly modify the textural and rheological properties of analyzed desserts. It was observed that preparation with a high degree of polymerization results in improved moduli, while preparation with a low degree of polymerization initially falls and then increases the quantities of both moduli. It was concluded that inulin may impart captivating ingredients in desserts with beneficial health properties (Witczak et al., 2020). Furthermore, a study was performed to evaluate the anti-platelet and anti-inflammatory characteristics of lipid bioactive substances from apple cider by-products. The functionality of lipid extracts and their high-performance liquid chromatography (HPLC) derived-based subclasses were evaluated in vitro against platelet aggregation influenced the by inflammatory and thrombotic platelet agonists adenosine di-phosphate and platelet activating factor. Overall, it was observed that apple cider's by-products are an excellent source of biologically active polar lipid substances with great anti-platelet and anti-inflammatory properties. These products can also be used in producing nutraceutical compounds or functional foods against cardiovascular diseases (Tsoupras, Moran, Byrne, et al., 2021). Likewise, apple cider vinegar has also possessed pharmacological activities toward Alzheimer's disease. Apple-based cider vinegar contains polyphenolic substances like chlorogenic acids, catechin, gallic acid, p-coumaric acid, and caffeic acid that have strong antioxidant potential. Regular consumption of these phenolic substances can help in reducing the stress that is the main cause of Alzheimer's disease (Tripathi & Mazumder, 2020). Polyphenolic substances and health properties of different apple products and by-products have been summarized in Table 18.1.

TABLE 18.1

Polyphenolic Substances and Health Properties of Different Apple Products and By-Products

Apple products and by-products	Polyphenolic substances	Health benefits	References
Ancient Italian apple	Gallic acid, Epicatechin, Procianidin, Cyanidin 3-glucoside, Kampferol-3-glucoside, p-Coumaric acid, Phloridzin, Phloretin, Oleanolic acid, Ursolic acid	antioxidation, anti-inflammation	Wandjou et al. (2020)
Apple peel	Chlorogenic acid, caffeic acid, epicatechin, catechin	antioxidation, anti-inflammation, anti-wrinkle, anti-diabetic	Lee et al. (2020)
Pulp of 'Himalayan' apple	Phloridzin and phloretin	Antioxidant potential	Dadwal et al. (2018)
Apple varieties	Total flavonoid compounds, ascorbic acid, total phenolic compounds, anthocyanins	Strong antioxidant potential	Sethi et al. (2020)
Red apple seeds	Coumarins compounds	Strong antimicrobial properties	Mohammed & Mustafa (2020)
Apple juice	Different polyphenol components	Antioxidant activity	Zielinski et al. (2019)
Cider and Irish apple juice	Bioactive polar lipid substances	Anti-inflammatory and anti-platelet properties	Tsoupras et al. (2021)
Apple fruit	Phloretin	Anti-cancer	Cho et al. (2019)
Korean apple cultivar	Different polyphenol substances	Anti-diabetic, antioxidant, anti-wrinkle properties	Cho et al. (2019)
Apple fruit pomace	Polyphenol compounds	Anti-carcinogenic, antioxidant, anti-inflammatory, anti-allergenic properties	Iqbal et al. (2021)

18.7 CONCLUSION

Fruits are consumed daily all over the world because they are a rich source of nutritional components and are also highly recommended in the food pyramid. Among all fruits, apples are the most nutritious and tasty fruit. In this chapter, different properties and health benefits of apple parts like seeds, pomace, pulp, juice, and peel along with apple products such as cider, vinegar, jams, jellies, juice, and concentrates have been evaluated. This manuscript comprehensively explains the latest extraction and spectroscopy techniques that have been used for the phytochemical and antioxidant analysis of apple and apple products. It has been observed that the Fourier transform infrared spectroscopy (FTIR), NMR, X-Ray diffraction, TEM, scanning electron microscopy (SEM), and ultraviolet spectroscopy techniques used to study the antioxidant, anti-inflammatory, anti-cancer, and antiproliferative properties of apple and apple products. Moreover, the main polyphenolic substances that have been evaluated in the different apple parts are catechin, quercetin glycosides, phloridzin, chlorogenic acid, procyanidins, and epicatechin.

REFERENCES

Ahmad, A., and Kaleem, M., 2018. β-Glucan as a food Ingredient. In: Grumezesu, A. M., and Holban, A. M. (Eds.), *Biopolymers for food design* (pp. 351–381), Elsevier, San Diego, CA.

Ahmad, A., and Khalid, S., 2018. Therapeutic aspects of probiotics and prebiotics. In: Grumezesu, A. M., and Holban, A. M. (Eds.), *Diet, microbiome and health* (pp. 53–91), Elsevier, San Diego, CA.

Ahmed, Z., and Ahmad, A., 2017. Biopolymer produced by the lactic acid bacteria: Production and practical application. In: Holban, A. M., and Grumezescu, A. M. (Eds.), *Microbial production of food Ingredients and additives* (pp. 217–257), Academic Press, San Diego, CA.

Alarcon-Rojo, A., Lucero, V., Carrillo-Lopez, L., and Janacua, H., 2019. Use of apple pomace in animal feed as an antioxidant of meat. *South African Journal of Animal Science*, 49(1): 131–139.

Alpaslan, D., Ersen Dudu, T., and Aktas, N. J. S. M., 2021. Synthesis of smart food packaging from poly (gelatin-co-dimethyl acrylamide)/citric acid-red apple peel extract. *Soft Material*, 19(1): 64–77.

Alvariño, R., Alonso, E., Alfonso, A., Botana, L. M. J. M. N., and Research, F., 2020. Neuroprotective effects of apple-derived drinks in a Mice model of Inflammation. *Molecular Nutrition Food Research*, 64(2): 1901017.

Bassi, M., Lubes, G., Bianchi, F., Agnolet, S., Ciesa, F., Brunner, K., Guerra, W., Robatscher, P., and Oberhuber, M., 2017. Ascorbic acid content in apple pulp, peel, and monovarietal cloudy juices of 64 different cultivars. *International Journal of Food Properties*, 20(sup3): S2626–S2634.

Bat, K. B., Vodopivec, B. M., Eler, K., Ogrinc, N., Mulič, I., Masuero, D., and Vrhovšek, U. J. L., 2018. Primary and secondary metabolites as a tool for differentiation of apple juice according to cultivar and geographical origin. *LWT-Food Science and Technology*, 90: 238–245.

Bolarinwa, I. F., Orfila, C., and Morgan, M. R., 2015. Determination of amygdalin in apple seeds, fresh apples and processed apple juices. *Food Chemistry*, 170: 437–442.

Bondonno, N. P., Bondonno, C. P., Ward, N. C., Hodgson, J. M., and Croft, K. D., 2017. The cardiovascular health benefits of apples: Whole fruit vs. isolated compounds. *Trends in Food Science Technology*, 69: 243–256.

Büker, M., Angın, P., Nurman, N., Rasouli Pirouzian, H., Akdeniz, E., Toker, O. S., Sagdic, O., and Tamtürk, F., 2021. Effects of apple pomace as a sucrose substitute on the quality characteristics of compound chocolate and spread. *Journal of Food Processing Preservation*, 45(10): e15773.

Chatzimarkou, A., Chatzimitakos, T. G., Kasouni, A., Sygellou, L., Avgeropoulos, A., Stalikas, C. D. J. S., and Chemical, A. B., 2018. Selective FRET-based sensing of 4-nitrophenol and cell imaging capitalizing on the fluorescent properties of carbon nanodots from apple seeds. *Sensors Actuators B: Chemical*, 258: 1152–1160.

Chen, C., Lu, Y., Yu, H., Chen, Z., and Tian, H. J. F. B., 2019. Influence of 4 lactic acid bacteria on the flavor profile of fermented apple juice. *Food Bioscience*, 27: 30–36.

Cho, Y.-J., Lee, E.-H., Yoo, J., Kwon, S.-I., Choi, H. W., and Kang, I.-K. J. H., 2019. Analysis of functional properties in the new Korean apple cultivar Arisoo. *Horticulture, Environment, Biotechnology*, 60(5): 787–795.

Dadwal, V., Agrawal, H., Sonkhla, K., Joshi, R., and Gupta, M., 2018. Characterization of phenolics, amino acids, fatty acids, and antioxidant activity in pulp and seeds of high altitude Himalayan crab apple fruits (*Malus baccata*). *Journal of Food Science Technology*, 55(6): 2160–2169.

Demirci, M. A., Ipek, Y., Gul, F., Ozen, T., and Demirtas, I. J. F. C., 2018. Extraction, isolation of heat-resistance phenolic compounds, antioxidant properties, characterization, and purification of 5-hydroxy-maltol from Turkish apple pulps. *Food Chemistry*, 269: 111–117.

de Oliveira Raphaelli, C., Azevedo, J. G., dos Santos Pereira, E., Vinholes, J. R., Camargo, T. M., Hoffmann, J. F., Ribeiro, J. A., Vizzotto, M., Rombaldi, C. V., and Wink, M. R. J. P. P., 2021. Phenolic-rich apple extracts have photoprotective and anti-cancer effect in dermal cells. *Phytomedicine Plus*, 1(4): 100112.

Dhalaria, R., Verma, R., Kumar, D., Puri, S., Tapwal, A., Kumar, V., Nepovimova, E., and Kuca, K., 2020. Bioactive compounds of edible fruits with their anti-aging properties: A comprehensive review to prolong human life. *Antioxidants*, 9(11): 1123.

Egüés, I., Hernandez-Ramos, F., Rivilla, I., and Labidi, J., 2021. Optimization of ultrasound assisted extraction of bioactive compounds from apple pomace. *Molecules*, 26(13): 3783.

El-Messery, T. M., El-Said, M. M., Demircan, E., and Ozçelik, B. J. A. S. P. T. A., 2019. Microencapsulation of natural polyphenolic compounds extracted from apple peel and its application in yoghurt. *Acta Scientiarum Polonorum Technologia Alimentaria*, 18(1): 25–34.

El-sayed, H., and El-Shemy, N. J. J. O. N. F., 2021. Utilization of red apple peel extract in dyeing of wool fabric. *Journal of Natural Fibers*: 1–14.

FAO. (2021). Retrieved from www.fao.org/faostat/en/

Gowman, A. C., Picard, M. C., Rodriguez-Uribe, A., Misra, M., Khalil, H., Thimmanagari, M., and Mohanty, A. K., 2019. Physicochemical analysis of apple and grape pomaces. *BioResources*, 14(2): 3210–3230.

Gunes, R., Palabiyik, I., Toker, O. S., Konar, N., and Kurultay, S. J. J. O. F. E., 2019. Incorporation of defatted apple seeds in chewing gum system and phloridzin dissolution kinetics. *Journal of Food Engineering*, 255: 9–14.

Han, M., Zhang, M., Wang, X., Bai, X., Yue, T., and Gao, Z., 2021. Cloudy apple juice fermented by Lactobacillus prevents obesity via modulating gut microbiota and protecting intestinal tract health. *Nutrients*, 13(3): 971.

Iqbal, A., Schulz, P., and Rizvi, S. S., 2021. Valorization of bioactive compounds in fruit pomace from agro-fruit industries: Present Insights and future challenges. *Food Bioscience*, 44: 101384.

Issar, K., Sharma, P., and Gupta, A., 2017. Utilization of apple pomace in the preparation of fiber-enriched acidophilus yoghurt. *Journal of Food Processing Preservation*, 41(4): e13098.

Ito, V. C., Zielinski, A. A. F., Avila, S., Spoto, M., Nogueira, A., Schnitzler, E., and Lacerda, L. G., 2017. Effects of gamma radiation on physicochemical, thermogravimetric, microstructural and microbiological properties during storage of apple pomace flour. *LWT-Food Science and Technology*, 78: 105–113.

Khalil, R. R., and Mustafa, Y. F. J. S. R. P., 2020. Phytochemical, antioxidant and antitumor studies of coumarins extracted from Granny Smith apple seeds by different methods. *Systematic Reviews in Pharmacy*, 11(2): 57–63.

Koutsos, A., Lima, M., Conterno, L., Gasperotti, M., Bianchi, M., Fava, F., Vrhovsek, U., Lovegrove, J. A., and Tuohy, K. M., 2017. Effects of commercial apple varieties on human gut microbiota composition and metabolic output using an in vitro colonic model. *Nutrients*, 9(6): 533.

Kruczek, M., Gumul, D., Kačániová, M., Ivanišhová, E., Mareček, J., and Gambuś, H., 2021. Industrial apple pomace by-products as a potential source of pro-health compounds in functional food. *Journal of Microbiology, Biotechnology Food Sciences*, 2021: 22–26.

Lan, W., Renard, C. M., Jaillais, B., Leca, A., and Bureau, S., 2020. Fresh, freeze-dried or cell wall samples: Which is the most appropriate to determine chemical, structural and rheological variations during apple processing using ATR-FTIR spectroscopy? *Food Chemistry*, 330: 127357.

Lee, E.-H., Cho, E.-B., Kim, B.-O., Jung, H.-Y., Lee, S.-Y., Yoo, J., Kang, I.-K., and Cho, Y.-J., 2020. Functional properties of newly-bred 'Summer King'apples. *Horticultural Science and Technology*: 405–417.

Lee, P. Y., Kebede, B. T., Lusk, K., Mirosa, M., Oey, I. J. I. J. O. F. S., and Technology. 2017. Investigating consumers' perception of apple juice as affected by novel and conventional processing technologies. *International Journal of Food Science Technology*, 52(12): 2564–2571.

Li, C. X., Zhao, X. H., Zuo, W. F., Zhang, T. L., Zhang, Z. Y., and Chen, X. S., 2020. Phytochemical profiles, antioxidant, and antiproliferative activities of four red-fleshed apple varieties in China. *Journal of Food Science*, 85(3): 718–726.

Li, Z., Teng, J., Lyu, Y., Hu, X., Zhao, Y., and Wang, M., 2019. Enhanced antioxidant activity for apple juice fermented with Lactobacillus plantarum ATCC14917. *Molecules*, 24(1): 51.

Lyu, F., Luiz, S. F., Azeredo, D. R. P., Cruz, A. G., Ajlouni, S., and Ranadheera, C. S., 2020. Apple pomace as a functional and healthy ingredient in food products: A review. *Processes*, 8(3): 319.

Madrera, R. R., Valles, B. S. J. E. F. R., and Technology. 2018. Characterization of apple seeds and their oils from the cider-making industry. *European Food Research Technology*, 244(10): 1821–1827.

Manzoor, M., Singh, J., and Gani, A. J. L., 2021. Characterization of apple (Malus domestica) seed flour for its structural and nutraceutical potential. *LWT-Food Science and Technology*, 151: 112138.

Mohammed, E. T., and Mustafa, Y. F. J. S. R. I. P., 2020. Coumarins from red delicious apple seeds: Extraction, phytochemical analysis, and evaluation as antimicrobial agents. *Systematic Reviews in Pharmacy*, 11(2): 64–70.

Montañés, F., Catchpole, O. J., Tallon, S., Mitchell, K. A., Scott, D., and Webby, R. F., 2018. Extraction of apple seed oil by supercritical carbon dioxide at pressures up to 1300 bar. *The Journal of Supercritical Fluids*, 141: 128–136.

Musacchi, S., and Serra, S., 2018. Apple fruit quality: Overview on pre-harvest factors. *Scientia Horticulturae*, 234: 409–430.

Nakov, G., Brandolini, A., Hidalgo, A., Ivanova, N., Jukić, M., Komlenić, D. K., Lukinac, J. J. F. S., and International, T., 2020. Influence of apple peel powder addition on the physico-chemical characteristics and nutritional quality of bread wheat cookies. *Food Science Technology International*, 26(7): 574–582.

Okokon, E. J., Okokon, E. O. J. G. J. O. P., and Sciences, A., 2019. Proximate analysis and sensory evaluation of freshly produced apple fruit juice stored at different temperatures and treated with natural and artificial preservatives. *Global Journal of Pure Applied Sciences*, 25(1): 31–37.

Opyd, P. M., Jurgoński, A., Juśkiewicz, J., Milala, J., Zduńczyk, Z., and Król, B. J. N., 2017. Nutritional and health-related effects of a diet containing apple seed meal in rats: The case of amygdalin. *Nutrients*, 9(10): 1091.

Paini, J., Benedetti, V., Ferrentino, G., Baratieri, M., Patuzzi, F. J. B. C., and Biorefinery. 2021. Thermochemical conversion of apple seeds before and after supercritical CO2 extraction: An assessment through evolved gas analysis. *Biomass Conversion Biorefinery*, 11(2): 473–488.

Potenza, L., Minutelli, M., Stocchi, V., and Fraternale, D., 2020. Biological potential of an ethanolic extract from 'Mela Rosa Marchigiana' pulp callus culture. *Journal of Functional Foods*, 75: 104269.

Purić, M., Rabrenović, B., Rac, V., Pezo, L., Tomašević, I., and Demin, M. J. L., 2020. Application of defatted apple seed cakes as a by-product for the enrichment of wheat bread. *LWT-Food Science and Technology*, 130: 109391.

Ramos-Aguilar, A. L., Victoria-Campos, C. I., Ochoa-Reyes, E., de Jesus Ornelas-Paz, J., Zamudio-Flores, P. B., Rios-Velasco, C., Reyes-Hernández, J., Pérez-Martínez, J. D., and Ibarra-Junquera, V., 2017. Physicochemical properties of apple juice during sequential steps of the industrial processing and functional properties of pectin fractions from the generated pomace. *LWT-Food Science and Technology*, 86: 465–472.

Rather, S. A., Masoodi, F., Akhter, R., Ganaie, T. A., and Rather, J. A., 2015. Utilization of apple pomace powder as an antioxidant fat replacer in meat emulsions. *Journal of Food Measurement and Characterization*, 9: 389–399.

Riaz, A., Lagnika, C., Abdin, M., Hashim, M. M., Ahmed, W. J. J. O. P., and Environment, T., 2020. Preparation and characterization of chitosan/gelatin-based active food packaging films containing apple Peel nanoparticles. *Journal of Polymers the Environment*, 28(2): 411–420.

Riaz, A., Lei, S., Akhtar, H. M. S., Wan, P., Chen, D., Jabbar, S., Abid, M., Hashim, M. M., and Zeng, X. J. I. J. O. B. M., 2018. Preparation and characterization of chitosan-based antimicrobial active food packaging film incorporated with apple peel polyphenols. *International Journal of Biological Macromolecules*, 114: 547–555.

Ryadinskaya, A., Ordina, N., Koschaev, I., Mezinova, K., Chuev, S., and Zakharova, D. (2021). *Study of technological properties of apples for jam production*. Paper presented at the IOP Conference Series: Earth and Environmental Science.

Sethi, S., Joshi, A., Arora, B., Bhowmik, A., Sharma, R., Kumar, P. J. E. F. R., and Technology. 2020. Significance of FRAP, DPPH, and CUPRAC assays for antioxidant activity determination in apple fruit extracts. *European Food Research Technology*, 246(3): 591–598.

Shehzadi, K., Rubab, Q., Asad, L., Ishfaq, M., Shafique, B., Ali Nawaz Ranjha, M., Mahmood, S., Mueen-Ud-Din, G., Javaid, T., and Sabtain, B., 2020. A critical review on presence of polyphenols in commercial varieties of apple peel, their extraction and health benefits. *Biogeneric Science and Research*, 6: 18.

Skinner, R. C., Gigliotti, J. C., Ku, K.-M., and Tou, J. C., 2018. A comprehensive analysis of the composition, health benefits, and safety of apple pomace. *Nutrition Reviews*, 76(12): 893–909.

Srivastava, P., David, J., Rajput, H., Laishram, S., and Chandra, R., 2017. Nutritional information of custard apple and strawberry fruit pulp. *Chemical Science Review Letters*, 6(24): 2337–2341.

Sudha, M., Dharmesh, S. M., Pynam, H., Bhimangouder, S. V., Eipson, S. W., Somasundaram, R., and Nanjarajurs, S. M., 2016. Antioxidant and cyto/DNA protective properties of apple pomace enriched bakery products. *Journal of Food Science Technology*, 53(4): 1909–1918.

Taher, M. A., Tadros, L. K., and Dawood, D. H., 2018. Phytochemical constituents, antioxidant activity and safety evaluation of Kei-apple fruit (*Dovyalis caffra*). *Food Chemistry*, 265: 144–151.

Tripathi, S., and Mazumder, P. M., 2020. Apple cider vinegar (ACV) and their pharmacological approach towards Alzheimer's disease (AD): A review. *Indian Journal of Pharmaceutical Education and Research*, 54: s67–s74.

Tsoupras, A., Moran, D., Byrne, T., Ryan, J., Barrett, L., Traas, C., and Zabetakis, I., 2021. Anti-inflammatory and anti-platelet properties of lipid bioactives from apple cider by-products. *Molecules*, 26(10): 2869.

Tsoupras, A., Moran, D., Pleskach, H., Durkin, M., Traas, C., and Zabetakis, I. J. F., 2021. Beneficial anti-platelet and anti-inflammatory properties of irish apple juice and cider bioactives. *Molecules*, 10(2): 412.

Vukušić, J. L., Millenautzki, T., Cieplik, R., Obst, V., Saaid, A. M., Clavijo, L., Zlatanovic, S., Hof, J., Mösche, M., and Barbe, S., 2021. Reshaping apple juice production into a zero discharge biorefinery process. *Waste Biomass Valorization*, 12(7): 3617–3627.

Waldbauer, K., McKinnon, R., and Kopp, B., 2017. Apple pomace as potential source of natural active compounds. *Planta Medica*, 83(12/13): 994–1010.

Wandjou, J. G. N., Lancioni, L., Barbalace, M. C., Hrelia, S., Papa, F., Sagratini, G., Vittori, S., Dall'Acqua, S., Caprioli, G., and Beghelli, D., 2020. Comprehensive characterization of phytochemicals and biological activities of the Italian ancient apple 'Mela Rosa dei Monti Sibillini'. *Food Research International*, 137: 109422.

Wilczyński, K., Kobus, Z., and Dziki, D. J. S., 2019. Effect of press construction on yield and quality of apple juice. *Sustainability*, 11(13): 3630.

Wilczyński, K., Kobus, Z., Nadulski, R., and Szmigielski, M., 2020. Assessment of the usefulness of the twin-screw press in terms of the pressing efficiency and antioxidant properties of apple juice. *Processes*, 8(1): 101.

Witczak, M., Jaworska, G., and Witczak, T., 2020. Rheological and colour properties of apple jellies supplemented with inulin with various degrees of polymerisation. *International Journal of Food Science Technology*, 55(5): 1980–1991.

Zhang, F., Wang, T., Wang, X., and Lü, X., 2021. Apple pomace as a potential valuable resource for full-components utilization: A review. *Journal of Cleaner Production*, 329: 129676.

Zheng, J., Li, H., Wang, D., Li, R., Wang, S., and Ling, B., 2021. Radio frequency assisted extraction of pectin from apple pomace: Process optimization and comparison with microwave and conventional methods. *Food Hydrocolloids*, 121: 107031.

Zielinska, D., Laparra-Llopis, J. M., Zielinski, H., Szawara-Nowak, D., and Giménez-Bastida, J. A., 2019. Role of apple phytochemicals, phloretin and phloridzin, in modulating processes related to intestinal inflammation. *Nutrients*, 11(5): 1173.

Zielinski, A. A., Zardo, D. M., Alberti, A., Bortolini, D. G., Benvenutti, L., Demiate, I. M., and Nogueira, A. 2019. Effect of cryoconcentration process on phenolic compounds and antioxidant activity in apple juice. *Journal of the Science of Food Agriculture*, 99(6): 2786–2792.

19 Utilization of By-Products
Peel, Seeds and Pomace

Farhan Saeed, Faqir Muhammad Anjum,
Muhammad Afzaal, Bushra Niaz,
Huda Ateeq and Muzzamal Hussain

CONTENTS

19.1 INTRODUCTION

The apple (*Malus sp.*) is an important and widely globally traded fruit. Apples belong to the *Rosaceae* family's Malus genus, and there are thousands of cultivars grown all over the world. Apples are one of the most significant economic fruit species (Mushtaq and Nayik, 2020). According to the Food and Agriculture Organization (FAO), apples are widely consumed in all countries throughout the world, and they are well-liked for their flavor, juiciness, color, texture and nutritional value (Anstalt, 2013). They also have a high preservation capacity, are accessible year-round in stores at reasonable rates, and are considered a healthy food. Apples are known for having a high concentration of bioactive chemicals that have health-promoting properties (Wojdyło et al., 2021). However, depending on meteorological, agronomic, harvest and postharvest conditions, as well as the food processing procedures and storage, the amounts and types of bioactive compounds found in apples can vary significantly by species and cultivars (Soppelsa et al., 2018; Rabetafika et al., 2014).

One of the key concerns of the food industry is waste control. The apple juice industry produces large amounts of waste material, approximately 12 million tons (Rabetafika et al., 2014). The significant amount of microbial decomposable by-product may cause environmental difficulties if disposed of, resulting in additional waste treatment expenditures for food makers. The large volume of the low-cost by-product, on the other hand, may provide economic implications for its potentially valuable components. It has been considered for use as feed; however, little saving are recovered as a cheaper price cattle feed. As a result, recovering commodities from by-products look to be beneficial, both economically and environmentally. Fresh fruits are essential sources of nutrients such as natural bioactive antioxidants, phytosterols (Dimou et al., 2019; Alonso-Salces et al., 2001), fiber, and others functional compounds that are widely advised for a healthy lifestyle. The transformation of a raw material into food results in a 'waste' and degraded throughout the food supply chain. Food 'waste' has long been thought of as an undesirable items that were disposed of in an expensive ways, such as through animal feed, but they are increasingly seen as health endorsing sources of important nutraceutical (Ordoudi et al., 2018; Zafar et al., 2005).

Apple by-product biotechnology processing has mostly been explored. Apple by-products have been used as a substrate for the manufacture of enzymatically catalyzed compounds including citric acid, flavor components, fuel and pigments (Rabetafika et al., 2014). Extractions of phenolic compounds like dietary fibers, phenolic acids and tannins are evaluated in a non-fermentative manner. Much research has shown that dietary fiber can be extracted on a massive scale. The fractionation of the main components, on the other hand, will result in increased importance of apple waste products. As a result, it will added to the current and unique application of the bio-refinery perception to food wastes for ideal usage and the generation of value-added products. Phytochemicals are another useful component found in apple by-products that have recently been studied (Aachary and Prapulla, 2011).

Fruits such as apples, bananas and mangos are abundant in Pakistan, India, and other Asian countries (Faqeerzada et al., 2018). Meredith et al. (1990) explained that in apple producing areas, millions of tons of apples are produced each year, of which 30–40 percent are damaged and therefore are instead kept in the trees until they decay and degrade. Depending on the demand for apple juice in different countries, 20%–40% of apples are processed in factories, and thousands of tons of pomace (remaining material from apple after juice extraction) are thrown away with no further purpose. Apples and their by-products provide nutritional value for animal feed. Fresh apples have a moderate energy value; however, processed apple by-products have an energy value that is either comparable to fresh apples or higher due to concentration, dehydration or sugar addition during processing. Many factors influence the chemical makeup of apples including cultivar, growing region, climate, cultural practices and processing.

19.2 APPLE BY-PRODUCTS

19.2.1 Apple Peels

Apples are an important and widely consumed fruit worldwide. Apple peels are comprised of a substantial number of important biomolecules, such as polyphenols (e.g., procyanidins and flavanols) (Wang et al., 2020). Polyphenols, such as those found in apples, are play a significant role in human health because of their powerful anti-inflammatory and antiradical properties (Joseph et al., 2016). Polyphenol intake and human health have been discussed in many studies, with a focus on diabetes, oncological, intestinal related syndromes, cardiovascular, metabolic disorder and obesity. Several experiments using ultrasound (Wang et al., 2019), microwave (Prothon et al., 2001), pulsed electric fields (PEF) (El Kantar et al., 2018), high voltage electrical discharges (Wang et al., 2020), supercritical fluids (Watson, 2018) and high hydrostatic pressure (Łata et al., 2009) to extract polyphenols from fruit and vegetable tissues have been published. PEF treatment allowed for the specific extraction of several biomolecules from apples (e.g., pectin, carbohydrates, proteins flavors and pigments). Here we discuss the different ways to utilize apple peels.

19.2.1.1 Utilization of Apple Peels

19.2.1.1.1 Water Purification with Apple Peels

Apples are high in catechin, procyanidins, chlorogenic acid, epicatechin, phlorizin and quercetin conjugates, among other flavonoids (Bondonno et al., 2017; Boyer and Liu, 2004; Wolfe and Liu, 2003). As an effective adsorbent, mostly Zr cations immobilized apple peels are used. Several hazardous anions such as phosphate and chromate ions were extracted. To further understand the adsorption mechanism, researchers conducted adsorption and desorption experiments. The adsorption process was validated using Langmuir and Freundlich isotherm models. To explore more about adsorption processes, which are controlled by pseudo-second-order kinetics, kinetic studies were conducted. The experimental adsorption capabilities (mg/g) of treated apple peels toward distinct anionic pollutants were 15.64 (AsO2), 15.68 (AsO4 3), 25.28 (Cr2O7 2) and 20.35 (PO4 3). The extraction efficiency is also influenced by experimental parameters such as the medium's pH and temperature. The Zr in apple peel is responsible for removing various pollutants from water. It is possible that using biowaste as a water treatment method is an easy, low cost and efficient method that may be applied on a large scale (Mallampati and Valiyaveettil, 2013).

19.2.1.1.2   Anti-Inflammatory Properties of Apple Peel Extract

Ursolic acid (UA), a pentacyclic triterpene acid naturally present in apple peels, has a wide range of medicinal properties. In several therapeutic plants and phytopharmaceutical preparations, the triterpenes (3b-hydroxy-olea-12-en-28-oic) UA plays a significant function as a chemical marker (Radhiga et al., 2012). This molecule has antioxidant and hepato-protective properties, also studied in conjunction with other triterpene acids that are similarly related. UA is also thought to be functional and nutraceutically beneficial for human body. UA has also been determined to have immune-modulatory and anti-inflammatory effects and to have the ability to trigger apoptosis in a variety of cell lines (Pádua et al., 2014). UA is present across the plant kingdom, including in the apple cuticle, where it reported (30% to 35%) of total lipophilic content (Szakiel et al., 2012). Ursolic acid generally leads around 10% of its isomer, oleanolic acid, in this specific fruit matrix (Bringe et al., 2006).

19.2.1.1.3   Apple Peels as a Food Additive

Apples are an important part of many people's diets (Wolfe and Liu, 2003). According to Hertog et al. (1995), apples constitute the third most important source of flavones. These are the leading contributors in Finland, alongside onions (Knekt et al., 1997). Apples account for 22% ingested in the United States making them the most abundant source of all fruit phenolic (Vinson et al., 2001). Several studies have connected apple consumption to the inhibition of different diseases. In previous two studies Le Marchand et al. (2000) and Vinson et al. (2001) reported that apple consumption is related to a lower risk of lung cancer. Its consumption is also effective against different heart diseases. Apple eating also related to lower rates of coronary and overall mortality (Hodgson et al., 2016; Knekt et al., 1996), as well as a lower risk of pulmonary disease (Kunadian et al., 2018; Tabak et al., 2001) and thrombotic stroke (Knekt et al., 2000).

Apples are good source of many phenolic compounds (Geană et al., 2021; Eberhardt et al., 2000). Total extractable phenolic contents of fresh apples were evaluated, and they range from 110–357 mg/100 g (Podsędek et al., 2000; Liu et al., 2001). Total phenolic contents (TPC) in the peel of an apple are substantially higher than in the flesh (Escarpa and Gonzalez, 1998; Burda et al., 1990; Ju et al., 1996). The apple flesh and peel have different types and concentrations of polyphenols. Catechins, caffeic acid, procyanidins, phloretin glycosides, phloridzin and chlorogenic acid are among the components found in the flesh; the peel includes all of these constituents as well as flavonoids (Golding et al., 2001; van der Sluis et al., 2001). The peels of several fruits and vegetables have been found to contain high levels of phytochemicals and/or antioxidant effects (Kubola and Siriamornpun, 2011; Talcott et al., 2000).

Chronic disorders including cancer and cardiovascular disease may be caused by oxidative stress according to some studies. Apple peels have high levels of phenolic components that have many functional and nutraceutical characteristics. Different research studies argue that the peels of these apples may be used to make functional ingredients if some processing modification are applied without losing a lot of phytochemicals (Helkar et al., 2016). Only blanching treatments maintained the phenolic components to a significant degree with peels blanched for 10 seconds having the highest total phenolic concentration. Researchers reported that apple peel powder component are useful and appealing addition to functional foods (Can-Cauich et al., 2017). A foods' phytochemical content and antioxidant activity could be considerably increased with just a tiny amount. Incorporating apple peels phenolic compounds into food products could help with chronic disease prevention and management (Wolfe and Liu, 2003).

19.2.1.1.4   Apple Peels Produce Plant Cell Wall Destroying Enzymes

The apple is a famous fruit with production of several million tons per year globally. The usage of apples in the food sector resulted in a large amount of waste in the form of peels and pomace. Apple peels are used as a source of substrate for enzymes that break down cell walls. Due to the high levels of lignocellulose in the trash, it can be used as a substrate for the manufacture of plant cell wall disintegrating enzymes (Jalis et al., 2014). The cell walls of plants, which are made up of cellulose, hemicelluloses and lignin are considered renewable chemical feedstock. The existence of a variety of components, on the other hand, makes it stable, and its biodegradation necessitates the action of several enzymes (Kumar et al., 2008).

Different studies have described the potential utility of apple pomace, but nothing has been written on the possibility of using its peels for the manufacture of fungal enzymes (Ghorai et al., 2009). In another study of apple peels, peels were obtained from fruit juice vendors, chopped into small pieces and sun dried for the fermentation trials. The peels' particle size was sustained by sifting them using a 100-mesh sieve (Mahawar et al., 2012).

19.2.1.1.5   Isolated Ursolic Acid From Apple Peels

This cuticular layer of ursolic acid is largely consistent on aliphatic wax materials with up to 30 carbon atom chains, as well as additional emulsifying compounds like fatty acids, free alcohols and triterpenes that help hydrophilic polymers work together (cutin, polymers, pectin, protein, etc.). The structure of cuticular wax researched in apples because it is important in preventing fruit damage caused by insects and other pathogenic agents (Frighetto et al., 2008). The wax protection has linked to antioxidant qualities that prevent fruit scald caused by α-farnesene oxidation that occurs naturally in the apple by-products. Variations in apple peel wax ingredients have been documented in relation to climatic circumstances, genetic variables and plant ontogenesis.

The major chemical naturally found in apple cuticular wax and leaves is ursolic acid [(3b)-3-hydroxyurs-12-en-28-oic acid]. Depending on the ontogenetic period, it can make up to 32% of the total wax. Ursolic acid is a widespread triterpenoid throughout the plant with evidence of its existence in the wax-like covering of a variety of different fruits and others numerous eminent medicinal plants. The epicuticular wax in the adaxial juvenile leaves of the apple tree comprised 26%–28% ursolic acid, which dropped to 25% through ontogenesis. Oleanolic acid accounts for approximately 7% of the overall wax composition (Saucedo-Pompa et al., 2007).

19.2.1.1.6   Utilization of Polyphenols Extracted from Apple Peels

Apples are a globally important and popular fruit, and they are a good source of phytochemicals and antioxidants. Apples have a diversified and stable constituent with a wide range of vitamins and a high level of fibers. Apple consumption is related to the prevention of numerous chronic diseases, particularly heart diseases, the key bioactive substances found in apples especially dietary fiber and phenolic compounds.

More than 50 polyphenols are found in apples. Hydroxycinnamic acids (of which chlorogenic acid is the most prevalent), dihydrochalcone and its derivatives (especially phloridzin), flavan-3-ols like monomers of catechin and flavonols are the four major phenolic groups (quercetin and quercetin glycosides). Polyphenol concentrations and distribution vary significantly between apple varieties, 70–160 mg/100 g TPC present in apple.

Polyphenols found in apple peels have also been demonstrated to reduce oxidative stress. In France, apples are consumed per inhabitant average of 20 kg, accounting for a market share of 20.4%, significantly higher than the second and third most popular fruits, bananas (14%) and oranges (14%) (12%). With 1.5 million tons produced in 2013, the apple is France's most popular fruit. About 30% of the crop is converted into juices, compotes and concentrates which results in an enormous amount of trash, including peels. A study investigated the usage of compressed CO_2 to improve polyphenols content from apple peels. Herrero et al. (2013) studied subcritical and supercritical extraction of useful substances from fruit sources. Hertog et al. (1995) explored the phenolic and flavonoids contents in different plant materials.

19.2.1.1.7 Production of Organic Acid and Pectin from Apple Peels

In the apple food industry, HCL is commonly employed as the main isolation of solvent for the separation of pectin from plant tissues, although HCL has no negative environmental impact. Pectin is a polysaccharide and comprises about 65% sequential galacturonic acids with monosaccharide side chains in its chemical composition. Pectin is classified as a dietary fiber that does not digest to the small intestine by digestive enzymes including -amylase and –glucosidases. Pectin has been proven in several studies to have health-related functions such as boosting immunity, lowering blood cholesterol, demonstrating anti-cancer activity and protecting the gastrointestinal tract (Cho et al., 2019).

Pectin is a useful thickening and gelling agent in many food industries, and it is also used as an emulsifier, binder and protein stabilizer in different food. Pectin is often used as basic materials for a food or medicine constituent because of these functional characteristics.

19.2.2 APPLE POMACE

According to prior reports, around 30% of produced apples are thrown away as trash by the apple processing industries globally. As a waste from the food industry, more than 1 million tons of apple pomace were traded in 2010 (Min et al., 2011). The waste materials of apple processing, which involves primarily 70%–75% moisture and agro-industrial waste including peels, flesh and pulps may pollute the environment through biological degradation and raise biological oxygen demand (BOD) and chemical oxygen demand (COD) levels. The goal of the study was to find a way to use the discarded and unused apple pomace as material for the production of pectin (Agrahari and Khurdiya, 2003).

19.2.2.1 Utilization of Apple Pomace

19.2.2.1.1 Production of Pectin from the Waste of Apple Pomace

Malic, citric and tartaric acids, all of which are food-grade organic acids, were used in a study to extract pectin instead of strong acids (e.g., HCl). Although an enzymatic procedure could be a possible solution to this problem, it is costly. The FDA has designed food-grade organic acids as generally recognized as safe (GRAS) compounds with no restrictions, and these are primarily used as food preservatives to inhibit microbial development. Furthermore, consumers have recently developed a strong interest in 'clean label' items which are frequently linked to health advantages and environmental concerns (Shalini and Gupta, 2010).

19.2.2.1.2 Using Pomace from the Apple Processing Industry Reduces Cost of Disposal

Apple fruits are commonly produced in hot regions of the world and are the most popular fruit globally (Shalini and Gupta, 2010). The huge amount of wastes generated after apple processing enterprises can be divided into two categories. The first category is belt rejection, which refers to fruit that is thrown into the sorting belt because it is partially bruised or ruined. The apple pomace attained after juice extraction is the second type (Fernandes et al., 2019). Apple pomace and belt rejection apples are also thrown away as garbage. To avoid pollution, it is critical to dispose of processing waste properly. Because apple pomace includes a lot of water, simply in a fermentable state, it has a lot of disposal issues. The disposal of such trash comes at a significant cost. Pomace is a good example of a discarded food resource (Bates and Roberts, 2001).

Because apple pomace is biodegradable in nature and has a high biochemical oxygen demand (BOD), dumping it into the environment pollutes the environment; this has forced researchers to attempt to find a suitable solution. The products' cost and the expense of waste disposal, together with pressure from environmental protection agencies in enforcing legislation, will ultimately determine the commercial use of pomace. Apple pomace has traditionally been used as cattle fodder, however due to the rapid deterioration of wet pomace, only a small portion of it is utilized. It is an excellence source of different nutrients because it is high in carbohydrate, pectin, fibers and minerals (Shah and Masoodi, 1994).

Because of the vast amount of apple pomace created during apple processing, it is unlikely that a single product would be commercially viable, hence the manufacture of all possible products should be investigated. The research work on the use of apple pomace from the apple processing businesses to produce diverse products, as well as the potential for small-scale industry development, is discussed in this chapter.

19.2.2.1.3   Apple Pomace for Fuel

Apple pomace used as a source of steam in processing plants after drying, contributing significantly to the energy budget. The economic viability of plant burning of apple processing wastes was investigated by Sargent et al. (1986). Apple processors might save money on fossil fuels and trash disposal by burning apple pomace in the plant, according to the researcher Duman et al. (2020).

19.2.2.1.4   Apple Pomace as an Ingredient in Different Food Products

Apple pomace has been used in the manufacturing of food goods such as apple pomace jam and sauce, as well as the production of citric acid (Quiles et al., 2018). Apple pomace has also been used to make different food products (Lyu et al., 2020). Rotova (1983) developed a method for making apple powder from apple press cake that entails crushing, drying, molding and fractionation. Some confectionary recipes incorporating this powder were also developed. The confectionery sector in Ukraine is said to consume 2000 tons of apple powder per year. In blends used in toffees apple pomace powder was replaced for soy meal without affecting their quality.

The fibers from apple pomace were studied by Walter et al. (1985). Fibers of apple pomace have mild alkaline degradation, which resulted in a cellulosic component of around 25% of the untreated dry matter. These extracted fibers used in several aqueous solvents, yielding water dispersible uronide fractions that contained 12%–20% of the raw material and had varying viscometrical properties depending on the isolated (aqueous solvents) utilized. Fibers exhibited the ability to add bulk to low-fiber meals as non-nutritive.

Shah and Masoodi (1994) investigated how trash from the apple processing companies may be used. They found that roughly 25% of apple pomace and 1.5% of apples are lost as processing wastes in huge apple processing unit. After autoclaving, industrially rejected apples were turned to homogeneous pulp. The pulp held well for over a year at 20°C when preserved with 1% potassium metabisulphite in pulp material. Pulp-based beverages were very well received. For the creation of pomace sauce, Joshi and Sandhu (1996) used apple pomace from three maturation stages and different sugar levels with total soluble solids (TSS) (per kg of pulp).

Sugars, brix/acid ratio, proteins, starch and fibers all increased from stage T1 to T3, however ascorbic acid levels declined. Based on sensory attributes, T2 apple pomace sauce was deemed to be the best. Except for pectin, starch and ascorbic acid, which decreased from T1 to T3, all quality indices increased. During the 6 months of storage, the TSS, sugars (reducing and total), titratable acidity and standard plate counts grew significantly, but non-reducing sugars, starch, pectin, crude fiber and ascorbic acid levels dropped. The alterations that occurred during storage were identical to those that occurred in sauces and were not unique to apple pomace sauce.

Kaushal et al. (2002) prepared cookies by mixing varying quantities of apple pomace powder (10%–50%) into the batter. Sensory examination of the produced cookies revealed that 30% apple pomace powder was used in the creation of high-quality cookies. Bates and Roberts (2001) investigated the use of apple pomace as a press assist in the production of fruit juice. The dynamics of drying apple pomace were also studied. The juice was compared to juice squeezed using rice hulls and paper in terms of yield and flavor. The pomace desorption isotherms were measured at 25°C, 35°C, and 45°C. The juice produced by pressing with apple pomace had a better flavor than the juice produced by pressing with hulls (Yan, 2012).

Extraction of pectin takes a long time, so apple pomace has been utilized to extract pectin. Attempts to extract pectin from apple pomace have been made in the past. Jewell and Cummings (1984) developed a multistep countercurrent approach to extract pectin from apple pomace and recommended that the processing be done without mixing raw material batches and after identifying the best processing conditions.

Countercurrent extraction of pectin, on the other hand, reduced its gelling strength. Ngadi and Correia (1992) investigated obtaining pectin from apple pomace (using 'Pippin') and used ethanol or 2%–4% AlCl3 to precipitate pectin from crude extract. With ethanol, pectin yields were 7.2%, 1.11% with 2% AlCl3, 2.38% with 3% AlCl3, and 2.8% with 4% AlCl3. The gel strength was best in samples precipitated with 3% AlCl3, while the residual aluminum concentration was considerable. Because of the large yield and acceptable characteristics of pectin, overall precipitation with ethanol was advised.

Extraction method of pectin from apples under acidic conditions was patented by Ezhov et al. (1993). Separation of pressings into solid and liquid phases, precipitation of pectin from the liquid phase, and the subsequent drying are all part of the technique. After the pectin is removed, the solid and liquid phases can both sustain microbial growth. Additional extraction with equimolar quantities of NH4 OH was observed to improve pectin output and microbial growth on the solid phase.

19.2.2.1.5 Apple Pomace Utilized as Cattle Feed

Traditionally, apple pomace was used as cow feed. An innovative strategy for recovering ethanol and producing animal feed at the same time has also been proposed by Joshi and Sandhu (1996). Based on body weight gains, live body measures and metabolic trials, Narang and Lal (1985) claim that apple pomace can be safely added to animal feeds. Bae (1994) evaluated a total mixed ration (AP-TMR) with standard feeds that contained 39% apple pomace (control). When comparing the cows fed the AP-TMR diet to the cows fed the control diet, they found that the AP-TMR diet boosted protein content while decreasing lactose content in milk. Both diets had identical amounts of milk fat and solid not fat (SNF). The body weight of AP-TMR-fed cows was also higher than that of the control-fed animals. The AP-TMR diet had a higher feed cost per kg milk output but a greater gross income than the control.

19.2.2.1.6 Biotransformation of Apple Pomace

Various microbial transformations of apple pomace are used to produce important products such as biogas, ethanol, butanol, citric acid and pectinases. Yeast can convert fermentable sugars in apple pomace including glucose, fructose and sucrose. Ethanol is being studied as a potential alternative fuel to supplement or completely replace fossil fuel. Hang and Woodams (1994) defined a solid-state fermentation method for producing ethanol from apple pomace using a *Saccharomyces cerevisiae* strain. Depending on the samples fermented, ethanol yield ranged from 30 g/kg –40 g/kg pomace. A rotary vacuum evaporator was used to separate up to 99% of ethanol from discarded apple pomace. The findings suggested that the apple pomace fermentation could be a cost-effective way to solve waste disposal issues while also producing ethanol.

Two processing technologies for transforming apple processing wastes material into a valuable products were identified by Jewell and Cumming (1984). The first step contains the anaerobic digestion of pomace, which produces energy in the form of biogas. According to a batch and pilot scale studies, almost 80% of the pomace material might be transformed into a natural gas replacement. Ultrafiltration and gel chromatography were used to purify crude-fructofuranosidase homogenate. Of the three species tested, *Aspergillus foetidus* had the highest -fructofuranosidase activity (Ngadi and Correia, 1992). According to Bhalla and Joshi (1994), solid-state and liquid state fermentation boosted the nutritional profile of apple pomace. Fermentation was said to be just marginally useful. The modified apple pomace had two times the amount of essential amino acids as the control.

19.2.2.1.7 Apple Pomace as a Fiber

Apple pomace, according to chemical research, is high in total dietary fiber but also comprises a considerable content of soluble dietary fiber, which includes pectin. Bielig et al. (1984) reported a method for producing bakery items that are lower in energy and higher in fiber. Up to 20% of the product weight was added in the form of dried and powdered residue from juice extraction. The most frequent method for extracting juice is to use an apple or a combination of apples and pears. To generate unique flavors, stone fruit or berry pulps may be utilized goods. The sugar concentration in the pulp expedited dough fermentation, resulting in a sour dough reduction. The bread's shelf life has been improved.

Patt et al. (1984) reduced the calorie content of bread (wheat, rye and mixed) by adding 5%–10% apple pomace powder. Baking studies revealed that apple powder, which is less expensive than flour, can be used to add fiber to bread and manage acidity as a sugar alternative in breads made with rye or rye-wheat flour blends (Walter et al., 1985). In another study, Chen et al. (1988) used chemical and physical methods to assess apple fiber and found, and excellent source dietary fibers. Apple fiber added to cookies and muffins at a rate of 4% without degrading the quality of the cookies or muffins. Sensory characteristics of muffins with 50% plain wheat bran replaced with powdered apple pomace were much better than control bran muffins, according to Wang and Thomas (1989). Pomace was utilized in pie fillings and oatmeal cookies by Carson et al. (1994).

19.2.2.1.8 Various Other Applications of Apple Pomace

Shalini and Gupta (2010) recognized apple pomace as a potential filler, extender, bulking agent and microcrystalline cellulose alternative. Hinsch and Simon (1992) patented a method for recycling apple peeling factory trash. Fresh apple peels and cores were comminuted into a pumpable pulp at apple peeling factories. This pulp can be kept refrigerated for up to 8 days. It was crushed to extract a juice that could be used to make apple concentrate and fragrance. When Ramm et al. (1994) investigated the extraction of waxes from dried apple pomace, they discovered that the yield of waxes was higher with solvents than with CO_2 and that this yield could be further increased by de-pectinization and grinding with 90% ethanol, but this required the removal of water soluble components. Triterpenoids were found in all waxes, regardless of extraction method.

19.2.3 APPLE SEEDS

The chemical profile of apple seed oil and the study of its in vitro biological activities revealed that it had intriguing features. The oil's biological activities reported that it is useful in the food and pharmaceutical industries. Apples are widely regarded as one of the most important fruit resources (Tian et al., 2010).

19.2.3.1 Utilization of Apple Seeds

19.2.3.1.1 Antioxidant and Antibacterial Properties of Apple Seed Oil

Fruit seed oil is used in a variety of products, including food, fragrances, cosmetics and chemical additives. Because of its powerful antibacterial and insecticidal properties, plant seed oil is an important and fascinating essential oil. Large regions of apple plantations can be found in northern China. Every year a substantial quantity of apple seeds is generated which is typically thrown away as agro-industrial waste but might be beneficially used for oil production and value addition. This suggests that a complete investigation to extract and describe apple seed oil is required. Most abundant were 50.7–51.4 g/100 g linoleic acid and 37.49–38.55 g/100 g oleic acids in apple seed oil. The chemical profile of apple seed oil and the study of its in vitro biological activities revealed that it had intriguing features. The oil's biological activities proposed that is useful in the food and medicinal industries (Sicuro et al., 2010).

19.2.3.1.2 Apple Seed Oils

Because of the growing tendency to substitute synthetic alternatives for their natural equivalents, altering customer demand and environmental concerns, several industries are reverting to the procedure of chemicals derived from raw materials. The fundamental concern for most commercial size businesses is the control of by-products and the reduction of escalating production costs. As a result, in many circumstances, proper by-product use allows not only for the acquisition of natural bio-components, but also for the acquisition of large economic advantages. The utilization of seed oil manufacturing is an excellent illustration of how by-products created by the fruit industry are handled (Górnaś et al., 2013).

The composition of lipophilic compounds of oils extracted from seeds of five dessert and six crabapple varieties were studied. By-products formed during the preparation of fruit salads and juice pressing were used to recover apple seeds. Apple juice production is one of the industries that produces a significant number of by-products in comparison to the original number of processed apples; it is estimated at 25% (Mahawar et al., 2012). Apple seeds can make up as much as 0.7% of the fresh fruit, depending on the variety (Górnaś et al., 2014).

Phytosterols are improved from cellulose manufacturing oil waste in the pharmaceutical and cosmetics industries and used to make medicinal steroids, lotions and lipsticks (Górnaś et al., 2014). Squalling is a natural isoprenoid chemical. It is responsible for protecting human mammary epithelial cells from oxidative DNA damage as an antioxidant (Górnaś et al., 2013; Warleta et al., 2010). However, there is a significant variation in oil yield as well as amounts of phytosterols and squalene and a lower variation in fatty acid composition in seed oils extracted from different apple species.

Furthermore, the amount of phytosterols and the content of fatty acids were more connected to oil output than to the division between dessert and crab apples. Nonetheless, efficient apple seed utilization would allow for the control of tons of by-product created during apple processing (Arain et al., 2012).

19.2.3.1.3 Therapeutic Potential of Apple and Its Waste Products

Seeds are a natural component of apple pomace, accounting for between 2% and 3% of the dry material. Apple seeds are high in proteins, dietary fiber and lipids, all of which are concentrated in the endosperm (Manzoor et al., 2021). They are good for oil extraction because of their high lipid content (about 29%). Apple seed oil is high in unsaturated fatty acids, particularly linoleic acid, which accounts for around 44% of the total fatty acids in the oil. Significant levels of protein and fiber can still be discovered in seed remnants after extraction for future use.

On the other hand, apple seeds comprise a considerable amount of toxigenic amygdalin (1.0–3.9 mg/g seeds), a cyanogenic glycoside categorized as a mandelonitrile gentiobioside (Madubuike and Okereke, 2009).

The hazardous activity of cyanide is mostly due to its affinity for the mitochondrial respiratory pathway's terminal cytochrome oxidase, which results in its blockage. Cells are unable to utilize oxygen as a result. As a result, prolonged amygdalin use is considered risky and can disrupt key physiological systems in the body. However, animal research demonstrating amygdalin's positive effects are also accessible in the literature. Intraperitoneally administered amygdalin, for example, reduced renal fibrosis in rats as their chronic kidney disease progressed (Bolarinwa et al., 2015; Madubuike et al., 2003). Finally, defatted apple seed is a high-protein, high-fiber food. According to a previous study, apple seed meal supplementation can promote satiety or decrease diet palatability in rats and has positive effects on the digestive tract, blood lipid profile and antioxidant status. In most circumstances, the presence of amygdalin has no influence on these consequences (Opyd et al., 2017).

19.3 CONCLUSION

Apples and their waste products have the potential to be converted into edible food products because these are a part of the fruit. These waste products are excellence source of nutritional components because they are high in carbohydrate, pectin, crude fiber and minerals. Utilization of apple by-products is beneficial for human being. Moreover, different bioactive components extracted from apple waste products are a cheap source of many functional and nutraceutical moieties. So there is dire need for complete utilization of apple industrial waste as food, medicine and cosmetics products.

REFERENCES

Aachary, A. A., and Prapulla, S. G. 2011. Xylooligosaccharides (XOS) as an emerging prebiotic: Microbial synthesis, utilization, structural characterization, bioactive properties, and applications. *Comprehensive Reviews in Food Science and Food Safety 10*(1): 2–16.

Agrahari, P. R., and Khurdiya, D. S. 2003. Studies on preparation and storage of RTS beverage from pulp of culled apple pomace. *Indian Food Packer 57*(2): 56–61.

Alonso-Salces, R. M., Korta, E., Barranco, A., Berrueta, L. A., Gallo, B., and Vicente, F. 2001. Determination of polyphenolic profiles of Basque cider apple varieties using accelerated solvent extraction. *Journal of Agricultural and Food Chemistry 49*(8): 3761–3767.

Anstalt, S. V. 2013. *Food and Agriculture Organization of the United Nations (FAOSTAT)*. Food and Agriculture Organization of the United Nations. http://faostat.fao.org/site/377/DesktopDefault.aspx? PageID=377#ancor

Arain, S., Sherazi, S. T. H., Bhanger, M. I., Memon, N., Mahesar, S. A., and Rajput, M. T. 2012. Prospects of fatty acid profile and bioactive composition from lipid seeds for the discrimination of apple varieties with the application of chemometrics. *Grasas y Aceites 63*(2): 175–183.

Bae, D. H. 1994. Effect of total mixed ration including apple pomace for lactating cows. *Korean Journal of Dairy Science 16*: 295–302.

Bates, A. W., and Roberts, J. S. 2001. The utilization of apple pomace as a press aid in fruit juicing. In *IFT Annual Meeting – New Orleans, Louisiana: Session E* (Vol. 88). www.ift.confex.com/ift/paper_8281. htm

Bhalla, T. C., and Joshi, M. 1994. Protein enrichment of apple pomace by co-culture of cellulolytic moulds and yeasts. *World Journal of Microbiology and Biotechnology 10*(1): 116–117.

Bielig, H., Fanghaelnel, K., Kuettner, D., Viehweger, T., and Strauss, J. 1984. Process for manufacture of bakery products and patisserie products. *German Democratic Republic Patent. DD*: 213587.

Bolarinwa, I. F., Orfila, C., and Morgan, M. R. 2015. Determination of amygdalin in apple seeds, fresh apples and processed apple juices. *Food Chemistry 170*: 437–442.

Bondonno, N. P., Bondonno, C. P., Ward, N. C., Hodgson, J. M., and Croft, K. D. 2017. The cardiovascular health benefits of apples: Whole fruit vs. isolated compounds. *Trends in Food Science & Technology 69*: 243–256.

Boyer, J., and Liu, R. H. 2004. Apple phytochemicals and their health benefits. *Nutrition Journal 3*(1): 1–15.

Bringe, K., Schumacher, C. F., Schmitz-Eiberger, M., Steiner, U., and Oerke, E. C. 2006. Ontogenetic variation in chemical and physical characteristics of adaxial apple leaf surfaces. *Phytochemistry 67*(2): 161–170.

Burda, S., Oleszek, W., and Lee, C. Y. 1990. Phenolic compounds and their changes in apples during maturation and cold storage. *Journal of Agricultural and Food Chemistry 38*(4): 945–948.

Can-Cauich, C. A., Sauri-Duch, E., Betancur-Ancona, D., Chel-Guerrero, L., González-Aguilar, G. A., Cuevas-Glory, L. F., . . . and Moo-Huchin, V. M. 2017. Tropical fruit peel powders as functional ingredients: Evaluation of their bioactive compounds and antioxidant activity. *Journal of Functional Foods 37*: 501–506.

Carson, K. J., Collins, J. L., and Penfield, M. P. 1994. Unrefined, dried apple pomace as a potential food ingredient. *Journal of Food Science 59*(6): 1213–1215.

Chen, H., Rubenthaler, G. L., Leung, H. K., and Baranowski, J. D. 1988. Chemical, physical, and baking properties of apple fiber compared with wheat and oat bran. *Cereal Chemistry 65*(3): 244–247.

Cho, E. H., Jung, H. T., Lee, B. H., Kim, H. S., Rhee, J. K., and Yoo, S. H. 2019. Green process development for apple-peel pectin production by organic acid extraction. *Carbohydrate Polymers 204*: 97–103.

Dimou, C., Karantonis, H. C., Skalkos, D., and Koutelidakis, A. E. 2019 Valorization of fruits by-products to unconventional sources of additives, oil, biomolecules and innovative functional foods. *Current Pharmaceutical Biotechnology 20*(10): 776–786.

Duman, A. K., Özgen, G. Ö., and Üçtuğ, F. G. 2020. Environmental life cycle assessment of olive pomace utilization in Turkey. *Sustainable Production and Consumption 22*: 126–137.

Eberhardt, M. V., Lee, C. Y., and Liu, R. H. 2000. Antioxidant activity of fresh apples. *Nature 405*(6789): 903–904.

El Kantar, S., Boussetta, N., Lebovka, N., Foucart, F., Rajha, H. N., Maroun, R. G., Louka, N., and Vorobiev, E. 2018. Pulsed electric field treatment of citrus fruits: Improvement of juice and polyphenols extraction. *Innovative Food Science & Emerging Technologies 46*: 153–161.

Ezhov, V. N., Lebedev, V. V., Lukanin, A. S., Petrushevskii, V. V., and Romanovskaya, T. I. 1993. Wasteless. *Processing of Apple Pressings. USSR Patent SU 1*(785): 639.

Faqeerzada, M. A., Rahman, A., Joshi, R., Park, E., and Cho, B. K. 2018. Postharvest technologies for fruits and vegetables in South Asian countries: A review. *Korean Journal of Agricultural Science 45*(3): 325–353.

Fernandes, P. A., Le Bourvellec, C., Renard, C. M., Nunes, F. M., Bastos, R., Coelho, E., Wessel, D. F., Coimbra, M. A., and Cardoso, S. M. 2019. Revisiting the chemistry of apple pomace polyphenols. *Food chemistry 294*: 9–18.

Frighetto, R. T., Welendorf, R. M., Nigro, E. N., Frighetto, N., and Siani, A. C. 2008. Isolation of ursolic acid from apple peels by high speed counter-current chromatography. *Food Chemistry 106*(2): 767–771.

Geană, E. I., Ciucure, C. T., Ionete, R. E., Ciocârlan, A., Aricu, A., Ficai, A., and Andronescu, E. 2021. Profiling of phenolic compounds and triterpene acids of twelve apple (Malus domestica borkh.) cultivars. *Foods 10*(2): 267.

Ghorai, S., Banik, S. P., Verma, D., Chowdhury, S., Mukherjee, S., and Khowala, S. 2009. Fungal biotechnology in food and feed processing. *Food Research International 42*(5–6): 577–587.

Golding, J. B., McGlasson, W. B., Wyllie, S. G., and Leach, D. N. 2001. Fate of apple peel phenolics during cool storage. *Journal of Agricultural and Food Chemistry 49*(5): 2283–2289.

Górnaś, P., Rudzińska, M., and Segliņa, D. 2014. Lipophilic composition of eleven apple seed oils: A promising source of unconventional oil from industry by-products. *Industrial Crops and Products 60*: 86–91.

Górnaś, P., Siger, A., and Segliņa, D. 2013. Physicochemical characteristics of the cold-pressed Japanese quince seed oil: New promising unconventional bio-oil from by-products for the pharmaceutical and cosmetic industry. *Industrial Crops and Products 48*: 178–182.

Hang, Y. D., and Woodams, E. E. 1994. Apple pomace: A potential substrate for production of β-glucosidase by Aspergillus foetidus. *LWT-Food Science and Technology 27*(6): 587–589.

Helkar, P. B., Sahoo, A. K., and Patil, N. J. 2016. Review: Food industry by-products used as a functional food ingredients. *International Journal of Waste Resources 6*(3): 1–6.

Herrero, M., Castro-Puyana, M., Mendiola, J. A., and Ibañez, E. 2013. Compressed fluids for the extraction of bioactive compounds. *TrAC Trends in Analytical Chemistry 43*: 67–83.

Hertog, M. G., Kromhout, D., Aravanis, C., Blackburn, H., Buzina, R., Fidanza, F., Giampaoli, S., Jansen, A., Menotti, A., Nedeljkovic, S., and Katan, M. B. 1995. Flavonoid intake and long-term risk of coronary heart disease and cancer in the seven countries study. *Archives of Internal Medicine 155*(4): 381–386.

Hinsch, D., and Simon, M. 1992. Process for utilization of wastes from apple peeling plants and the products. *German Federal Republic Patent DE 4115*(162): C1.

Hodgson, J. M., Prince, R. L., Woodman, R. J., Bondonno, C. P., Ivey, K. L., Bondonno, N., . . . and Lewis, J. R. 2016. Apple intake is inversely associated with all-cause and disease-specific mortality in elderly women. *British Journal of Nutrition 115*(5): 860–867.

Jalis, H., Ahmad, A., Khan, S. A., and Sohail, M. 2014. Utilization of apple peels for the production of plant cell-wall degrading enzymes by Aspergillus fumigatus MS16. *Journal of Animal and Plant Sciences 24*(2): 64–67.

Jewell, W. J., and Cummings, R. J. 1984. Apple pomace energy and solids recovery. *Journal of Food Science 49*(2): 407–410.

Joseph, S. V., Edirisinghe, I., and Burton-Freeman, B. M. 2016. Fruit polyphenols: A review of anti-inflammatory effects in humans. *Critical Reviews in Food Science and Nutrition 56*(3): 419–444.

Joshi, V. K., and Sandhu, D. K. 1996. Preparation and evaluation of an animal feed byproduct produced by solid-state fermentation of apple pomace. *Bioresource Technology 56*(2–3): 251–255.

Ju, Z., Yuan, Y., Liu, C., Zhan, S., and Wang, M. 1996. Relationships among simple phenol, flavonoid and anthocyanin in apple fruit peel at harvest and scald susceptibility. *Postharvest Biology and Technology* 8(2): 83–93.

Kaushal, N. K., Joshi, V. K., and Sharma, R. C. 2002. Effect of stage of apple pomace collection and the treatment on the physico-chemical and sensory qualities of pomace Papad (fruit cloth). *Journal of Food Science and Technology (Mysore)* 39(4): 388–393.

Knekt, P., Isotupa, S., Rissanen, H., Heliövaara, M., Järvinen, R., Häkkinen, S., Aromaa, A., and Reunanen, A. 2000. Quercetin intake and the incidence of cerebrovascular disease. *European Journal of Clinical Nutrition* 54(5): 415–417.

Knekt, P., Järvinen, R., Reunanen, A., and Maatela, J. 1996. Flavonoid intake and coronary mortality in Finland: A cohort study. *BMJ* 312(7029): 478–481.

Knekt, P., Järvinen, R., Seppänen, R., Heliövaara, M., Teppo, L., Pukkala, E., and Aromaa, A. 1997. Dietary flavonoids and the risk of lung cancer and other malignant neoplasms. *American Journal of Epidemiology* 146(3): 223–230.

Kubola, J., and Siriamornpun, S. 2011. Phytochemicals and antioxidant activity of different fruit fractions (peel, pulp, aril and seed) of Thai gac (Momordica cochinchinensis Spreng). *Food Chemistry* 127(3): 1138–1145.

Kumar, R., Singh, S., and Singh, O. V. 2008. Bioconversion of lignocellulosic biomass: Biochemical and molecular perspectives. *Journal of Industrial Microbiology and Biotechnology* 35(5): 377–391.

Kunadian, V., Chan, D., Ali, H., Wilkinson, N., Howe, N., McColl, E., ... and De Soyza, A. 2018. Antiplatelet therapy in the primary prevention of cardiovascular disease in patients with chronic obstructive pulmonary disease: Protocol of a randomised controlled proof-of-concept trial (APPLE COPD-ICON 2). *BMJ Open* 8(5): e020713.

Łata, B., Trampczynska, A., and Paczesna, J. 2009. Cultivar variation in apple peel and whole fruit phenolic composition. *Scientia Horticulturae* 121(2): 176–181.

Le Marchand, L., Murphy, S. P., Hankin, J. H., Wilkens, L. R., and Kolonel, L. N. 2000. Intake of flavonoids and lung cancer. *Journal of the National Cancer Institute* 92(2): 154–160.

Liu, R. H., Eberhardt, M. V., and Lee, C. Y. 2001. Antioxidant and antiproliferative activities of selected New York apple cultivars. *New York Fruit Quarterly* 9(2): 15–17.

Lyu, F., Luiz, S. F., Azeredo, D. R. P., Cruz, A. G., Ajlouni, S., and Ranadheera, C. S. 2020. Apple pomace as a functional and healthy ingredient in food products: A Review. *Processes* 8(3): 319.

Madubuike, F. N., Agiang, E. A., Ekenyem, B. U., and Ahaotu, E. O. 2003. Replacement value of rubber seed cake on performance of starter broiler. *Journal of Agriculture and Food Science* 1: 21–27.

Madubuike, F. N., and Okereke, C. O. 2009. The potential use of white star apple seeds (Chrysophyllum albidum) and physic nut (Jatropha curcas) as feed ingredients for rats. *Nigeria Agricultural Journal* 40(1–2).

Mahawar, M., Singh, A., and Jalgaonkar, K. 2012. RETRACTED: Utility of apple pomace as a substrate for various products: A review. *Food and Bioproducts Processing* 90(4): 597–605.

Mallampati, R., and Valiyaveettil, S. 2013. Apple Peels: A versatile biomass for water purification? *ACS Applied Materials & Interfaces* 5(10): 4443–4449.

Manzoor, M., Singh, J., and Gani, A. 2021. Assessment of physical, microstructural, thermal, techno-functional and rheological characteristics of apple (Malus domestica) seeds of Northern Himalayas. *Scientific Reports* 11(1): 1–10.

Meredith, F. I., Leffler, R. G., and Lyon, C. E. 1990. Detection of firmness in peaches by impact force response. *Transactions of the ASAE* 33(1): 186–188.

Min, B., Lim, J., Ko, S., Lee, K. G., Lee, S. H., and Lee, S. 2011. Environmentally friendly preparation of pectins from agricultural byproducts and their structural/rheological characterization. *Bioresource Technology* 102(4): 3855–3860.

Mushtaq, R., and Nayik, G. A. 2020. Apple. In *Antioxidants in Fruits: Properties and Health Benefits* (pp. 507–521). Springer, Singapore.

Narang, M. P., and Lal, R. 1985. Evaluation of some agro-industrial wastes in the feed of Jersey calves. *Agricultural Wastes* 13(1): 15–21.

Ngadi, M. O., and Correia, L. R. 1992. Kinetics of solid-state ethanol fermentation from apple pomace. *Journal of Food Engineering* 17(2): 97–116.

Opyd, P. M., Jurgoński, A., Juśkiewicz, J., Milala, J., Zduńczyk, Z., and Król, B. 2017. Nutritional and health-related effects of a diet containing apple seed meal in rats: The case of amygdalin. *Nutrients* 9(10) 1091.

Ordoudi, S. A., Bakirtzi, C., and Tsimidou, M. Z. 2018. The potential of tree fruit stone and seed wastes in Greece as sources of bioactive ingredients. *Recycling 3*(1): 9.

Pádua, T. A., de Abreu, B. S., Costa, T. E., Nakamura, M. J., Valente, L. M., das Graças Henriques, M., Siani, A. C., and Rosas, E. C. 2014. Anti-inflammatory effects of methyl ursolate obtained from a chemically derived crude extract of apple peels: Potential use in rheumatoid arthritis. *Archives of Pharmacal Research 37*(11): 1487–1495.

Patt, V. A., Vasin, M. I., Shcherbatenko, V. V., Petrash, I. P., Kramynina, A. A., and Kuzetsova, N. V. 1984. Use of powdered fruit preparation in bread making. *Khlebopekarnaya i Konditerskaya Promyshlennost 1*: 18–20.

Podsędek, A., Wilska-Jeszka, J., Anders, B., and Markowski, J. 2000. Compositional characterisation of some apple varieties. *European Food Research and Technology 210*(4): 268–272.

Prothon, F., Ahrné, L. M., Funebo, T., Kidman, S., Langton, M., and Sjöholm, I. 2001. Effects of combined osmotic and microwave dehydration of apple on texture, microstructure and rehydration characteristics. *LWT-Food Science and Technology 34*(2): 95–101.

Quiles, A., Campbell, G. M., Struck, S., Rohm, H., and Hernando, I. 2018. Fiber from fruit pomace: A review of applications in cereal-based products. *Food Reviews International 34*(2): 162–181.

Rabetafika, H. N., Bchir, B., Blecker, C., and Richel, A. 2014. Fractionation of apple by-products as source of new ingredients: Current situation and perspectives. *Trends in Food Science & Technology 40*(1): 99–114.

Radhiga, T., Rajamanickam, C., Senthil, S., and Pugalendi, K. V. 2012. Effect of ursolic acid on cardiac marker enzymes, lipid profile and macroscopic enzyme mapping assay in isoproterenol-induced myocardial ischemic rats. *Food and Chemical Toxicology 50*(11): 3971–3977.

Ramm, A., Baumann, G., and Gierschner, K. 1994. Waxes (including tripernoids) in dried apple pomace and in dried residues after the extraction of pectin. *Industrielle Obst-und Gemeseverwertung (Germany) 79*: 2–9.

Rotova, G. P. 1983. Waste free technology for apple processing. *Pishchevaya Promyschlennost 1*: 49–50.

Sargent, S. A., Steffe, J. F., and Pierson, T. R. 1986. The economic feasibility of in-plant combustion of apple processing wastes. *Agricultural Wastes 15*(2): 85–96.

Saucedo-Pompa, S. A. U. L., Jasso-Cantu, D. I. A. N. A., Ventura-Sobrevilla, J. A. N. E. T. H., Sáenz-Galindo, A. I. D. É., Rodríguez-Herrera, R. A. U. L., and Aguilar, C. N. 2007. Effect of candelilla wax with natural antioxidants on the shelf life quality of fresh-cut fruits. *Journal of Food Quality 30*(5): 823–836.

Shah, G. H., and Masoodi, F. A. 1994. Studies on the utilization of wastes from apple processing plants. *Indian Food Packer 48*: 47–47.

Shalini, R., and Gupta, D. K. 2010. Utilization of pomace from apple processing industries: A review. *Journal of Food Science and Technology 47*(4): 365–371.

Sicuro, B., Dapra, F., Gai, F., Palmegiano, G. B., Schiavone, R., Zilli, L., and Vilella, S. 2010. Olive oil by-product as a natural antioxidant in gilthead sea bream (Sparus aurata) nutrition. *Aquaculture International 18*(4): 511–522.

Soppelsa, S., Kelderer, M., Casera, C., Bassi, M., Robatscher, P., and Andreotti, C. 2018. Use of biostimulants for organic apple production: Effects on tree growth, yield, and fruit quality at harvest and during storage. *Frontiers in Plant Science 9*: 1342.

Szakiel, A., Pączkowski, C., Pensec, F., and Bertsch, C. 2012. Fruit cuticular waxes as a source of biologically active triterpenoids. *Phytochemistry Reviews 11*(2–3): 263–284.

Tabak, C., ARTS, I. C., Smit, H. A., Heederik, D., and Kromhout, D. 2001. Chronic obstructive pulmonary disease and intake of catechins, flavonols, and flavones: The MORGEN Study. *American Journal of Respiratory and Critical Care Medicine 164*(1): 61–64.

Talcott, S. T., Howard, L. R., and Brenes, C. H. 2000. Contribution of periderm material and blanching time to the quality of pasteurized peach puree. *Journal of Agricultural and Food Chemistry 48*(10): 4590–4596.

Tian, H. L., Zhan, P., and Li, K. X. 2010. Analysis of components and study on antioxidant and antimicrobial activities of oil in apple seeds. *International Journal of Food Sciences and Nutrition 61*(4): 395–403.

van der Sluis, A. A., Dekker, M., de Jager, A., and Jongen, W. M. 2001. Activity and concentration of polyphenolic antioxidants in apple: Effect of cultivar, harvest year, and storage conditions. *Journal of Agricultural and Food Chemistry 49*(8): 3606–3613.

Vinson, J. A., Su, X., Zubik, L., and Bose, P. 2001. Phenol antioxidant quantity and quality in foods: Fruits. *Journal of Agricultural and Food Chemistry 49*(11): 5315–5321.

Walter, R. H., Rao, M. A., Sherman, R. M., and Cooley, H. J. 1985. Edible fibers from apple pomace. *Journal of Food Science 50*(3): 747–749.

Wang, H. J., and Thomas, R. L. 1989. Direct use of apple pomace in bakery products. *Journal of Food Science 54*(3): 618–620.

Wang, L., Boussetta, N., Lebovka, N., Lefebvre, C., and Vorobiev, E. 2019. Correlations between disintegration degree of fruit skin cells induced by ultrasound and efficiency of bio-compounds extraction. *Ultrasonics Sonochemistry 52*: 280–285.

Wang, L., Boussetta, N., Lebovka, N., and Vorobiev, E. 2020. Cell disintegration of apple peels induced by pulsed electric field and efficiency of bio-compound extraction. *Food and Bioproducts Processing 122*: 13–21.

Warleta, F., Campos, M., Allouche, Y., Sánchez-Quesada, C., Ruiz-Mora, J., Beltrán, G., and Gaforio, J. J. 2010. Squalene protects against oxidative DNA damage in MCF10A human mammary epithelial cells but not in MCF7 and MDA-MB-231 human breast cancer cells. *Food and Chemical Toxicology 48*(4): 1092–1100.

Watson, R. R. (Ed.). 2018. *Polyphenols in Plants: Isolation, Purification and Extract Preparation.* Academic Press. https://www.elsevier.com/books/polyphenols-in-plants/watson/978-0-12-813768-0

Wojdyło, A., Nowicka, P., Turkiewicz, I. P., Tkacz, K., and Hernandez, F. 2021. Comparison of bioactive compounds and health promoting properties of fruits and leaves of apple, pear and quince. *Scientific Reports 11*(1): 1–17.

Wolfe, K. L., and Liu, R. H. 2003. Apple peels as a value-added food ingredient. *Journal of Agricultural and Food Chemistry 51*(6): 1676–1683.

Yan, H. 2012. *Vacuum Belt Dried Apple Pomace Powder as a Value-added Food Ingredient* (Doctoral dissertation, University of Georgia).

Zafar, F., Idrees, M., and Ahmed, Z. 2005. Use of apple by-products in poultry rations of broiler chicks in Karachi. *Pakistan Journal of Physiology 1*(1–2).

20 Current World Trade and Market Trends

S. A. Wani, Farheen Naqash and Fehim J. Wani

CONTENTS

20.1 INTRODUCTION

During the first few 5-year-plans, priority was assigned to achieve self-sufficiency in food-grains production. Over the years, horticulture emerged as an important and growing subsector of agriculture, offering a wide range of choices to the farmers for crop diversification. The horticulture sector has emerged as an important sector for diversification of agriculture and has established its credibility in improving farm income through increased productivity, generating employment and in enhancing exports in addition to providing household nutritional security. The focused attention on investment in horticulture during past few decades has been rewarding in terms of increased production and productivity of horticultural crops with manifold export potential.

DOI: 10.1201/9781003239925-20

Horticulture sector contributes about 24.5 percent of the GDP from about 8 percent of the area under the sector. A large variety of fruits are grown in India and account for an area of 6.3 million ha under fruit crops with a production of 90.2 million metric tons and 10 percent of the total world production.

Apple is the fourth most widely produced fruit in the world after bananas, oranges and grapes with a growing demand. About 87 million tons of apples are produced worldwide with China contributing to the maximum in production followed by the United States. The area, production and productivity of apples at the world level has recorded a compound growth rate of 1.2 percent, 2.4 percent and 1.2 percent per annum from 1973–74 to 2014–15 respectively. During the same period the area, production and productivity in Asia recorded a growth rate of 3.9 percent, 6.3 percent and 2.4 percent per annum respectively. India recorded a growth rate of 2.4 percent, 3.5 percent and 1.1 percent, respectively. India isthe fifth largest apple producer in the world with productivity of 9.46 tons per hectare,

Apples are an ideal value chain in the temperate horticulture sector of India, and it is profitable for all players. Apples in India are mainly grown in three mountainous states of northern India, Jammu and Kashmir, Himachal Pradesh and Uttaranchal at an altitude of 4000 to 11000 feet. Jammu and Kashmir and Himachal Pradesh have roughly equal acreage under apples, but Jammu and Kashmir has the highest average yield and accounts for 77 percent of total apple production; it is important for the economic growth of the region (NHB, 2018). Jammu and Kashmir state, being endowed with natural advantages of topography, climate and enormous diversity of agro-climatic niches, has immense scope for apple cultivation. A productivity of 12 metric tons per hectare was also achieved in Jammu and Kashmir, which is the highest in India and even better than in China.

Although apple production under the horticulture sector is increasing year by year in the state, there is not a significant growth in exports. The improvements in the production are quite important, but marketing also has an equal importance to develop a commercial crop, which is purely produced to be processed and sold in the market. Apple marketing is a complex phenomenon that requires special treatment and the utmost care at present in the Kashmir Valley. Due to powerful intermediaries in the marketing system, present marketing has an inherent tendency to give more benefits to these intermediaries at the cost of the small apple growers (Naqash *et al.*, 2017). Value chain is a wider concept and deserves careful attention toward preharvest and postharvest technologies and operations. It is in this context that the chapter is conceived to analyze the market of the apple value chain, its effectiveness and inherent lacunae which mar the prospects of small apple growers to realize better returns for their produce. The information generated through this chapter can provide a guideline to flag agendas for policy planners, practitioners, development departments and financial institutions to better serve the interests of small holder apple growers with efficient delivery system.

20.2   WORLD PRODUCTION SCENARIO OF APPLES

About 87 million tons of apples are produced worldwide with China producing more than 48 percent of this total production. The United States, with more than 5 percent of the world production is the second leading producer. The other important global producers are Turkey, Poland, India, Italy, Iran, Russia, France and Chile. India's apple production is 23.16 million tons on 0.28 million hectare making it the fifth largest apple producer in the world with productivity of 9.46 tons per hectare, which is much lower than that of countries like Belgium (46.22 t/ha), Denmark (41.87 t/ha) and Netherlands (40.40 t/ha) (FAO, 2019–20). Substantial progress has been recorded in previous plan periods for apple fruit cultivation in terms of area coverage, production and productivity and still a vast potential exists for both vertical and horizontal expansion. Presently, a small quantity of apple produced in India is mainly exported to Bangladesh and Sri Lanka (Figure 20.1).

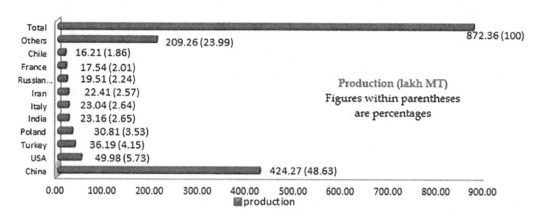

FIGURE 20.1 Leading apple producing countries globally.

Source: FAO (2019)

TABLE 20.1

Region Wise Compound Growth Rates in Apple (1973–74 to 2014–15)

Regions	Area	Production	Yield
India	2.4** (0.044)	3.5* (0.090)	1.1 (0.088)
Asia	3.9* (0.181)	6.3* (0.093)	2.4* (0.133)
World	1.2 (0.088)	2.4* (0.044)	1.2 (0.088)

Source: Wani *et al.*, 2021

Note: * Significant at 1 percent level of significance; ** Significant at 5 percent level of significance

20.3 TRENDS IN AREA, PRODUCTION AND PRODUCTIVITY OF APPLES

The area, production and yield of apple at the world level has recorded a compound growth rate of 1.2 percent, 2.4 percent and 1.2 percent per annum from 1973–74 to 2014–15 respectively. During the same period the area, the production and yield of apples in Asia recorded a growth rate of 3.9 percent, 6.3 percent and 2.4 percent per annum respectively. India recorded a growth rate of 2.4 percent, 3.5 percent and 1.1 percent respectively (Wani *et al.*, 2021). The region wise compound growth rates can be seen in Table 20.1.

20.4 NATIONAL SCENARIO OF APPLES

Jammu and Kashmir, Himachal Pradesh, Uttarakhand and Arunachal Pradesh are the major apple producing states of India. The two important states, namely Jammu and Kashmir and Himachal Pradesh, account for 97 percent of the total production and about 89 percent of the total area under apples in India. As far as productivity of apple is concerned, Jammu and Kashmir has the highest productivity (11.43 tons/hectare) followed by Himachal Pradesh (3.96 tons/hectare) and Uttarakhand (2.31 tons/hectare) (Figure 20.2).

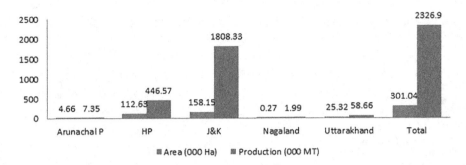

FIGURE 20.2 State-wise area and production of apples during 2018–19.

Source: National Horticulture Board (NHB) New Delhi, 2018

TABLE 20.2
Area, Production and Yield of Apples in Jammu and Kashmir (1980–81 to 2019–20)

Year	Area (ha)	Production (MT)	Yield (MT/ha)
1980–81	60286	536300	8.9
1985–86	63796	760666	11.9
1990–91	68723	658165	9.6
1995–96	78007	714834	9.1
2000–01	88149	757610	8.6
2005–06	111879	1151341	10.3
2010–11	141717	1852413	13.1
2016–17	143501	1726834	12.0
2019–20	164854	2026472	12.3

Source: Directorate of Horticulture, Govt. of Jammu and Kashmir

20.5   STATE SCENARIO OF APPLES

The dynamics of the apple industry in the state (Table 20.2) reveal that area increased by 1.4 per cent from 2010–11 to 2016–17, while as production decreased by 0.2 percent during the same period. In spite of many fold increase in area and production, yield has remained almost stagnant at around 9 metric tons during the past two decades. However, by concerted efforts of farmers, yield has picked up and during the year 2010–11 it was recorded more than 13 metric tons per hectare. Area under apple has witnessed a continuous increase since 1980s. During 1980–81, area under this crop was 60,286 hectares, which increased to 164854 hectare in 2019–20. Production of this fruit crop also exhibited the same pattern. However, during the last decade it has witnessed a marginal increase.

Apples are the principle fruit crop of Jammu and Kashmir and account for 51 percent of total area of 2.72 lac hectare under all temperate fruits grown in this state. The annual apple production in the state ranges between 12–18 lac M tons. The average yield of commercially important apple cultivars per unit area is the highest in the country ranging between 10–13 tons/ha, but it compares poorly to the yields of > 40 tons/ha in horticulturally advanced countries around the world. Climatic and other agro-ecological factors of Kashmir are ideally suited to the cultivation of many apple varieties. Alternate bearing, defective pruning and training, use of seedling rootstock of unknown performance, lack of proper nutrients and water management, deficiency of suitable

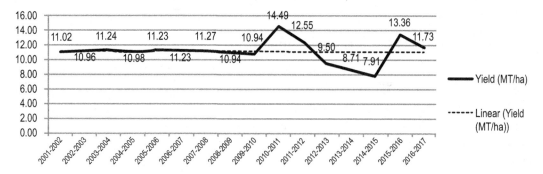

FIGURE 20.3 Area, production and yield of apples in Jammu and Kashmir (2001–2018).

Source: Department of Horticulture, Govt. of Jammu and Kashmir

TABLE 20.3

Exponential Growth Rates of Area, Production and Yield of Apples in Jammu and Kashmir (1981–82 to 2018–19)

Time period	Area	Production	Yield
1981–82 to 1990–91	1.5*	2.7***	1.2***
1991–92 to 2000–01	2.8*	3.0*	0.2***
2001–02 to 2018–19	5.2*	6.5*	1.3***
1981–82 to 2018–19	2.8*	4.0*	1.2*

* Significant at 1 percent; ** Significant at 5 percent; *** Significant at 10 percent

pollinizers and ineffective control of pests and diseases are the main causes of low productivity (Figure 20.3).

Table 20.3 summarizes the exponential growth rate of area, production and yield of apples in Jammu and Kashmir from 1981–82 to 2018–19. The data has been divided into three subperiods from 1981–82 to 2018–19 (1981–82 to 1990–91, 1991–92 to 2000–01 and 2001–02 to 2018–19) as well as the overall period from 1981–82 to 2018–19. This division of periods is considered appropriate to estimate the structural changes that have taken place in respect to the area, production and productivity. The area, production and yield witnessed growth momentum of 2.8 percent, 4 percent and 1.2 percent respectively during the overall period. During the previous decade the state apple industry showed an overriding performance by achieving higher a trajectory in area expansion as well as production and productivity.

20.6 DISTRICT SCENARIO OF APPLES

Apple production in the state extends across several districts regardless of the agro-climatic suitability and productivity. Given the current levels of productivity, districts such as Bandipora, Budgam and Pulwama do not seem well-suited for inclusion in the apple value chain on account of low productivity. In the first stage the focus should be on districts having a higher productivity so that investments have a probability of producing better returns. The average yield across the state was about 12.3 tons/ha, which was less than 30 percent of the level of the top apple producing countries. The yield gap is large and the potential for increases in yield makes the investments in the sector attractive (Table 20.4).

TABLE 20.4
Area, Production, Yield across Districts (2019–20)

Serial No.	District	Area (Ha)	Production (MT)	Productivity (MT/Ha)
1	Baramula	25307	479570	18.95
2	Shopian	21676	276226	12.74
3	Kupwara	19441	283460	14.58
4	Kulgam	18144	207873	11.46
5	Anantnag	18426	259959	14.11
6	Pulwama	15785	172112	10.90
7	Budgam	14061	153356	10.91
8	Bandipore	5088	50385	9.90
9	Ganderbal	7465	91955	12.32
10	Srinagar	1738	20205	11.63

Source: Directorate of Horticulture, Govt. of Jammu and Kashmir

TABLE 20.5
Demand Estimates (2018–19)

20 percent of Indian population	240 million	240 million	240 million
No. of days consumption/year	60	75	90
Per capita consumption/year (150 grams per day)	9 kg	11.25 kg	13.5 kg
Total consumption	2.16 mn tons	2.70 mn tons	3.24 mn tons
Production of apple in Jammu and Kashmir	1.80 mn MT		
Production in other states	0.57 mn MT		
Imports	0.35 mn MT		
Total supply	2.72 mn MT		

20.7   DEMAND ESTIMATES OF APPLES FOR JAMMU AND KASHMIR

The demand for apple table fruit is likely to be strong. The changing food habits as captured by National Sample Survey Office (NSSO) in its monthly per capita expenditure (MPCE) surveys show that the consumption basket is changing all over the country. There is increased consumption of fruits and processed fruit products, and the increases have been secular over the last 10 years. Imported apples that are expensive by about two fold more than domestic apples find a ready market. The per capita availability even at the increased production level of 2017–18 is about 15 fruits in a year. If 20 percent of the population wants to consume an apple for 75 days in a year, the production will not be adequate (Table 20.5).

20.8   TRENDS IN DISTRIBUTION OF APPLES

20.8.1   1950 TO 1990

During the four decade period of 1950 to 1990, almost the entire (100 percent) production of apple from Kashmir valley had its destination as the Delhi/Azadpur market. This pattern of distribution to the sole destination of the Delhi/Azadpur market involved 15 percent quantity supplied by free growers directly and 85 percent supplied through preharvest contractors (PHC). But the ultimate destination was Delhi, from where it was further distributed to the rest of India (ROI) markets. Within this period of 40 years, there were gradual changes taking place in the distribution pattern of apple. Marketing began during 1950s with 100 percent quantity supplied through preharvest contractors but slowly some free growers emerged by 1990 to control 15 percent of production, which nonetheless found its destination in Delhi but without PHC as an intermediary.

20.8.2   During 2017, 2018

The pattern of distribution predominated by preharvest contractors (PHC) had continued during 1990s and 2000s; however, after 2010, there were drastic and more visible changes in the distribution pattern of apples from Kashmir (Figure 20.4). By 2017, the share of free growers in apple distribution had increased from 15 percent to 25 percent, but the apples are now headed straight to local Kashmir markets covering all ten districts in Kashmir division. From Kashmir markets, the apples are forwarded to Delhi and rest of India markets either through local commission agents, forwarding agents or both. Sopore is the biggest local Kashmir market. The share of apple supplied through preharvest contractors (PHC) has come down to 75 percent as compared to 85 percent earlier. At the same time, now these preharvest contractors supply apples not only to Delhi-based commission agents (CA) in Azadpur market but also to the CAs of Kashmir and rest of India markets. The respective shares of the commission agents in different markets of the country getting captive supply of apples through preharvest contractors (75 percent) whereas under 10 percent in Kashmir local markets; 40 percent in the Azadpur Delhi market; and 25 percent in rest of India markets. An important change happening is that now only exportable 'A' and 'B' grades apple are considered for distribution to the various markets. This is a big change compared to the earlier era of 1950 to 1990 when the entire apple production (A, B, C grades) was considered for supply to these markets through whatever channels. Therefore, over a period of time, not only did the respective shares of free grower and preharvest contractor change in movement of apples from the Kashmir Valley, but apple grades considered for marketing also underwent a change. The third change was destination, where apples reached from Kashmir.

The following types of changes in apple distribution and movement have taken place, which are comparable between the 1950–1990 and 2017–2018 periods:

(i)   Shares of free grower and preharvest contractor (departure points)
(ii) Grades of apple
(iii) Destination markets.

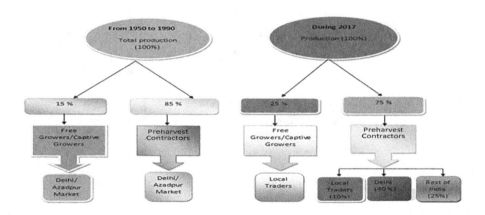

FIGURE 20.4   Trends in distribution of apples.

20.8.2.1 Allocation of Apple Production in 'A', 'B', 'C' Grades

There is visible variation in the ratio of various grades of apple (A, B and C grades) produced in the different districts of the Kashmir valley. Distribution of apples through market channels for various uses depends on ratios and grades which pass through such channels. Various assumptions are made by different agencies for relative proportions of grades of apple in Jammu and Kashmir, which are indicated in Table 20.6.

Out of the total apple production of 17.27 lakh MT in the Kashmir division, A grade, B grade and C grade output was 7.94 lakh MT, 6.22 lakh MT and 3.12 lakh MT, respectively. Exportable (fine variety sold in domestic markets within country) production was 14.16 lakh MT (A and B grades). The 'A' and 'B' grade production of 14.16 lakh MT was disposed of as 5.66 lakh Mt (40 percent) to Delhi markets and the rest 8.50 lakh Mt (60 percent) to the rest of the Indian markets. The 'C' grade production of 3.12 lakh was disposed of as under 4.5 percent (0.78 lakh MT) sold in Kashmir markets for local consumption by low/middle income groups; 4.5 percent (0.78 lakh MT) retained by households for their own consumption; 4.5 percent (0.78 lakh MT) was available for processing within the Kashmir division (big/small food processing units making juice, pulp and concentrate, mainly) in/around Srinagar/Ganderbal/Budgam/Baramulla districts; and 4.5 percent (0.78 lakh MT) was transported and sold in the markets of the Jammu division (ten districts) diverted through Agricultural Produce Market Committee (APMC) Jammu and Udhampur APMC market, for local consumption in Jammu division (Figure 20.5).

TABLE 20.6
Grades of Apple Production in Kashmir (%)

Grade	Horticulture department	Horticulture (P&M) department	Jammu and Kashmir HPMC	High density (HD) cultivation	Average	Assumption for present Study*
A	30	35	50	50	41	40
B	40	35	30	30	34	30
C	30	30	20	20	25	30

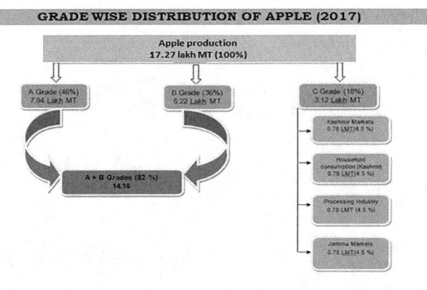

FIGURE 20.5 Grade wise distribution of apple (2017).

20.9 MARKETING SCENARIO OF APPLES

20.9.1 EXISTING MARKETING INFRASTRUCTURE AND FACILITIES

In order to provide a marketing facility for agriculture/horticulture produce at the doorsteps of the farmers/growers, various programs have been initiated to establish three terminal fruit and vegetable markets mainly for apples such as Parimpore (Srinagar), Sopore (Baramulla) and Narwal (Jammu) which are functional, and the trade is carried on in these markets.

A new terminal fruit and vegetable market is being established at Jablipora Anantnag for which 441 Kanals of land has been acquired. In addition, the Department of Horticulture is also developing 21 Satellite fruit and vegetable markets in the state out of which seven are completed and functional, six are likely to be completed and made functional, five are in progress and the land acquisition process for three fruit and vegetable markets is in progress. Furthermore, 11 Apni *Mandis* are also under the process of development out of which three are completed, six are in progress and the land acquisition for two Apni *Mandis* is in progress (Table 20.7). Most of the existing satellite and terminal markets were set up during late 1990s and are being strengthened on piecemeal basis from time to time. These markets lack basic infrastructure in terms of adequate auction platforms, rest houses for farmers/outside buyers, parking areas for trucks, toilets, proper drainage systems, electrification, sanitation, fencing, adequate land for shops, etc. For example, in Sopore Mandi the total handling capacity is about 320 trucks per day, whereas outflow touches 500 trucks/day in peak season leading to traffic jams and temporary closures of the market for fresh arrivals. The traders in all the markets are ready to pay the *Mandi* tax provided the basic infrastructure is created and improved.

20.9.2 MARKET PRACTICE

The price discovery process in the local markets within the state is not transparent. The markets do not have a price dissemination mechanism, and it is difficult to know the prevailing prices on any given day. While there is some understanding of the fees payable to the market intermediaries, there are no norms and enforcement of such norms. High commissions and fees payable to intermediaries tend to get blurred with other fees and charges; they are often adjusted in the price thereby making the realization uncertain. While the APMC law is passed in the State, it has not been implemented. The law needs amendments in line with the changes suggested by the Central Government and adopted by several states. This has led to the proliferation of unlicensed traders, agents acting in the

TABLE 20.7

Fruit and Vegetable Markets in Jammu and Kashmir State

Serial No	Market category	Market location	Total markets
A	Terminal markets	Narwal* Parimpora* Sopore* Jablipora[1]	4
B	Satellite markets	Shopian*, Pulwama*, Kulgam*, Baramulla* Handwara*, Udhampur*, Chari Sharief*, Kupwara[1], Kathua[1], Rajouri[1], Bisnah[1], Akhnoor[1], Batingoo[1], Zazna[1], Leh[1], Kargil[1], Pryote Doda[1], Poonch[1], Samba[1], Sunderbani[1], Aglar(shopain)[1]	21
C	Apni *Mandi*	Pachhar*, Nunmai, Kulgam[1] Mandi, Poonch[1], Pouria, Reasi[1], Gharian (Udh)[1], Tapyal[1], Raya Bagala[1], Dyala Chak[1] (Chadwal), Bhaderwah Bani[1]	11
	TOTAL		37

Source: Directorate of Horticulture, Govt. of Jammu and Kashmir.
* functional markets; [1] markets under process to be made functional

market with nontransparent auction procedures. The APMCs do not earn revenues through collection of *Mandi* tax and are unable to improve the conditions in the *Mandi*.

The major destination of Kashmir apples is the Azadpur *Mandi*, which is a buyers' market and designed to be so. Manipulation of prices by traders in the *Mandi* is resorted through stopping apple trucks at the border of entry into Delhi, use of cold stores to alter supply of apples in the *Mandi*, keeping away small buyers with artificially high price quotes and later reducing prices to low levels to benefit preferred buyers and the use of proxies in auctions. The markets within the state are comparatively better in price determination and transparency. Growers with preharvest contracts (PHC) access the markets easier, but they lose out on the full benefit of market prices on account of taking money in advance. Free growers find it difficult to enter markets even when the demand is brisk, and the commission agents prefer their 'captive growers' with PHC. Setting up satellite markets has helped growers (especially the free ones) in marketing. Farmers who market apples through cooperatives realize higher prices. Trade margins range from 42 percent to 73 percent in the different channels of marketing. Price discovery by grower would be more realistic and effective if they are able to hold back and store their produce for some time. The farmer needs to have conditions (local storage and financial capacity to hold) under which distress selling can be checked. To improve and reform market practice, some measures are required. The first is to organize farmers into producer collectives or cooperatives. These organizations of farmers will help in aggregation of inputs and outputs and improve their bargaining power

20.9.3 MARKETING OF APPLES

The marketing of apples continues to be a difficult task in the whole process of fruit growing. In this section an attempt has been made to study the important parameters of apple marketing like market functionaries, marketing costs and margins, issues of efficiency, price realized by the growers and price spread in the study region.

20.9.3.1 Marketing Costs of Apples (Per Box)

The various stages of marketing and the cost incurred at each stage are presented in Table 20.8. The results revealed a total marketing cost of Rs. 233.5 per box of apple. The expenditure incurred on different marketing stages is discussed in the following sections.

20.9.3.2 Picking, Assembling and Grading Cost

The apples are picked manually by skilled laborers and assembled at a plain place matted with paddy straw. The apples are then graded again by skilled laborers according to any of the prevailing grades. The results revealed an expenditure of Rs. 25 (10.70 percent) was incurred on picking, assembling and grading of one box of apple. Operation wise, the total marketing cost of the share of expenditure on picking, assembling and grading was 4.28 percent (Rs. 10), 2.14 percent (Rs. 5) and 4.28 percent (Rs. 10) respectively.

20.9.3.3 Packing Cost

An efficient packing aims at arranging the fruit in suitable compact containers to avoid spoilage, breakage and pilferage during transit in order to deliver good quality fruit to the consumers. The analysis of information on packing cost (Table 20.8) revealed that an amount of Rs. 93 (39.82 percent) was incurred on packing out of the total marketing cost of Rs.233.5 per box. Packing boxes alone costs Rs.65 (27.83 percent) when the grower used wooden boxes as packing material and Rs. 35 when they used the cardboard boxes. The packing cost included other packing material costs like packing labor, wrapping paper, paddy straw, nails, assembling and closing of boxes and labeling and stenciling, which worked out to be Rs. 15, Rs. 4, Rs.4, Re.1, Rs.3 and Rs.1 per box respectively.

TABLE 20.8

Marketing Cost of Apple (Rs. per box)

Serial No.	Cost components	Amount (Rs.)	Percentage of total cost
1	**Prepacking cost**		
(a)	Picking charges	10.00	4.28
(b)	Assembling charges	5.00	2.14
(c)	Grading charges	10.00	4
	Total	**25.00**	**10.70**
2	**Packing cost**		
(a)	Cost of wooden packing box	65.00	27.83
(b)	Cost of wrapping paper	4.00	1.71
(c)	Cost of Paddy straw	4.00	1.71
(d)	Cost of nails	1.00	0.42
(e)	Cost of packing	15.00	6.42
(f)	Closing and assembling of boxes	3.00	1.28
(g)	Labeling and stenciling	1.00	0.42
	Total	**93.00**	**39.82**
3	**Transportation cost**		
(a)	Warehouse, loading and unloading charges	5.00	2.14
(b)	Orchard to road head	15.00	6.42
(c)	Forwarding charges	30.00	12.84
(d)	Freight to Delhi	55.00	23.55
(e)	Loading at road head	3.00	1.28
(f)	Unloading at destination	2.00	0.85
(g)	Communication, etc.	0.50	0.21
	Total	**110.5**	**47.32**
4	**Miscellaneous costs**	5.00	2.14
	Grand Total**	**(233.5)**	**(100)**

* Standard wooden box contains 18 kg of apple.

The orchardist has to bear a huge amount of money (27.83 percent of the total marketing cost) to meet the cost of wooden packing, which, if minimized by some innovative packing material, would considerably reduce the total marketing cost of apples. Therefore, efforts need to be devised to find alternative means for the prevailing wooden packing boxes. The other packing boxes such as carton boxes that cost about Rs.35 per box are still not accepted by the majority of farmers due to the lack of infrastructure in the orchards. The growing concern about the dwindling timber plantation and what it means for wooden box manufacturing also aggravates the need to seek an alternative to such packings.

20.9.3.4 Transportation Cost

Like other perishable agricultural products, apples require an efficient transportation system for quick disposal. The transportation cost of apple have been taken into account with the freight incurred to lift the fruit from the orchards to the Delhi market, and also the loading, unloading and forwarding charges. The transportation cost per box of fruit was found to be Rs.110.5 which accounted for 47.32 percent of the total marketing cost. Among the components of the transportation cost, a maximum expenditure of Rs.55 was incurred on freight from road head to Delhi which accounted for 23.55 percent of the total marketing cost (Table 20.10). The forwarding charges amounted to Rs.30 and constituted 12.84 percent of the total marketing cost. However the other costs that prevailed previously like a state tax and *octroi* at the destination have now been abolished. A critical insight of the findings revealed that a very high expenditure was incurred on transporting the fruit from road head to the Delhi market, the main terminal hub which digests the major quantity of apple production. Efforts, therefore, need to be made to identify transportation means, routes and models which would economize on this expenditure. Establishment of cooperative societies for transportation of the fruit to distant markets could be a healthy step in the right direction. Route scheduling is also a viable option.

20.9.3.5 Miscellaneous Cost

Miscellaneous costs included the expenditures incurred on watch and ward, incidentals and advertisements. This accounted for Rs.5 per box of fruit and constituted 2.14 percent of the total marketing cost.

20.9.4 Marketing Channels and Price Spread

20.9.4.1 Marketing Channels

Apples are produced by a large number of small farmers scattered around the valley whereas the consumers are located throughout the country. The marketing system for apples is highly complex and comprises different marketing channels for distribution of apples in different markets. In each channel a number of functionaries are involved who perform numerous business activities called marketing functions. The following are some commonly encountered channels of apple distribution in the sampled area.

Apple distribution was comprised of movement of produce from producer to the ultimate consumer. In this process the fruit has to pass through more than one functionary, except when it is directly sold to the consumer by the producer. The chain involves various intermediaries like growers, preharvest contactors, wholesalers, retailers, etc. and is called the marketing channel. The following channels were identified as important channels in the sampled area for the marketing of produce (Table 20.9).

The price spread consists of marketing costs and margins of intermediaries involved in the marketing process. It explains the variance in the price received by the producer and the price paid by the consumer. The study of price spread is essential from the stand point of efficiency in the marketing system. The channel-wise price spread in terms of percent of consumers price is given in Table 20.10.

A cursory glance of the Table 20.10 reveals that, net price received by farmer (NPRF) is more in channel I (68.17 percent of the consumer's price) followed by channels III and IV because farmers directly sold their produce to the wholesaler. To sum up, NPRF is more in the channel where the numbers of intermediaries are less.

As far as the price spread of apples was concerned, preharvest contractor's (PHC) margin was more than the commission agent because of more bargaining power. Retailer margin is more in

TABLE 20.9
Marketing Channels of Apples

Channel I	Producer–Wholesaler/Commission agent–Retailer–Consumer
Channel II	Producer–Preharvest contractor–Wholesaler/Commission agent–Retailer–Consumer
Channel III	Producer–Commission agent–Wholesaler–Retailer–Consumer
Channel IV	Producer–Preharvest contractor–Commission agent–Wholesaler–Retailer–Consumer
Channel V	Producer–Postharvest contractor/potential growers–Commission agent– Wholesaler–Retailer–Consumer

** This does not include commission and market fee paid at 10 percent of the total bill

TABLE 20.10

Marketing Costs, Margins and Price Spread in Different Channels of Apple

Functionary	Marketing channels				
(Values in terms of percent of consumers price)	I	II	III	IV	V
Expenses incurred by producer	22.24	-	35.92	11.61	-
Producers net margin	45.93	43.35	37.19	27.85	28.92
Commission agents margin	-	-	7.31	7.31	7.31
Expenses incurred by preharvest contractor	-	22.24	-	-	35.92
Sale price of preharvest contractor	-	90.07	-	-	73.86
Preharvest contractors margin	-	4.09	-	-	9.02
Expenses incurred by postharvest contractor	-	-	-	17.02	-
Sale price of postharvest contractor	-	-	-	74.17	-
Postharvest contractors margin	-	-	-	10.29	-
Expense incurred by whole seller	0.48	0.48	3.06	3.06	3.06
Sale price of whole seller	72.54	72.54	78.92	78.92	78.92
Whole sellers margin	3.88	3.88	2.75	1.80	2.12
Expenses incurred by retailer	12.08	12.08	12.57	12.57	12.57
Retailer margin	15.38	15.38	8.51	8.51	8.51
Sale price of retailer/consumers price	**100**	**100**	**100**	**100**	**100**

channel I and channel II than other channels because they sold the produce in much small quantities and furnishes it before consumers on relatively high prices.

It could be concluded that producers received higher proportion of consumer's price as net return in channels with lower number of intermediaries. It was seen that net price received by farmer decreased considerably with increased in number intermediaries in marketing chain of apples. In order to improve net profit of producer/farmer and to provide a competitive price to consumers, it is necessary to reduce the number of intermediaries in the marketing supply chain.

20.9.4.2 Marketing Efficiency

Marketing efficiency essentially reflects the degree of market performance. An efficient marketing system is an effective agent of change and an important means of the raising income level of orchardists and the satisfaction level of consumers. It can be harnessed to improve the quality of life of the masses. The existence of competitive conditions and a desire to maximize profit are the main forces which induce firms to operate efficiently. In this section an attempt has been made to measure the marketing efficiency, particularly for comparing the efficiency of alternate markets/channels in the apple trade. The marketing efficiency of different channels as presented in Table 20.11 revealed that channel I (0.68) turns out to be economically more efficient, followed by channels III and channels IV (0.43) and least efficient is channel II (0.28). It was observed that producers got maximum shares of consumers rupee in the channel where, produce was directly marketed to wholesaler. The contractor, in turn, trades their produce to the wholesaler at a higher prices than the producer because of higher bargaining power. An orchardist could earn maximum share of the consumers price in the channel where he sells his produce directly to the wholesaler. However, lack of liquidity potential, ignorance of market demand, etc. capitalizes into a distress sale. Liberal cheap credit facility along with other incentives to apple growers would definitely increase their bargaining power.

The study on the marketing of apples revealed the following significant features related to market practice that need attention as part of the value chain improvement effort. The major destination of Kashmir apples is the Azadpur *Mandi*, which is a buyer's market and designed to be so.

TABLE 20.11

Marketing Efficiency of Apple in Different Channels (in percent)

Serial No.	Particulars	Channels				
		I	II	III	IV	V
1	Net price received by the farmer	68.17	28.92	43.35	39.46	28.92
2	Total marketing cost	34.8	34.8	51.55	44.26	51.55
3	Total marketing margin	65.19	66.7	48.45	48.45	48.57
4	Marketing efficiency	0.68	0.28	0.43	0.43	0.29

Manipulation of prices by traders in the *Mandi* is resorted through stopping apple trucks at the border of entry into Delhi, the use of cold stores to alter supply of apples in the *mandi*, keeping away small buyers with artificially high price quotes and later reducing prices to low levels to benefit preferred buyers and the use of proxies in auctions. The markets within the state are comparatively better in price determination and transparency. Growers with preharvest contracts (PHC) access the markets easier but lose out on the full benefit of the market prices on account of taking money in advance. Free growers find it difficult to enter markets even when the demand is brisk, and the commission agents prefer their 'captive growers' with PHC. Setting up satellite markets has helped growers (especially the free ones) in marketing. Farmers who market apples through cooperatives realize higher prices. Trade margins range from 42 percent to 73 percent in the different channels of marketing.

Price discovery by grower would be more realistic and effective if they are able to hold back and store their produce for some time. The farmer needs to have conditions (local storage and financial capacity to hold) under which distress selling can be checked.

20.10   PRICE BEHAVIOR OF APPLES

Consumers in India have taste preferences for distinct, specific varieties of apple coming from Kashmir. Therefore, imported varieties of apple cannot be a perfect substitute or near replacement for local Kashmiri varieties. The Kashmir apple is likely to have a secure market for its segmented consumers, not only within India but also in Bangladesh, Sri Lanka and Nepal. Therefore, the idea of cutthroat competition from imported apples against the local varieties is a pure myth. The local 'American' variety (small and sweet) is preferred in West Bengal, Bangladesh and locally within Jammu and Kashir State by the poorest classes. The imported 'Washington' apple cannot replace it. 'Maharaji' (sweet and sour) is exported to South India mainly because the people there prefer the sour taste of the apple. This is the cheapest but good quality apple from Kashmir; its shelf life is 6 months without storage because of its hardness and 'pressure'. The red 'Delicious' variety is mainly exported to the metropolitan cities of Delhi, Mumbai and Ahmedabad.

Demand and supply factors are powerful determinants of market price of Kashmir apples. But market prices does not remain in a stable equilibrium over a period of time as can be seen from the Table 20.12. The growth rates for different apple varieties in the different markets (Figure 20.6) also show a fluctuating trend with the highest growth rates for the 'Delicious' varieties in Delhi, Kolkatta and Bangalore markets, i.e., 10.84 percent, 10.24 percent and 10.18 percent respectively. For Kashmir apples, generally demand for consumption (including processing demand) is more than supply. Production/supply of Kashmir apple is increasing, but the demand is also increasing, often at faster rate. This dynamic situation is expected to destabilize the market into disequilibrium to push up the market price until supply is reduced or managed in such a way (say by commission agents/self buyers who own cold stores) that high market price again falls back to an equilibrium

TABLE 20.12
Variety Wise Wholesale Rates of Apples over the Years (Rs per box)

Year	Delhi	Ahmedabad	Kolkatta	Bangalore	Mumbai	Amritsar	Sopore	Parimpora
2008	451	758	442	469	476	367	355	340
2009	664	783	581	739	611	401	404	459
2010	535	708	610	731	547	429	489	436
2011	713	785	690	824	813	528	515	538
2012	820	783	725	913	839	532	554	611
2013	718	857	648	855	625	430	465	507
2014	634	824	748	948	783	500	512	558
2015	787	306	785	900	775	600	486	517
2016	743	441	858	912	805	652	516	541
2017	787	580	933	936	857	657	548	514

Source: APMC Azadpur Delhi (2016–17)

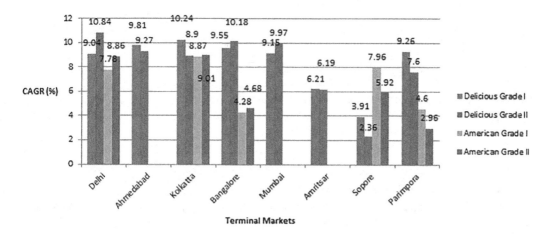

FIGURE 20.6 Compound growth rates of the price of apple varieties in different markets.

level. This process of determination and adjustment of market prices of Kashmir apples takes a very short time, is dynamic, ongoing, and happens on a day-to-day basis.

20.11 EXPORT AND IMPORT SCENARIOS OF APPLES

India's share in the total world apple production is merely 2.05 percent. Only around 1.6 percent of the country's production gets exported. The country wise export of apples from India during 2016–17 is given in Figure 20.7 that shows a maximum quantity of apples is exported to Bangladesh and Nepal followed by others.

The apple has been traded worldwide, and it shows continuous increase from last the 10 years. The global export of apples was 5308225 metric tons in 1999 and has grown continuously; it was 7756841 metric tons in 2009 which shows an increase of 40 percent. Similarly the global import of apples was 4879292 metric tons in 1999 and has grown continuously to 7,395,772 metric tons in 2009. That is an increase of 51 percent (Figure 20.8).

The apple has been traded worldwide, and it shows continuous a increase over the last 10 years. The global export of apples was 5000 metric tons in 2001 and has grown continuously to 20000

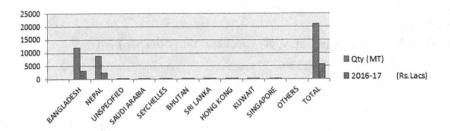

FIGURE 20.7 Country wise export of apples from India during 2016–17.

Source: APEDA

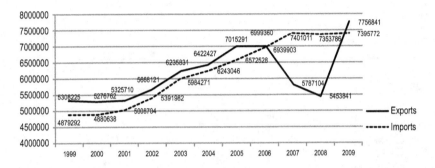

FIGURE 20.8 Global import and export of apples (in metric tons).

Source: FAO STAT

TABLE 20.13
India's Share in Global Trade of Apples (metric tons)

Market Year	Imports (MT)	Growth rate	Exports (MT)	Growth rate
2001	20000	NA	5000	NA
2005	33000	8.20%	30000	29.87%
2010	144000	10.51%	47200	76.78%
2011	207600	44.17%	30200	−36.02%
2012	197100	−5.06%	27300	−9.60%
2013	196800	−0.15%	32600	19.41%
2014	204300	3.81%	20700	−36.50%
2015	201700	−1.27%	20500	−0.97%
2016	369800	83.34%	20800	1.46%
2017	350000	−5.35%	20000	−3.85%
CAGR		**22.05%**		**4.88%**

Source: Export Import Data Bank, Dept. of Commerce

metric tons in 2001 which is an increase of 4.88 percent. Similarly, the global import of apples was 20000 metric tons in 2001 and has grown to 350000 metric tons in 2017. That is an increase of 22.05 percent. Overall, in recent years the export of apples to other countries has been almost negligible because of stiff competition from other countries. However, some meager quantities of apple have been traded with Pakistan in a barter system through the LOC trade agreement between the two countries (Table 20.13).

TABLE 20.14
Price Analysis of Imported Apple in India

Particulars	Price in US $ per 20 Kg box	Price in INR per 20 Kg box	Price in INR per Kg
Import price of 'Washington' apples	25	1650	83
Expenses incurred by importer	17.1	1128	56
Importer's margin	3.3	217	11
Price at wholesale market	45.4	2996	150
Expenses incurred by trader	1.1	72	3.5
Trader's margin	2.2	145	7.0
Retailer's purchase price	48.7	3214	160
Retailer's expenses	4.4	290	14.5
Retailer's margin	8.9	587	29
Consumer price	62	4092	200

Source: NABCONS-2014; $ = 66 INR

India imported approximately 197100 metric tons of apples during 2012, and the import price of the 'Washington' apple was 83 INR per kg. Apples accounted for a major chunk of fruit imports despite the high import tariff rate of Rs 56, and India's customs duty on apples is high (NABCONS, 2014). The 'Delicious' group constitutes approximately 80 percent of the apples imported into India. The remaining portion included varieties such as 'Fuji', 'Royal Gala', and 'Granny Smith'. Generally imported apples are priced approximately double the domestically produced apples. Higher import tariff and transportation cost along with higher margins charged by importers are important contributing factors. Important qualities by which imported apples fetch a premium price and give a tough competition to domestic apples include a tempting bright red, gleaming surface and a consistent taste year round. Indian growers have made little effort in the value chain to improve the quality and yields to better compete with imported apples overall (Table 20.14).

20.12 CONCLUSION

Apple production and marketing is an important economic pursuit and a source of livelihood to 35 lakh people of the union territory of Jammu and Kashmir. In recent years the state has given lots of attention to the development process of the apple industry. However, there exists a wide and marked gap in the productivity of apples as compared to the major apple producing countries around the world. There is a growing concern that with technological change in production, variability has increased. Demand and supply factors are powerful determinants of the market price of Kashmir apples. But market prices does not remain in a stable equilibrium over a period of time. For Kashmir apples, generally demand for consumption (including processing demand) is more than supply. Production/supply of Kashmir apples is increasing but demand is also increasing, often at faster rate. This dynamic situation is expected to destabilize the market into disequilibrium to push up market price until supply is reduced or managed in such a way (say by commission agents/self buyers who own cold stores) that high market price again falls back to an equilibrium level. The relative peace in Jammu and Kashmir has made it possible for farmers to focus on improving their livelihoods. The apple sector has the potential to influence several households and improve their economic prospects. New market players have to be invited in, resources found for investments, changes in policies and support systems from the government and capacities built in individuals and institutions for effective and remunerative participation in the value chain. The market for imported apples in India has strong segmentation by variety and origin. Apples from the United States and

China are the backbone that facilitates the development of this market which has a low commercial concentration. India is a complex model for the marketing of imported apples, with small structural change at the country level between 2014 and 2015 against a background of strong growth in world exports. Important qualities by which imported apples fetch premium price and give a tough competition to domestic apples include a tempting, bright red, gleaming surface and consistent taste all year round. The complexity is the result of the interrelations of the monthly income of the importation, the varieties, the quality and the origin of the apples. However, all importers of apples show changes, which overlap with each other. This explains that the general view of the country is of little change in contrast to what occurs at the level of consignees or importers. Therefore, an interesting case develops to teach about commercial decisions and the need for intense and deep research on business movements. The revamp of the apple sector has to be planned with a mix of investments, capacity building, innovations and committed institutional leadership.

REFERENCES

Directorate of Horticulture *Kashmir, Government of Jammu & Kashmir in the World Wide Web.* www.hortikashmir.gov.in.

Export Import Data Bank, Dept. of Commerce 2017. *Export Import Data Bank, Ministry of Commerce & Industry, Government of India in the World Wide Web.* www.tradestat.commerce.gov.in.

Food and Agriculture Organization (FAO) 2019–20. *Production Year Book. Food and Agricultural Organization of the United Nations, Rome in the World Wide Web.* www.faostat.com.

NABARD Consultancy Services (NABCONS) 2014. *A Proposal for Strengthening the Apple Value Chain in Jammu & Kashmir.* NABCONS, Corporate Office, Bandra Kurla, Mumbai, 29 pp.

Naqash, Farheen, Hamid, Naveed and Kapila, Dinesh K. 2017. A case study on economic analysis of marketing and price spread of apple fruit in Kashmir Valley of J&K state. *International Research Journal of Agricultural Economics and Statistics* 8(2):440–447.

NHB 2018–19. *Indian Horticulture Database-2019, National Horticulture Board (NHB), Ministry of Agriculture, Government of India, in the World Wide Web.* www.nhb.gov.in.

Wani, S. A., Kumar, Shiv, Naqash, Farheen, Shaheen, F. A., Wani, Fehim J. and Rehman, Haseeb Ur, 2021. Potential of apple cultivation in doubling farmer's income through technological and market interventions: An empirical study in Jammu & Kashmir. *Indian Journal of Agricultural Economics* 76(2):278–291.

Index

Note: Page numbers in *italics* indicate a figure and page numbers in **bold** indicate a table on the corresponding page.